Field Guide to the Forest Trees of Uganda

For Identification and Conservation

Field Guide to the Forest Trees of Uganda

For Identification and Conservation

James Kalema (PhD)

Assoc. Professor of Botany
Makerere University Herbarium
Department of Plant Sciences Microbiology and Biotechnology
College of Natural Sciences
PO Box 7062, Kampala, UGANDA

Alan Hamilton (PhD)

Honorary Professor
Kunming Institute of Botany (Chinese Academy of Sciences)
Heilongtan
Kunming 650204
Yunnan, CHINA

Published exclusively in Uganda by Alan Hamilton
128 Busbridge Lane
Godalming, Surrey GU7 1QJ, UK

ISBN: 978 0 95414 963 5 (softback)

Title: Field Guide to the Forest Trees of Uganda
Subtitle: For Identification and Conservation
Authors: James Kalema and Alan Hamilton

This book is also published for worldwide distribution by CABI, which is a trading name of CAB International.

CABI
Noseworthy Way
Wallingford
Oxfordshire OX10 8DE
UK

Tel: + 44 (0)1491 832111
Fax: +44 (0)1491 833508
E-mail: info@cabi.org
Website: www.cabi.org

CABI
WeWork
One Lincoln St
24th Floor
Boston, MA 02111
USA

Tel: +1 (617)682-9015
E-mail: cabi-nao@cabi.org

CABI ISBNs:

978 1 78924 527 1 (hardback)
978 1 78924 528 8 (ePDF)
978 1 78924 529 5 (ePub)

Commissioning editor (CABI): Ward Cooper
Editorial assistant (CABI): Lauren Davies
Production editor (CABI): James Bishop

Typeset: Alan Hamilton
Photographs: James Kalema, Alan Hamilton, William Olupot, John Kalule
Figures and plates: Alan Hamilton, Naomi Hamilton (Figure 5.1)

Printed in the UK by CPI Group (UK) Ltd, Croydon, CR0 4YY

Available from Gustro Limited, Gustro Book Centre, Plot No. 355, Sir Apollo Kaggwa Road, P.O. Box 9997, Kampala. Tel: +256 778 282323; E-mail: info@gustro.com; Website: www.gustro.com

Cover photo: Moist lower montane forest, Bwindi Impenetrable National Park, southwest Uganda. The prominent umbrella-shaped tree is *Newtonia buchananii*.

Contents

Figures

Plates (all in Part 4)

The number ranges of the species illustrated are given in Column 3

Tables

Authors

James Kalema

James is an Associate Professor of Botany at Makerere University, Uganda. He is a member of the Global Tree Specialist Group of IUCN, serves on the East African Plant Red List Authority and is lead author of *Conservation Checklist of the Trees of Uganda.* He has extensive experience in field botany, including undertaking plant surveys and biodiversity assessments. He has assisted in management planning for some of Uganda's forested national parks. He is currently involved in the production of a conservation atlas of East African Acanthaceae.

Alan Hamilton

Alan is a botanist who has undertaken research into the environmental history of tropical Africa and written on the history of forests in Uganda. A lecturer in the UK and Uganda between 1972 and 1989, he later worked for the conservation groups WWF and Plantlife, including mounting programmes to enhance community involvement in conservation. He is a Fellow of the Linnean Society, a Doctor of Science of the University of Cambridge and an Honorary Professor in the Kunming Institute of Botany, Chinese Academy of Sciences.

Preface

This book provides an up-to-date list of indigenous tree species known to grow in the tropical forest of Uganda and a guide to their identification in the field. It draws on *A Field Guide to Uganda Forest Trees* (UFT), which one of us (AH) wrote in 1971 (but was published only in 1981). The original field guide referred extensively to *The Indigenous Trees of the Uganda Protectorate* (ITU) (Eggeling and Dale 1951). Use was also made of Dawkins (1951), the first published field guide to the indigenous forest trees of Uganda. It was presented in graphic form.

The authoritative regional flora for East Africa, the *Flora of Tropical East Africa* (FTEA 1952-2012), has been used as the basis for general taxonomic treatment. Parts of this flora were published after 1971 and today some of them need substantial revision. There have been major advances in the scientific understanding of how plants are related to one another during recent decades based on studies of their DNA. This has contributed to substantial changes to the delimitation of some plant families (APG IV 2016), as well as of genera and species. As much as possible, we have followed APG IV for classification and nomenclature of flowering plants.

Those familiar with forests in Uganda know that flowers and fruits are rarely seen and, even when present, can be high on the trees and nearly impossible to collect. Field workers therefore have to rely mainly on vegetative characters to identify trees. However, the vegetative parts of plants tend to be much more variable morphologically than their reproductive organs, varying for instance according to their position on the plant and its age, and the environment in which it has grown. Uncertainties in identification can therefore arise. Users of this field guide should note that the keys and descriptions are based on the normal vegetative characteristics, as seen on mature trees, unless otherwise stated. Average leaf sizes, rather than total ranges, are normally given, because this is usually more useful information for the field worker.

Leaves provide many characters useful for identification. Therefore, their features are widely used in the keys and descriptions, as well as for determining how the species are ordered in the descriptions and illustrations. However, leaves can be difficult to obtain for close examination in some cases, especially with taller trees. It can even be difficult to determine which leaves high above in the canopy are attached to which of the trunks among which one is standing on the ground or perhaps even to a large forest climber. A special key to tall trees is provided.

Apart from nomenclature, taxonomy and new records, the main substantive changes with respect to UFT are the inclusion of information on conservation status (from Kalema and Beentje 2012, IUCN 2019 and WCS 2016) and on cultivation and propagation (from Meunier *et al.* 2010). The indexing of vernacular names is now by language. The keys, descriptions and illustrations remain much the same as with UFT, except that a number of new keys have been added for the identification of species within selected genera.

Foreword

Anyone moving through a tropical high forest will quickly appreciate the difficulty of accurately identifying the tree species because of their diversity. This challenge is further compounded by inability to access some of the most useful parts, including leaves, but also the low likelihood of finding flowers and fruits.

As an undergraduate student at Makerere University in the late 1980s and early 1990s, the book that was in vogue and which we used for identification of forest trees of Uganda was mainly Hamilton (1981). This guide has been used by tree lovers and students, and has been instrumental in making researchers and other stakeholders know more about the trees that grow in forest habitat. The book was based on scientific principles but was also handy to non-professional users. However, Science is a very dynamic realm. From the time of its publication, many aspects of plant taxonomy and systematics have changed and so have the approaches to conservation. But these changes have happened as the cover and ecological integrity of forests in Uganda have continued to decline, yet these ecosystems harbour the bulk of our biodiversity.

Publication of a new guide is timely and it delights me to see that it has been produced by two well experienced Professors of Botany, who love nature and care to help our nation develop and prosper by providing the required knowledge for sustainable use of our natural resources. Field researchers and students in the areas of Botany, Forestry, Agriculture, and other biological science-related disciplines as well as conservation scientists, naturalists and environmentalists should find it useful in determining the identity of forest trees.

This book has been prepared not only for identification of the 451 forest tree species but the authors have gone an extra mile by providing useful information about propagation techniques for some of the species. The National Forestry Authority, in partnership with some stakeholders, intends to restore forests that have been degraded over time, as part of the organization's strategic action plans. This book has hence been published at a time when we needed it most to set us into motion, now that more information about propagation is provided.

Besides, this book draws attention to those species most critically in need of conservation attention. These have been highlighted, and their status provided based on acceptable standards of conservation assessment. As much as possible, the authors have provided names in local languages, something very useful to non-professional users of the guide.

Natural tropical high forests are, by far, the most important for conservation of biodiversity but also for providing ecosystem services for community livelihoods. The authors have provided the range of forest types in Uganda and laboured to explain why conservation of indigenous forests and indigenous species is important for this nation and the wider community. They have strengthened and ably backed up this reasoning with a rich and very current literature drawn from various examples around the world. Particular attention has been given to the influence of forest loss on climate change and impact of climate change on forest cover and its services. The underlying causes of forest loss and degradation, as explained in this book, need to be addressed if we are to continue getting the ecosystem services from forests and their support to different sectors of the economy, such as agriculture.

We need an enabling environment for implementation of programmes at national and all levels aimed at salvaging our forest estate, especially the natural high forest. To this end, the necessary policy framework has been reviewed in this book. It is my hope and wish that the up-to-date methods for identifying forest trees used in this book will enhance the understanding and appreciation of our forest tree species, our heritage and our pride.

Tom O. Okello
Executive Director
National Forestry Authority
January 2020

Acknowledgements

Many expert botanists and others have willingly provided information for use in this field guide. Dr Henk Beentje (Royal Botanic Gardens, Kew) helped in verifying some of the scientific names. The following assisted with the names of organizations in Uganda relevant to the purposes of this book: Assoc. Professor Gerald Eilu (Department of Forestry, Bio-Diversity and Tourism, Makerere University), Tom Forrest, Dr Robert Nabanyumya (Green Approaches Ltd.) and Professor Derek Pomeroy (Department of Zoology, Entomology and Fisheries Sciences, Makerere University).

Those who kindly edited the list of vernacular names in local languages were: (for Ateso) Nelson Omagor (JBN Consults and Planners); (for Luganda) Olivia Wanyana Maganyi and earlier Christine Kabuye (both c/o Makerere University Herbarium); (for Acholi, Lango and Luo) Dr Perpetra Akite (Department of Zoology, Entomology and Fisheries Sciences, Makerere University); (for Lusoga) Ben Kirunda (WCS, Kampala); (for Rufumbira) Samuel Nsutiyayesu (RICA Faculty-Conservation Ecology, Rwanda Institute for Conservation Agriculture); (for Rukiga) Robert Barigyira and Dr Dennis Babaasa (both Institute of Tropical Forest Conservation); (for Runyoro) Assoc. Professor Patrick Mucunguzi (Department of Plant Sciences, Microbiology and Biotechnology, Makerere University).

We thank Rudi Lemmens for helping us in the early stages of revision and for providing links to a website that he has created containing on-line information on some of the tree species.

We appreciate the support of Ward Cooper, James Bishop and Lauren Davies at CABI.

AH would like to thank his wife Naomi for the help she has given him in the preparation of this field guide, including tolerance for his unsociability (the writers' disease). He wishes to thank his son Dr Patrick Hamilton and Martin Foran for help with the design of the cover and Mike Lagan and Alan Brignell for assistance with computing.

Part 1

The Forests and Their Trees

What is a Forest Tree?

With rare exceptions, a tree is considered here to be a perennial, self-supporting woody plant, typically with a single main stem or trunk, a distinct crown and capable of growing to a height of at least five metres. Trees grade physiognomically into large shrubs. A few species on the borderline between trees and shrubs have been included in this field guide, providing useful information for readers trying to distinguish them from those similar-looking plants that are clearly small trees.

The term 'forest', as used here, refers to a type of vegetation that typically has a continuous stand of trees, a tall canopy (10 to 50 m or more) and usually several layers of trees with crowns interdigitating with one another or overlapping (Fig. 1.1) (McElhinny *et al.* 2005; Obua *et al.* 2010; Côte *et al.* 2018; FAO 2018a, 2018d). It is a type of vegetation that regenerates naturally to maintain a complex structure (Kalema and Kasenene 2007; FAO 2018a, 2018d). Also known as tropical rainforest or, in Uganda, as Tropical High Forest, forest contrasts with certain other types of vegetation that are similarly dominated by trees and that are normally known by scientists concerned specifically with Ugandan vegetation as woodland (Langdale-Brown *et al.* 1964; White 1983). Woodland can usually be distinguished from forest in having only a single tree layer, an abundance of narrow-leaved grasses in the herbaceous layer (not the broad-leaved grasses common in some forests) and in being subject (and adapted) to burning. Also excluded are those other types of vegetation found in Uganda dominated by woody plants known scientifically as thicket and evergreen scrub. Forest becomes reduced in stature at high altitudes, with fewer tree layers, and, above the limit of broad-leaved trees, can grade into vegetation dominated by microphyllous trees (typical of the Ericaceous Belt) and giant groundsels (typical of the Afroalpine Belt). We have included trees found in these two vegetation belts here.

Confusingly, the term 'forest' is sometimes applied to other types of Ugandan vegetation apart from forest as we understand it. This means that reports on the state of Ugandan forests need to be read cautiously. The definition of forest, as used by the Secretariat of the Convention on Biological Diversity (2001) defines forest as including types of 'ecosystems in which trees are the predominant life forms', which is a very broad definition meant to cover global variations.

Vegetation on the boundary between forest and other forms of vegetation is in an intermediate situation and its flora can be distinctive (Fig. 1.2) (Marfo *et al.* 2019). This zone with its abrupt to gradual change in species composition is associated with changes in other aspects of the environment, such as climate, soils and human use of natural resources (Liautaud *et al.* 2019), but the extent to which the position of the boundary is a consequence of these other factors or these other factors are responsible for the position of the forest can be difficult to judge (Brownstein *et al.* 2015). The type of ecosystem found on one side of the boundary can have profound influence on that on the other and the boundary itself (an ecozone) can be more biologically diverse in terms of numbers of species than the areas on either side (Hufkens *et al.* 2009; Marfo *et al.* 2018). We include here the commoner species of trees found in boundary zones and on forest edges, but omit those seen less frequently.

Fig. 1.1. Forest structure of Mpanga Central Forest Reserve, a lowland semi-deciduous forest of the Lake Victoria forest belt. Photo: Alan Hamilton (2019).

In an earlier field guide to Ugandan forest trees (Hamilton 1981), which was based on reports and observations made prior to 1972, it was stated that 'there is rarely any difficulty in determining whether or not a certain type of vegetation is forest, since marginal types of vegetation have been almost completely eliminated by burning, grazing and agriculture over a long period of time. Indeed, the boundaries of the great majority of forests are artificial and, in many cases *follow Forest Department demarcation lines* [our emphasis]'. This is no longer so true.

Many forests have been and continue to be degraded through human activities (Kalema *et al.* 2010), even in protected areas (Sassen and Sheil 2013). There has been widespread land use and land cover change (Kalema and Bukenya-Ziraba 2005; Kyarikunda *et al.* 2017). Extensive areas of ground now contain a mixture of forest and non-forest species. Sometimes, the clearance of forest to plant crops results in the leaving behind of impoverished ecosystems with only a few scattered tall trees (Fig. 1.3). It can be predicated that many of these trees, now abandoned to the elements, will soon die. This degradation and loss continue despite enactment of a National Forestry and Tree Planting Act (Government of Uganda 2003) and new institutional arrangements. The latter include establishment of a Forest Sector Support Department, the National Forestry Authority (NFA) and District Forestry Services (Tumushabe and Mugyenyi 2017; Josephat 2018).

Fig. 1.2. Boundary of Mpanga Central Forest Reserve. Photo: John Kalule (2019).

Forest Distribution and Types in Uganda

Figure 1.4 shows the distribution of larger areas of forest in the 1950s, based on a map in Government of Uganda (1967). Also shown are those parts of Uganda that would naturally have carried forest before its clearance by people (estimated from a combination of climatic parameters and the presence of forest remnants). It can be seen that many of the forests lie in two regions, both of which are characterized by relatively high and well distributed rainfall. One is to the north of Lake Victoria (the lake belt) and the other, lying on or close to the border with D.R. Congo, is associated with the Albertine Rift. The forests shown in north-eastern Uganda (Kadam, Timu, etc.) are on mountains. Groundwater sometimes sustains forest in climatically dry areas, as along river banks (riverine forest).

The floristic composition of the forests, which in turn contributes to their structure, is greatly influenced by temperature, which reduces with altitude (FAO 2017; Mau *et al.* 2018; Mujawamariya *et al.* 2018; Cabrera *et al.* 2019), as well as climatic moistness and environmental history (Hamilton 1989; Tang 2019). A standard system of classification used for forests in Uganda recognizes two principal altitudinal types, High Altitude (or montane) Forest above 5000 ft (1525 m) and Mid Altitude Moist Forest below (Langdale-Brown *et al.* 1964). Mid Altitude Moist Forest, especially that below 1400 m, is floristically akin to forests at much lower altitudes (towards sea level) elsewhere in tropical Africa (Hamilton 1989; White 1983) and can be alternatively referred to as lowland. All the lake-belt forests are lowland, while both lowland and montane forests can be found along the Albertine Rift. Lowland forest varies in species complement and physiognomy between wetter and drier areas, an increased proportion of deciduous trees being found in the latter *(semi-deciduous forest).

Fig. 1.3. Matiri Central Forest Reserve, severely damaged by encroachment for agriculture and felling trees for charcoal. Photo: James Kalema (2009).

If the total altitudinal ranges of tree species in the country as a whole are considered, then lowland and montane forests grade gradually into one another without an abrupt transition (Hamilton 1989). However, there are some species that can assume great abundance over particular altitudinal ranges, providing handy ways to classify the forests further. Mountain bamboo (*Sinarundinaria alpina*) tends to form extensive stands in climatically wetter areas at high elevation (normally 2450-3050 m), thereby enabling recognition of a moist lower altitude montane forest zone below (1500-2450 m) (also known as *Pygeum* [= *Prunus*] Moist Montane Forest) and an upper montane forest zone above (3050-3300 m) (also known as *Hagenia-Rapanea* Moist Montane Forest) (Langdale-Brown *et al.* 1964). *Cynometra alexandri* and *Parinari excelsa* can be locally abundant in some of the Albertine Rift forests at altitudes of 700-1200 m and 1400-1500 m respectively.

Forest was restricted in distribution during the last global ice age, which was marked by a dry climate across much of tropical Africa (Hamilton *et al.* 2016). The climate became wetter 12,000 years ago, allowing many forest species to expand their ranges away from dry period forest refugia, including one in Kivu Province (eastern D.R. Congo). Species had different abilities to spread, the net result being for Uganda the creation of gradients of decreasing numbers of forest species away from the border with D.R. Congo, especially away from the south-west. This pattern is superimposed on other patterns considered to be caused by modern environmental factors, such as temperature and rainfall (Hamilton 1989; Howard 1991; Brack 2019; Tang 2019). It is predicted that modern anthropogenic climate change will further affect the forests (Lewis 2006). There are indications that tropical trees may be more vulnerable to continued warming than temperate species, as tropical trees have shown greater declines in growth and photosynthesis at elevated temperatures (Mau *et al.* 2018).

The richest forests in Uganda in terms of biodiversity, as measured by species scores for four taxonomic groups (one being forest trees), are Bwindi (Fig. 1.7) and Semliki (Howard 1991). Ishasha Gorge in Kayonza Forest (northern part of Bwindi) has a particularly diverse and unusual flora and could possibly have been the site of a minor forest refugium during the time of ice age aridity.

Some idea of the botanical diversity of the forests may be gauged from the numbers of tree and shrub species encountered in transect surveys through five of Uganda's forests carried out for comparative biodiversity purposes (Howard 1991). The first number for each forest in the following list is the number of tree and shrub species classified as 'belonging to the forest interior' and the second for a wider ecological group of tree and shrub species 'deemed to be forest dependent': Budongo 123/233, Bwindi 106/188; Kalinzu 121/236; Kasyoha-Kitomi 120/226; Kibale 110/204; Semliki 108/199 (Davenport and Howard 1996; Davenport *et al.* 1996; Howard *et al.* 1996a-d).

Fig. 1.4. Distribution of forest in Uganda during the 1950s (Atlas of Uganda 1967).

There is much variation in the floristic composition of forest at the local level. Forests lying close to Lake Victoria within the lake belt (often developed on sandy soils) tend to have a distinctive tree flora, for example with an abundance of the large tree *Piptadeniastrum africanum*. They are known as lake-shore forests. Forests inland from Sango Bay (on the edge of Lake Victoria), some standing on swampy ground, are particularly unusual floristically, containing a number of typically montane trees, such as *Afrocarpus dawei* and *Podocarpus latifolius*. This may be the site of a minor forest refugium during the last ice age, a time when temperatures as well as rainfall were depressed. Possibly the forest was sustained by high levels of groundwater fed by a still active River Kagera.

More generally, forest composition varies everywhere according to position on slope, responding to catenary variations in soils and other environmental variables along gradients extending from hilltops to valley bottoms. Swampy ground has its particular trees. Both human activities and natural processes influence forest composition at the very local level. Forests are dynamic living systems, individual trees passing along pathways of establishment, growth, maturity and death. The falls of large trees create gaps in the forest canopy, triggering phases of new tree establishment and spurts of rapid growth on the part of trees already present. The dynamics of forest systems, such as this, have intimate influences on the exact positioning of individual trees on the ground.

Fig. 1.5. Making charcoal from indigenous forest trees in Mabira Central Forest Reserve. Photo: William Olupot (2018).

History of Human Influence on the Forests

Small-scale shifting agriculture within a forested environment started to have a significant influence on the local floristic composition of Ituri Forest (D.R. Congo) from the beginning of the first millennium CE (Hart *et al.* 1996) and the same is likely to have been the case in nearby Uganda. Shifting cultivation changes primary forest to secondary forest (Spracklen *et al.* 2018), which tends to be less diverse and structurally less complex. Probably all forests in Uganda have been influenced to at least some extent by the human hand, especially through previous clearance for agriculture (Hamilton *et al.* 2016). A widespread phase of forest reduction in Uganda at c. 1000 CE may have been associated with some major socio-economic developments, notably the establishment of more hierarchical societies (such as the interlacustrine kingdoms), the onset of large-scale cattle-herding and the adoption of a perennially productive type of garden centred on the banana (known as *lusuku* in Luganda). Little is known about forest management practices before the first written records were made, which was during the second half of the 19th century.

Fig. 1.6. Illegal felling of *Funtumia* for drum-making, Mpanga Central Forest Reserve. Certain types of trees are favoured for this use. Photo: Alan Hamilton (2016).

The colonial era (1894-1962) saw the introduction of new concepts of land ownership and management. Many larger forests became Central Forest Reserves (CFRs) under the administration of a Forest Department (part of central government), while many smaller areas of indigenous forest, as well as the numerous small plantations of conifers and eucalyptus that became established, fell under the local governments as Local Forest Reserves (LFRs).

Forest (Tropical High Forest) covered about 4% of the land area of Uganda in the 1950s (Langdale-Brown *et al.* 1964), since when its extent has become seriously reduced. Details about how some of the individual forests have become lost or degraded are given in Hamilton *et al.* (2016) and, for the period up to 1982, in Hamilton (1984). The rate of deforestation during recent years (2.72% per annum) has become one of the highest in the world (FAO 2010a, 2010b). It has been concluded from a study of NFA records that the total forest area of Uganda ('forest' being taken to include other woody types of vegetation, not just 'forest' as understood in the present field guide) decreased from 4.9 million to less than 2.0 million ha between 1990 and 2015, a reduction of about 60% (IUCN 2018). Over the same period the cover of 'Natural Tropical High Forest' (equivalent to forest as used here) decreased from 850,693 to 567,168 ha, a reduction of about 33.3% (Ministry of Water and Environment 2016).

The key drivers of forest loss and degradation in Uganda are reported to be expansion of agriculture (subsistence and commercial), the unsustainable harvesting of tree products, mainly charcoal (Fig. 1.5), firewood and timber, expansion of human settlement (including to house a growing numbers of refugees), livestock grazing, wild fires and artisanal mining operations (Ministry of Water and Environment 2017). Various factors have been mentioned as underlying some of these immediate influences. The rate of growth of the population is one of the highest in the world (3.4% per annum between 1991 and 2002), there are inappropriate systems of land tenure, there is a high rate of economic dependency on subsistence agriculture (which covers a larger total area than commercial agriculture), there are weaknesses in governance (including in the implementation of forestry extension services) and climate change is having adverse effects (Banana *et al.* 2007; FAO 2017; Ministry of Water and Environment 2011, 2017).

The governance structure of forestry was changed in 2003, when the Forest Department was closed and replaced by a new National Forestry Authority (NFA), taking over responsibility for the CFRs. Six of the larger forests, Bwindi (Fig. 1.7), Elgon, Kibale, Mgahinga, Rwenzori and Semliki, were transferred from the Forest Department to Uganda National Parks (now Uganda Wildlife Authority) during the 1990s, resulting in an increased rigour in law enforcement according to anecdotal evidence.

Fig. 1.7. Sharp boundary between forest in Bwindi-Impenetrable National Park and surrounding farmland, south-west Uganda. Photo: James Kalema (2010).

Fig. 1.8. Pine seedlings being raised by the National Forestry Authority in their nursery at Banda. Photo: Alan Hamilton (2019).

A high priority in government forestry over recent decades has been the establishment of plantations of the exotic trees eucalyptus and pines, both of which can be fast-growing on suitable sites (Fig. 1.8) (Tumushabe and Mugyenyi 2017). The immediate aim has been to produce high volumes of general grade wood for use in construction or as wood fuel (firewood and charcoal) (Kaboggoza 2011). A subsidiary intention has been to reduce pressure on indigenous trees and natural forests by providing alternatives as sources of these products. In practice, this policy has led to relatively few resources being devoted to the management of natural forest and the planting of indigenous trees. Moreover, there are products, such as drums and medicines, that require the use of specific indigenous trees, so eucalyptus and pines cannot serve as alternative sources of supply in these cases (Fig. 1.6).

Carbon credit schemes, usually linked in Uganda to the planting of eucalyptus and pines (rather than indigenous species), have been introduced to incentivize tree planting to sequester carbon, hence contributing to combating global climate change. The theory is that planting trees in Africa will compensate for the use of fossil fuels in richer countries (especially European ones in Uganda's case) (Nabunya 2017; de Oliveira *et al.* 2018; Mujawamariya *et al.* 2018; Brack 2019; FAO 2018c; Lee *et al.* 2019; van Goor and Snoep 2019).

A change in government policy in 2001 emphasized a greater role for the private sector in forestry operations. NFA, which became responsible for CFRs in 2003 (when it took over this responsibility from the old Forest Department), started to issue permits to private operators for the development and utilization of particular forests (Tumushabe and Mugyenyi 2017). At the same time, the number of government forestry staff was reduced. The implementation of these new management arrangements has proved extremely poor, attributed to a combination of lack of adequate capacity at NFA (including inadequate management systems), poor cross-sectoral coordination and weak forest laws (which, in practice, have been unevenly enforced) (Ministry of Water and Environment 2017).

There is general recognition globally that more attention should be devoted to the care and expansion of forests, as stated in Goal 1 of the UN's Strategic Plan for Forests 2017–2030 (Ellison 2018; Nakamura 2019): 'To reverse the loss of forest cover through sustainable forest management, including protection, restoration, afforestation and reforestation, and increase efforts to prevent forest degradation and contribute to the global effort of addressing climate change'. If achieved, this will assist in meeting the Millennium Development Goals (Garrity 2004). Uganda itself declared a commitment to restore 2.5 million hectares of degraded forest by 2020 under the Bonn Challenge process (IUCN 2018). Inadequate funding of the forestry sector is a major issue (Tumushabe and Mugyenyi 2017). It has been recommended that a commitment to tree planting should be mainstreamed into development plans, including support for agroforestry through credit schemes and expansion of forestry extension services (NatureUganda 2011; Basamba *et al.* 2016; FAO 2017; Hillbrand *et al.* 2017; Kyarikunda *et al.* 2017; Nabunya 2017; de Oliveira *et al.* 2018). Tree planting at household level should be promoted (Kabiru *et al.* 2018). Traditional conservation practices should be encouraged (Fig. 1.9).

Fig. 1.9. Sacred forest near Mpigi, 40 km west of Kampala. The large tree is *Antiaris toxicaria*. This is an example of a traditional conservation practice.
Photo: Alan Hamilton (2019).

The Values of Indigenous Forest Trees and Natural Forests

The diverse forest tree flora of Uganda has many known uses (Egeru *et al.* 2015; Mulugo *et al.* 2019) and doubtless others await discovery. They yield many types of timber and other construction materials, cloth, fibres, latex, resins, fuels, fruits, edible greens, medicines and more (Akwatulira *et al.* 2011; Ojelel *et al.* 2019; Tugume *et al.* 2016). Forestry in Uganda currently places a strong emphasis on planting the exotic trees eucalyptus and pines (Kaboggoza 2011; Ministry of Water and Environment 2016), mainly with the aim of producing large quantities of a few types of product, notably medium-grade wood, poles and wood fuel (FAO 2018b). It has been recommended that more emphasis be given to protection and restoration of natural forest across the landscape (Kazoora 2001; FAO 2017; Hillbrand *et al.* 2017; Lee *et al.* 2019; Schulte *et al.* 2019) and this has been accepted according to commitments made in the National Development Plan of Uganda and the National Forest Plan 2011/12 - 2021/22 (Ministry of Water and Environment 2013). If achieved, this will enable more people to enjoy local access to many types of forest produce. It will also better ensure future livelihood security (Egeru *et al.* 2015; HLPE 2017). Monocultures of plants, as increasingly represented by plantations of eucalyptus and pines, can be susceptible to attack by newly introduced or evolved pests and diseases (Nyeko and Nakabonge 2008; Liu *et al.* 2018).

To illustrate one of the diverse benefits that indigenous trees and associated knowledge can bring, there are many types of plants (some forest trees) that are used to this day in Uganda as anti-malarials (Adia *et al.* 2014; Anywar *et al.* 2016; Ssegawa and Kasenene 2007a, 2007b). The context is a high degree of reliance on local herbal medicine on the part of many people and the tendency of the available pharmaceutical drugs to lose effectiveness over time. Use of herbal medicine and indigenous knowledge in malaria treatment is not new. Many of the existing anti-malarial drugs owe their origin to research on ethnobotanical knowledge associated with the quinine tree (*Cinchona*) of South America and the Chinese annual mugwort (*Artemisia annua*). It is predictable that Uganda will benefit greatly if it manages to retain, in combination, its traditional healthcare knowledge and indigenous flora.

Additional to the particular benefits brought by each individual type of tree, natural forest collectively (as a form of vegetation) delivers a range of environmental services (Hillbrand *et al.* 2017), often with greater effectiveness than do plantations of eucalyptus and pines (Liu *et al.* 2018). Among them are:

* Carbon storage (Gallery 2014), thus contributing to reducing the amount of the greenhouse gas carbon dioxide in the atmosphere and so helping to combat global climate change (FAO 2017; Josephat 2018; Mujawamariya *et al.* 2018; Brack 2019; van Goor and Snoep 2019).
* Moderation of the local climate, making it more favourable for agriculture (Nabunya 2017; Spracklen *et al.* 2018; Brack 2019; Lee *et al.* 2019). This is related to the ways that natural forest influences the local environmental energy budget and how water circulates between air, soils and plants.
* Maintenance of perennial flows of water in streams and springs and into wells. The majority of households in Uganda, even those in Kampala, rely on natural sources of water, lacking the financial ability to pay for piped supplies.
* Reduction in soil erosion and risk of landslides. The dense canopy of natural forest reduces the erosive risk associated with the exposure of bare soil to heavy rain. The binding power of its tangle of roots resists loss of soil through surface wash and gully formation.
* Increasing the efficacy and maintenance of pollination systems that benefit agricultural crops (Nabunya 2017; van Goor and Snoep 2019). Forest plants provide additional sources of nectar and pollen that contribute to these systems. The forests provide places for the some of the pollinators to breed and to live during the earlier parts of their lives.
* Conservation of biodiversity. Trees and other forest plants capture solar radiation and make its energy available for the use of other organisms. The diverse forest flora supports a diverse forest fauna and the complex forest structure provides a wide range of physical habitats. Natural forest is the most biologically diverse type of ecosystem in Uganda.

The passing on to future generations of a world as diverse in natural wealth as the one that we ourselves have inherited will require widespread support across society. This responsibility is already recognized by many faiths, including the principal religions found in Uganda. Some political leaders in Uganda urge the planting of trees (we urge specific mention of indigenous species). However, there are others who continue to propose the replacement of substantial areas of indigenous forest with sugarcane, oil palm and other cash crops. Announcement of such schemes has sometimes resulted in public protest and expressions of concern by conservation groups (Birdlife International 2008; Nakkazi 2011; Tenywa 2005, 2013; van Schaik and Tickell 2015; Veit 2010).

Conservation Status of the Species

Three evaluations are available showing the conservation status of tree species indigenous to Uganda's forests. All refer to the same categories of threat (Table 4.2) and use the same criteria for assigning species to them (IUCN 2012, 2019). Two are for global conservation status (not necessarily the same as national conservation status for species that occur in other countries), one of them covering all 452 forest species known to occur in the forests at the time (Kalema and Beentje 2012) and the other, based on an on-going process, has so far covered 172 (IUCN 2019). The third evaluation referred to Uganda only, considered in geographical isolation (WCS 2016). WCS evaluated 42 species of forest trees, apparently pre-selected for consideration by an assembled group of experts. This evaluation made use of only three assessment categories (Vulnerable, Endangered, Critically Endangered), but helpfully provided the principal types of threat facing many of the species. The three sources are referred to below as TOU (Kalema and Beentje 2012), IUCN and WCS.

The total size of the indigenous forest tree flora of Uganda is 451 species by our own count. The slight difference with TOU is mainly due to a difference in assignment to life form of species close to the tree/shrub borderline. A total of 54 species of forest trees were identified as threatened taking all three evaluations together, but only three of these were common to all three lists, namely *Diospyros katendei*, *Encephalartos whitelockii* and *Uvariodendron magnificum*. The seven other species that were evaluated as globally threatened by TOU were *Cnestis mildbraedii*, *Desplatsia mildbraedii*, *Dicranolepis incisa*, *Ficus katendei*, *Idertia mildbraedii*, *Pandanus chiliocarpus* and *Vepris eggelingii*. All these are species with small total range or population sizes, and with their entire global distributions confined, or almost confined, to Uganda. Many of the species on the IUCN and WCS lists, but not on that of TOU, are species in high commercial demand for timber or harvested for medicinal properties (Galabuzi *et al.* 2015).

Regardless of what is happening globally, there is no doubt that many species of forest tree are in grave danger of extinction at Ugandan national level (Darbyshire *et al.* 2017; Deb *et al.* 2018; Kalema 2006; Kalema *et al.* 2010; Tumushabe and Mugyenyi 2017). TOU and WCS provide the following summary statements, clearly signalling great concern:

> *"... many species [of trees] are under threat in Uganda through habitat loss or habitat erosion, over-harvesting, and through other reasons."* (TOU, page 4);

> *"Very few of the plant species considered could be given a near threatened or least concern status. Forest or woodland clearance, degradation and grazing combined with the rate of extraction for many of the plant species, particularly those harvested for timber, medicinal use, building or other materials and as a source for fuel, means that many of the species are under threat. With the longevity of tree species, this puts the rate of loss over 3 generations of almost all species at 30% which might otherwise be considered widespread."* (WCS, page 62).

Part 2

How to Identify the Trees

How the Species are Classified and Named

Most species in the present work (numbers 8-447) are flowering plants (Angiosperms). For their classification and naming we follow the recommendations of the Angiosperm Phylogeny group (APG IV 2016), augmented by information in the African Plants Database (2019), the Flora of Tropical East Africa (1952-2012), Flora of Ethiopia and Eritrea (1980-2009), Flore du Congo, du Rwanda et du Burundi (1958-), the Flora of West Tropical Africa (1954-1963), Flora Zambesiaca (1960-), POWO (2019), TROPICOS (2019) and Clark *et al.* (2015). Explanations of botanical and other technical terms follow Harris and Harris (1994), Beentje and Cheek (2003) and Beentje (2010). The descriptions of some of the tree characters were obtained from Beentje (1994), Tor-Anyiin and Yakumbur (2012), Omino (1996), Okai (2012) and Christopher *et al.* (2009).

Identification Tools Provided

These include two keys in Part 3 (one for all species and the other for tall trees on which the leaves cannot be clearly seen), descriptions and illustrations of the species in Part 4 (plus additional keys, mainly for families and genera), a table showing the logic used for ordering the species (Table 4.1) and lists of vernacular names by language (Part 9). Construction of the keys has benefitted from consultation with Hamilton (1981) and Dawkins (1951), the latter the first vegetative key constructed for the identification of Ugandan forest trees (using a graphic form of presentation).

The keys and descriptions (Parts 3 and 4) are based on the typical features of mature trees. Young, damaged and heavily shaded individuals can differ in several ways. 'Young' in this context refers to small individuals that retain juvenile features. Note that small trees may not necessarily be young in calendar years, since young plants of many canopy species can linger under a heavily shading forest canopy for a long time. It is only when light breaks through, for instance following a large treefall, that they are able to accelerate their growth and development.

Many terms used to describe the appearance and structures of plants have proved their worth over the course of time. However, nature is diverse and the vegetative parts of plants are particularly liable to variation. The available vocabulary is inevitably not always completely adequate for the task of describing the huge amount of variety to be found in nature. Users should bear this in mind when comparing the specimens that they have before them in the forest with the descriptions and keys.

Recognize the Value of Local Knowledge

Local residents can be very knowledgeable about their local forests and trees, as well as the local cultural, socio-economic and political systems that influence them. People belonging to social groups that have long been present in an area or whose livelihoods are closely bound to the local natural world are liable to be particularly knowledgeable. Communicating with local residents about forests and trees arms researchers with additional knowledge and, more than that, can provide them with opportunities to establish personal relationships. These could serve them well should they become involved in practical attempts to improve the ways that the forests and trees are managed.

Know the Geographical Position, Name and Status of the Locality

Knowing where a locality is geographically, its altitude and its name facilitates comparisons with existing information. The names of some of the major forests are shown in Fig. 1.4, the divisions of the country traditionally used for botanical recording in Fig. 4.1 and the home areas of some of the indigenous languages in Fig. 7.1. The precise position of a locality in terms of longitude and latitude is important to note in botanical recording. Also, useful to know are who owns or has rights over the land and its resources, for instance whether the locality lies within a Forest Reserve, and who is responsible for its management.

Observe the Habitat

Aspects to note include the substrate and position along the catena, whether there are rocky outcrops nearby, whether a tree of interest is on or near a forest edge, and whether there are any signs of human or animal influence. How mature is the forest? Does the forest in the vicinity of the tree look as if it has regrown after a major disturbance event, such as a phase of intensive logging? What stage is the tree and its immediate surroundings in, in terms of the natural life stages of trees and the cycles associated with forest dynamics?

Observe the Tree

Take time to examine the specimen carefully. Is it a young or mature tree, or best classified as a sapling? Look at nearby trees and seedlings, since they may belong to the same species and give extra clues about its identity. Examine the litter on the forest floor, since it may contain leaves or other parts that have fallen or been broken off the tree. Leaves provide so many characters valuable for identification that every effort should be made to study them. Binoculars are useful for examining high-up foliage and a catapult or thrown stick for felling leaves to scrutinize. A long-forked pole can sometimes be used to twist off low-hanging branchlets to study them in the hand.

Consider Stratification

Species of trees naturally tend to reach maturity at different heights above ground level and are often described (for example) as canopy, second storey or understorey trees. Between two and five tree strata can commonly be visually discerned in many forests, but they frequently merge into one another and cannot be distinctly separated. Stratification of individuals, as actually seen in a forest, is different from stratification of species, since all plants must pass through smaller stages to reach the layers in which they reach maturity. To be able to accurately assess to which stratum a tree is liable to belong at maturity is a useful skill for the field worker.

Some Features of Trees

Refer to the Glossary (Part 5) for technical terms and Fig. 5.1 for illustrations of some leaf and lamina types.

Growth Habit

Some species of tree can transition into or out of other plant growth habits, such as shrubs, scramblers, climbers, stranglers and epiphytes. Some species of *Ficus* and *Schefflera* start life as epiphytes, then become stranglers and finally end up as free-standing trees. *Connarus longistipitatus* and *Uvaria angolensis*, which are sometimes found as small trees, can also be lofty climbers. Some species are sometimes found dwarfed, much smaller than normal, but flowering and fruiting just as they do normally when much taller.

Height

The height of tall trees can be accurately measured by triangulation, provided that their tops are clearly seen and the necessary equipment is available. Otherwise, one way of estimating the height of a tree is to stand back at a distance from its trunk and imagine the number of units of known length (for instance the height of a person) that can be stacked up on top of one another to reach the top of the crown.

Shape

This is a composite feature, referring to size, properties of the trunk, branches, branchlets and branching, the position and shape of the crown, and other characteristics. Trees that have been growing in more open conditions can differ significantly in shape from those that have always lived in dense forest.

Trunk

Features to note include its angle with reference to the vertical, the length of bole (the lower part with no branches), its shape in cross-section (round, fluted, etc.), whether there are spines or conical woody bosses, and the nature of the base. There may be buttresses, roots spurs, stilt roots or aerial roots of other kinds.

Some species appear always to have buttresses, even when quite young (e.g. *Newtonia buchananii*), and others typically only when more mature. *Uapaca paludosa* seems always to have stilt roots from a very early age, while *Musanga cecropioides* seems to only produce them when growing on damp land.

Bark

Features to note include colour, thickness, roughness, the method of accommodating expansion of the trunk (e.g. fissuring, exfoliation) and the presence of lenticels or ring marks. It is sometimes advisable to check the colour of the bark by shallowly scraping it with a knife, in so doing revealing its true colour beneath dirt and epiphytes. The bark of young trees is often thinner, smoother and lighter coloured than that of adults.

Slash

Slash refers to the appearance of that part of a trunk's interior immediately under the bark. It is revealed by making a shallow cut, only a small incision using a pocket knife being normally needed. Avoid using a machete so as to minimize damage to the tree. A cut angle of 30° from the vertical is usually satisfactory, but a shallower cut may be needed to clearly reveal patterns of fibres, for instance the reticulate patterning typical of some species of Malvaceae. The main features to note are colour, change in colour on exposure, layering (including whether a prominent phellogen layer is present very immediately under the bark), texture (such as fibrous and granular), exudate produced (if any) and its type (sap, latex or resin), quantity, colour and change in colour on exposure, and smell.

The slashes of young trees, and on buttresses or parts of trunks that have been damaged can be abnormal. The slashes of young trees are often less intensively coloured and have reduced quantities of exudate compared with older individuals.

Leaves

Several features of leaves can differ on young trees compared with those that are older, such as their arrangement, type, size and hairiness. Shaded leaves can differ in their properties from those that are exposed to full sun. Cotyledons are generally quite different from the leaves that follow later and the first few leaves after the cotyledons can also be distinctive.

Arrangement

This refers to whether the leaves are alternate or opposite and, if opposite, in pairs or whorls. Alternate leaves can be arranged in different ways along the stem, for instance they may be spirally arranged or clustered. Some species have more than one type of leaf arrangement, even on single mature trees. For instance, *Elaeodendron buchananii*, which normally has opposite leaves, can bear some leaves alternately.

Type

This refers to whether the leaves are simple or compound and, if compound, the type of compound (trifoliolate, paripinnate, etc. – see Fig. 5.1). It can sometimes be difficult with some species to determine whether the leaf is simple or compound, for instance with *Paropsia guineensis* and *Phyllanthus inflatus* (both of which have simple leaves, but which at first sight may appear to be pinnate). The best guide is provided by noting where the buds are placed, since these are only found in the axils of leaves and not in those of leaflets. Some species can have more than one type of leaf. For instance, *Ritchiea albersii* can bear simple and 2 to 5-foliolate leaves all on the same tree.

Size

The sizes of leaves and leaflets are usually given here as length x breadth (e.g. 15 x 5 cm). Length for a simple leaf is the distance between the base of the petiole and the tip of the lamina and, for a compound leaf, the distance between the base of the petiole and the tip of the most distant leaflet. These measurements refer to typical leaves, which is a more useful point of reference for field workers in most cases than being provided with the full range of known sizes. Users of this field guide should bear in mind that the sizes given provide only rough guides. Species vary in the amount of variation that they show in leaf size. There is a tendency for the leaves of many understorey species adapted to deep shade to be particularly constant, showing little variation.

Leaves of abnormal size are found particularly on young trees, epicormic shoots and on branchlets bearing flowers. Young plants often have larger leaves than those that are older. This is markedly the case with the saplings of light-requiring canopy species, such as species of *Entandrophragma*. Leaves fully exposed to the sun tend to be smaller than those that are shaded.

Shape

Some of the main terms used for the shapes of leaves are illustrated on Fig. 5.1. Take note of the apices and bases of leaves, as they often have characters useful for field identification.

Venation

This refers to the arrangement of the veins on the lamina. Aspects of interest include the degree of distinctiveness of the veins (which may be different on the two surfaces), the number, positions and shapes of the main lateral veins on either side of the midrib, and the colour and hairiness of the veins.

Colour

Leaves are usually green on both surfaces, though sometimes green of different shades. The undersurface is sometimes other coloured, such as yellow or brown, related to the presence of scales or hairs. Many members of the Annonaceae have a glaucous sheen, blushed with a greyish-green or blue colour. Sometimes young or old leaves are differently coloured, for instance in the case of the young leaves of *Carapa grandiflora* and the old leaves of *Shirakiopsis elliptica*.

Margin

The terms entire and non-entire are used here to describe the margins of laminas that are straight (smooth, uninterrupted) versus those that are toothed, crenate or otherwise broken or irregular. Leaves with undulating (wavy) margins are regarded as entire. Some species have variable margins. For instance, the margins of many leaves of *Ilex mitis* are entire, but there seem to be always some leaves bearing one or a few sharp teeth.

Hairs and scales

Some leaves lack hairs or scales, while others carry them, at least on occasion or in some places. Hairs may be confined only to the undersurface of the leaf, or to its veins, or to the axils between the main veins and the midrib. The leaves of some species are hairier when young or when exposed to full sun, or when the plants are growing at higher altitudes. Stellate (star-shaped) hairs are found in some species.

Petiole

Some of the points to note are length, whether grooved on the upper surface, and the presence of glands.

Stipules

Points to note are whether or not there are stipules, their shape, whether they are persistent or deciduous, and where they are placed in relation to the petioles.

Reproductive Parts

Flowers, fruits and seeds are generally only mentioned in the keys and descriptions for species on which they are frequently seen in the forest or particularly conspicuous. Their positions on the trees can be characteristic. For example, some species of *Ficus* bear their fruits (figs) on their trunks, while with others they on branchlets, as is normal with trees.

Confirmation of Identification

Herbaria, such as that of Makerere University, Kampala, are places where preserved specimens of plants are studied, classified and stored. It may be possible to visit herbaria to compare specimens of unknown plants with existing collections.

Contributing to Scientific Knowledge

Records of species, accurately identified, are useful to report, since this contributes to the sum total of information about them. It may assist in conservation and development efforts. Herbarium workers and other botanical scientists should be able to inform you about how this can be done. Also, contact the National Biodiversity Data Bank (Part 6).

Part 3

Keys to the species

What is a Key?

Keys in this field guide are lists of numbered points, users being required to choose between alternatives as they progress through the options at each stage. The starting point is the first set of options, numbered 1. In most cases there are two alternatives (couplets), so these are mainly dichotomous keys. Where there is more than one option, this is clearly signalled. Once the name of a species, genus, family or other set of trees is reached, then the descriptions and illustrations of the species given should be studied to see if they match the specimen at hand. Note that the keys and descriptions are based on the typical features of mature individuals.

The numbers given are those of species in Part 4 or (for species not allocated their own numbers in Part 4) those of the species in which they are mentioned or of the next numbered species.

Two keys are provided in this section, a Main Key (with 12 Sub-Keys, each containing a group of species sharing certain characteristics) and a Special Key for Tall Trees (those over about 20 m tall) and on which the leaves cannot be clearly seen. Other keys, mainly for families and genera, are embedded within the descriptions in Part 4 (see Table 4.1 for their locations).

Main Key to the Species (with 12 Sub-Keys)

1. Unusual-looking tree (refer to Table 4.1 and Plate 1 for the types of plants included). 2
   Normal-looking tree. 3
2. CHOOSE FROM ONE OF THE THESE 10 OPTIONS
   Tree fern (see illustration). ....**1-3. *Cyathea***
   Conifer. Leaves less than 1.5 cm broad. Above 2000 m altitude, except for near Sango Bay, Masaka, in the case of *Afrocarpus* and *Podocarpus*. ....**4-7. Conifers**
   Cycad (short stocky palm-like plant with huge spiny leaves). .....**7a. *Encephalartos***
   Palm. ....**8-10. Arecaceae**
   Screw palm (with simple, spiny, very large leaves borne spirally). .....**11. *Pandanus***
   Dracaena (little branched trees, with long, strap-shaped, leaves). ...**12-14. *Dracaena***
   Bamboo. ....**17-18. Poaceae**
   Giant groundsel; leaves over 30 cm long; above 2750 m altitude. ....**19-21. *Dendrosenecio***
   Ericaceous tree; leaves very small (under 1 cm long); above 2000 m altitude. ....**24. *Seriphium*; 25-29. *Erica***
   Tree a succulent (with thick fleshy leaves) and abundant white latex. ....**31-34. *Euphorbia***

3. Slash with white, off-white, yellow, orange or red latex or with conspicuously coloured (e.g. red) sap. The latex or sap may be produced in very small quantities, so allow time for it to appear. Latex will generally be present in young stems as well as in the slash, though not all trees with latex in branchlets also have latex in the slash. ....**Sub-Key 1**
Slash lacking latex or conspicuously coloured sap. Some species included in this part of the couplet produce large quantities of yellow or brown watery sap. 4

4. Leaves simple. 5
Leaves compound (or simple and very deeply palmately lobed). 18

5. Leaves alternate. 6
Leaves opposite or whorled. ....**Sub-Key 10**

6. Spines present on branchlets and/or trunk. (Note: all species that have spines on branchlets, but not on the trunk, are small to medium-sized trees). .....**Sub-Key 2**
Thorns absent. 7

7. Leaves with more than 3 main veins (including the midrib) from, or from near to, the base of the lamina. ....**Sub-Key 3**
Not as above. 8

8. Leaves with prominent, upwards extending, lateral veins from, or from near to, the base of the lamina and reaching more than a quarter of the way up the lamina. ....**Sub-Key 4**
Not as above. 9

9. Underside of lamina white, yellow or brown (i.e. not, or only slightly, green) due to a covering of hairs or scales. ....**Sub-Key 5**
Lamina about the same colour above and below, or of different shades of green. ...10

10. Main lateral leaf veins parallel to one another and continuing to the margin where they fuse with a marginal vein (see Plate 19). Small trees. ....**243-244. *Bridelia***
Not as above. 11

11. Leaf margin non-entire. ....**Sub-Key 6**
Leaf margin entire (leaves with wavy margins are included here). 12

12. Main lateral leaf veins prominent and easy to count and fewer than 13 on each side of the midrib. ....**Sub-Key 7**
Main lateral leaf veins not prominent, not easily distinguishable from secondary lateral veins or, if prominent, more than 12 on each side of the midrib. 13

13. Lamina on vegetative shoots normally over 22 cm long. 14
Lamina on vegetative shoots normally less than 22 cm long. 17

14. Petiole normally over 4 cm long. ....**240-240a. *Uapaca*; 241. *Spondianthus***
Petiole normally less than 4 cm long. 15

15. Young shoots and leaves covered with orange-brown hairs. ....**202. *Pycnanthus***
Not as above. 16

16. Branchlets hollow. A tree to 8 m. ....**236. *Barteria***
Branchlets not hollow. ....**219. *Monodora myristica*; 221. *Uvariodendron***

17. Lamina clearly broadest in upper half. ....**Sub-Key 8**
Lamina only indistinctly broadest in upper half, or broadest in lower half or near the centre, or margins of leaf more or less parallel. ....**Sub-Key 9**

18. CHOOSE FROM ONE OF THE THESE 3 OPTIONS
Leaves bifoliolate, trifoliolate, or digitate, or else simple and very deeply palmately lobed. ....**Sub-Key 11**
Leaves pinnate ....**Sub-Key 12**
Leaves bipinnate. ....**Fabaceae (see key before 421)**

Sub-Key 1

Slash with white, off-white, yellow, orange or red latex, or with conspicuously coloured (e.g. red) sap.

1. Exudate white or off-white in colour. 2
Exudate some other colour apart from white or off-white. 8

2. Plant a succulent. Thorns present. ....**31-33. *Euphorbia***
Plant not a succulent. Thorns absent. 3

3. Leaves pinnate. ....**392-393. *Trichilia dregeana*, *T. martineaui***
Leaves simple. 4

4. Leaves opposite or whorled. Small trees, except for *Alstonia*, *Funtumia* and *Rauvolfia caffra*. ....**Apocynaceae (see key before 267)**
Leaves alternate. 5

5. Forest-edge tree with old leaves turning red. Bark very rough, fibrous. Slash yellow, fibrous. This species only rarely exudes latex from the slash. ....**179. *Shirakiopsis elliptica***
Old leaves not turning red or, if so, then bark not very rough and fibrous. 6

6. Latex discolouring within about 15 seconds. ....**Moraceae (see key before 56)**
Latex not discolouring within about 15 seconds. 7

7. THE FOLLOWING TWO FAMILIES MAY NOT BE EASY TO DISTINGUISH FROM ONE ANOTHER. IF IN DOUBT, RUN THROUGH THE KEYS FOR BOTH FAMILIES
Latex usually not abundant. Bark smooth or, more often, moderately to very rough. Trunk often fluted or buttressed. Leaves sometimes markedly different in colour on the two surfaces. Stipules absent or, if present, then not folded around the young leaves. ....**Sapotaceae (see key before 35)**
Latex usually abundant. Bark thin and smooth (except in old *Milicia* and *Ficus vallis-choudae*), sometimes with prominent lenticels. Aerial roots sometimes present. Buttresses often present. Leaves of about the same colour above and below. Stipules claw-shaped, folded around the young leaves. ....**Moraceae (see key before 56)**

8. Exudate yellow or orange. 9
Exudate some other colour. 13

9. Leaves pinnate. ....**430-432. *Millettia***
Leaves simple. 10

10. Leaves opposite. ....**279-283. Clusiaceae, Hypericaceae, Calophyllaceae**
Leaves alternate. 11

11. Leaves not toothed. ....***Ficus* (see key before 56)**
Leaves toothed. 12

12. Leaves coarsely toothed. ....**164. *Maesa***
Leaves with many small teeth. ....**187. *Ficalhoa***

13. Exudate red. 14
Exudate another colour apart from red. 18

14. CHOOSE FROM ONE OF THESE 4 OPTIONS
Leaves simple. 15
Leaves pinnate. ....**385-386. *Pseudospondias, Trichoscypha***
Leaves digitate. ....**362. *Ricinodendron***
Leaves bipinnate. Leaflets very small. ....**435. *Newtonia***

15. Leaves with more than 3 main veins from base of lamina.
....**94. *Macaranga schweinfurthii***
Leaves not as above. 16

16. Trunk straight with branches at right angles. ....**202-203. *Pycnanthus, Staudtia***
Branches not obviously at right angles. 17

17. Bark thick and verticularly fissured (particularly on older trees). **241. *Spondianthus***
Bark thin, flaking. Only known from Ishasha Gorge. ....**124. *Strombosiopsis***

18. Slash smelling strongly of incense. ....**389. *Canarium***
Slash not as above. 19

19. Slash slowly exudes pale yellow-brown resinous droplets.
....**435. *Newtonia*; 440. *Parkia***
Not as above. 20

20. Exudate of large quantities of brown to yellow-brown sap.
....**73. *Ficus ingens*; 242. *Tetrorchidium***
Exudate brown and produced in small quantities (especially from small broken stems and petioles). ....**164. *Maesa***

Sub-Key 2

Leaves simple, alternate; trees with spines; no latex.

1. Leaves gland-dotted. Small tree to 7 m on forest edges. ....**167. *Aeglopsis***
Leaves not gland-dotted. 2

2. Leaves small (c. 7 x 3.5 cm or less), normally distinctly obovate. ……………………………………………………………**162. *Maytenus heterophylla***
   Leaves not both small and markedly obovate. ……………………………………………3

3. Leaves usually with a well-marked mucronate tip. Leaf margin entire. A small tree. ……………………………………………………………**252. *Chaetachme***
   Leaves not mucronate or, if so, then mucronate tip very small and leaf margin not entire. ……………………………………………………………4

4. Main lateral veins parallel to each other and continuing to the margin where they fuse with a marginal vein. Small trees. ……………………………**243-244b. *Bridelia***
   Not as above. ……………………………………………………………5

5. Petiole 2.5 cm long or less. ……………………………………………………6
   Petiole normally over 2.5 cm long. If about 2.5 cm long, then undersurface of lamina yellowish. ……………………………………………………………7

6. Very large tree with thick leathery leaves, with numerous narrow lateral veins. ……………………………………………………………**209. *Klainedoxa***
   Venation not as above. Trees generally less than 20 m tall. Note: *Phyllanthus inflatus* (248) may key out here. ……**Salicaceae, Achariaceae, etc. (see key before 132)**

7. Leaves normally 3-lobed. ……………**94-95. *Macaranga schweinfurthii, M. angolensis***
   Leaves not normally 3-lobed. ……………………………………………………8

8. Leaf margin with conspicuous teeth or crenations. ……………………………9
   Leaf margin entire or with a few scattered teeth. ……………………………10

9. Leaf base cordate. ……………………………………**102. *Alchornea cordifolia***
   Leaf base not, or only very slightly, cordate. ……………**96. *Macaranga monandra***

10. Petiole about 7 cm long or more. ……………………………………**132. *Caloncoba***
    Petiole normally less than 7 cm long. ……………………………………………11

11. Leaf base rounded to slightly cordate. ……………………**97. *Macaranga spinosa***
    Leaf base wedge-shaped. ……………………………………**98. *Macaranga barteri***

Sub-Key 3

Leaves simple, alternate, with more than 3 main veins (including the midrib) from, or from near to, the base of the lamina; no latex or spines.

1. At least some leaves conspicuously lobed. ……………………………………2
   Mature leaves not conspicuously lobed. ……………………………………3

2. Slash with a more or less distinct fibrous network (best seen with slashes made at a narrow angle). ……………………………**Malvaceae (see key before 115)**
   Slash without a fibrous network.
   …..**94-95. *Macaranga angolensis, M. schweinfurthii*; 106. *Croton macrostachyus***

3. Leaves often peltate (look very carefully at several leaves). Tree to 20 m, common above 1400 m in wetter montane areas. ....**99. *Macaranga capensis***
Leaves not peltate. 4

4. Leaf margin entire or very nearly so (the margin may be wavy or, rarely, have minute teeth). 5
Leaf margin toothed or crenate. 16

5. At least one of the basal, or near basal (e.g. *Cordia*), main lateral veins reaching half way or more up the lamina. 6
Basal lateral veins reaching less than half way up the lamina. 11

6. Slash white to yellow, rapidly turning dark green with vertical brown lines. ....**112. *Cordia millenii***
Slash not as above. 7

7. Leaves about 1.5 times as long as broad or less. 8
Leaves over 1.5 times as long as broad. 10

8. Slash with some shade of red as the dominant colour. ....**116. *Cola gigantea***
Slash white, yellow or brown. 9

9. Leaf base asymmetric. ....**111. *Alangium***
Leaf base symmetric. ....**100-101. *Neoboutonia*; 115. *Pterygota***

10. Slash with brown or black rings, dots or lines. ....**91. *Celtis adolfi-fridericii***
Slash not as above. ....**93. *Holoptelea*; 111. *Alangium***

11. Slash yellow, not turning rapidly darker, producing little or no sap. A tall tree, often with buttresses. ....**93. *Holoptelea***
Major colour of slash other not yellow or, if yellow, then either slash turning rapidly darker or producing abundant brown sap. 12

12. Tree to 35 m tall with comparatively large leaves (normally over 17 cm long and/or 8 cm wide). 13
Leaves less than 17 x 8 cm. Tree often (but not always) smaller than above. 14

13. Bark flaking in large pieces. ....**123. *Strombosia***
Bark flaking in small pieces or with vertical fissures. ....**118. *Sterculia***

14. Petiole less than 1 cm long. ....**121. *Leptonychia***
Petiole usually over 1 cm long. 15

15. Bark smooth, light-coloured. ....**73. *Ficus ingens*; 97. *Macaranga spinosa***
Bark brown or dark-coloured, fibrous. ....**122. *Nesogordonia***

16. Petiole c. 2 cm long or less (or up to 3 cm long and then covered with red-brown hairs). 17
Petiole usually more than 2 cm long. 21

17. Slash white or off-white, with brown dots or other markings. ....**89. *Celtis africana***
Slash not as above. 18

18. Mature leaves over 15 cm long. ....**127-129a. *Desplatsia, Glyphaea***
Mature leaves normally less than 15 cm long. 19

19. Trunk straight with branches at right angles and curving upwards. ....**92. *Trema***
Shape not as above. 20

20. Young leaves covered with stellate hairs. ....**107. *Croton sylvaticus***
Young stems not covered with stellate hairs. ....**141. *Flacourtia*; 147. *Trimeria***

21. Leaf base usually distinctly cordate. 22
Leaf base not or only slightly cordate. Note: occasionally *Alchornea laxiflora*, *Macaranga monandra* and *M. spinosa* (included on this side of the dichotomy) are slightly cordate. 25

22. Leaves not heart-shaped or rounded. ....**128-129. *Desplatsia***
Leaves more or less heart-shaped or rounded. 23

23. Shrubby species, with spreading main stems. Note: *Acalypha* (94) also keys out here. ....**102. *Alchornea cordifolia*; 105. *Acalypha ornata***
Trees, usually with a single main stem. 24

24. Slash without a fibrous network. Upper surface of leaves covered with small, but conspicuous and regularly arranged, stellate hairs. Main leaf veins not red. ....**100-101. *Neoboutonia*; 106. *Croton macrostachyus*; 107. *Croton sylvaticus***
Slash with a fibrous network (best seen if slashed shallowly). Upper surface of leaves sometimes with stellate hairs, but these not so regularly arranged or as conspicuous as above. Main veins sometimes red. ....**119-120. *Dombeya***

25. Leaves with comparatively few, large teeth (see illustration). ....**96. *Macaranga monandra***
Leaves not as above. 26

26. Slash white to yellow, rapidly turning dark green with brown lines. ....**112. *Cordia millenii***
Slash not as above. 27

27. Mature leaves large (often c. 20 x 7.5 cm). Uncommon small tree recorded from Budongo and Mabira forests. ....**176. *Discoclaoxylon***
Mature leaves smaller than above. 28

28. Green young shoots contrasting sharply in colour with brown older shoots. Small tree to 10 m. ....**103. *Alchornea laxiflora***
Not as above. 29

29. Old leaves turning conspicuously yellow/orange. ....**106. *Croton macrostachyus***
Old leaves not turning conspicuously yellow/orange. 30

30. Slash with some shade of red as the dominant colour. ....**97. *Macaranga spinosa*; 131. *Grewia mildbraedii***
Slash not red. ....**107. *Croton sylvaticus***

Sub-Key 4

Leaves simple, alternate, with prominent, upwards extending, lateral veins from, or from near to, the base of the lamina and reaching more than a quarter of the way up the lamina; no latex, spines or more than 3 main veins (including the midrib) from, or from near to, the base of the lamina.

1. Leaves very rough (like sandpaper) on both surfaces. ....**63. *Ficus exasperata*; 64. *F. gnaphalocarpa*; 68. *F. asperifolia***
 Leaves not as above. 2

2. At least one of the basal lateral veins extending more than 3/4 way up the lamina. 3
 Venation not as above. 6

3. Slash yellow or white, with brown or black rings, dots or other markings. ....***Celtis* (see key before 86)**
 Slash not as above. 4

4. Lamina comparatively small and comparatively wide in proportion to length (c. 7 x 4.5 cm). Tall tree, only known from Budongo Forest. ....**110. *Discoglypremna***
 Not as above. Lamina often over twice as long as wide. 5

5. Leaf venation characteristic (see Plate 8). ....**90. *Celtis philippensis***
 Leaf venation not as above. ....**127-131d. *Desplatsia, Glyphaea, Grewia***

6. Leaves heart-shaped. 7
 Leaves not heart-shaped. 15

7. Slash soft, white, yellow or yellow-brown (sometimes with darker markings), very rapidly turning darker. Note: the slashes of *Leptonychia* and some Malvaceae also turn rapidly darker, but they are not particularly soft and are not included in this part of the dichotomy. 8
 Slash not as above. 9

8. Trunk with branches at right angles and curving upwards. .... **92. *Trema***
 Shape not as above. ....**112-114. *Cordia, Ehretia***

9, Slash whitish, with brown spots or other brown markings. ....**89. *Celtis africana***
 Slash not as above. 10

10. Leaves often peltate (look carefully at several leaves). Tree to 20 m, common above 1400 m in wetter areas. ....**99. *Macaranga capensis***
 No leaves peltate. 11

11. Lamina with comparatively few, large and widely-spaced teeth (see illustration). ....**96. *Macaranga monandra***
 Not as above. 12

12. Lamina and young parts more or less glabrous. ....**102-103. *Alchornea cordifolia, A. laxiflora***
 Lamina and/or young parts hairy. 13

13. Trunk straight, branches at right angles and curving upwards. .... **92. *Trema***
Shape not as above. 14

14. Shrub with spreading stems. ....**102. *Alchornea cordifolia***
Tree, usually with a single straight trunk. ....**107. *Croton sylvaticus***

15. Slash quite hard, with brown or black rings or dots on a white or yellow background. ....***Celtis* (see key before 86)**
Slash not as above. If brown or black markings present, then slash soft. 16

16. Slash very soft, white, yellow or yellow-brown, sometimes with darker markings, very rapidly turning darker. Note: the slashes of *Leptonychia* and some Malvaceae also turn rapidly darker, but they are not particularly soft and are not included in this part of the dichotomy. 17
Slash not as above. 18

17 Trunk straight, branches at right angles and curving upwards. ....**92. *Trema***
Shape not as above. ....**112-114. *Cordia, Ehretia***

18. Leaf margin with easily distinguishable teeth or crenations. 19
Leaf margin not as above (the margin may be wavy). 29

19. Leaves with relatively few teeth which are large, regularly arranged and spaced well apart (see Plate 9). ....**96 *Macaranga monandra***
Not as above. 20

20. CHOOSE FROM ONE OF THESE 3 OPTIONS
Leaves commonly over 20 cm long. ....**127-129. *Desplatsia, Glyphaea***
Leaves small, commonly under 6 cm long. 21
Leaves commonly 6-20 cm long. 22

21. Leaves normally distinctly obovate. ....**162. *Gymnosporia heterophylla***
Leaves not obovate. ....**160. *Mystroxylon*; 163. *Maytenus acuminata***

22. Petiole often 3 cm long or more. 23
Petiole less than 3 cm long. 24

23. Base of leaf wedge-shaped. ....**98. *Macaranga barteri***
Base of leaf not wedge-shaped. ....**103. *Alchornea laxiflora*; 107. *Croton sylvaticus***

24. Slash smelling of cold cooked chicken. Leaves characteristically distantly toothed (see Plants 16). ....**189. *Maesopsis***
Slash not smelling of cold cooked chicken. 25

25. Leaves normally very hairy on lower surface. ....**92. *Trema*; 130. *Grewia pubescens***
Lamina glabrous below or with a few to a moderate number of hairs. 26

26. Slash smelling strongly of pepper. ....**107. *Croton sylvaticus***
Slash not smelling strongly of pepper. 27

27. Bark smooth and thin. Slash granular orange-brown on a white background, not turning darker. Tall tree to 45 m, only known from Budongo Forest. ....**110. *Discoglypremna***
Not as above. Tree shorter. 28

28. Venation characteristic, the second main lateral veins on each side of the midrib arising some distance above the first (see Plate 12). Slash often turning darker. ....**127-131. *Desplatsia, Glyphaea, Grewia***
Venation not as above. ....**Achariaceae, Salicaceae, etc. (see key before 132)**

29. Petiole often c. 7 cm long or more. ....**132. *Caloncoba***
Petiole shorter than above. 30

30. Mature leaves normally over 17 cm long. ....**123. *Strombosia***
Mature leaves often 15 cm long or less. 31

31. Leaves sessile or almost so. Only recorded from Ishasha Gorge. ....**126. *Brazzeia***
Leaves not sessile or almost sessile. 32

32. At least some leaves with mucronate tips. Small tree, only recorded from Budongo, Kibale and Rwoho forests. ....**251. *Microdesmis***
Leaves not mucronate. 33

33. Leaves comparatively long and thin (often 10-15 x 3-5 cm), with conspicuous acumens. Petiole less than 1.25 cm long. ....**121. *Leptonychia***
Combination of leaf shape and petiole length not as above. 34

34. Base of leaf rounded to slightly cordate. .. **93. *Holoptelea*; 97. *Macaranga spinosa***
Base of leaf wedge-shaped. ....**98. *Macaranga barteri***

Sub-Key 5

Leaves simple, alternate; leaf colour different above and below (which may be white, yellow or brown); no latex, spines or prominent, upwards extending, lateral veins from the base of the lamina.

1. Leaves large (over 18 cm long, sometimes much bigger). ....**170-171. *Vernonia***
Mature leaves less than 18 cm long. 2

2. Slash yellow to white, smelling of pepper. Old leaves turning conspicuously yellow. ....**109. *Croton megalocarpus***
Slash with some shade of red as the dominant colour (at least on larger trees). Old leaves not turning conspicuously yellow. 3

3. Slash fragrant. Leaves with c. 4-9 main lateral veins on each side of the midrib. A rare tree, commonest in Kayonza Forest. ....**205. *Ocotea usambarensis***
Slash not fragrant. 4

4. Tall tree to 45m. Leaves small when mature (c. 6.5 x 2 cm), with c. 20 or more main lateral veins on each side of the midrib. ....**200. *Parinari***
Leaves not as above. Tree smaller. 5

5. Petiole often over 2 cm long. ....**97-98. *Macaranga barteri, M. spinosa***
Petiole less than 2 cm long. 6

6. Young parts covered with brown hairs. ....**235. *Trichocladus***
Young parts not covered with brown hairs. ....**161. *Maytenus undata***

Sub-Key 6

Leaves simple, alternate, margins non-entire (i.e. toothed, crenate or lobed); no latex, spines or prominent veins from the base of the lamina; leaf colour similar above and below.

1. Teeth on leaf margin sharp and stiff; margin therefore spiny (see Plate 14). ....139. ***Rawsonia***; **148.** ***Rinorea ilicifolia***
   Leaf margin not spiny (large teeth may be present). 2
2. Leaves small (usually c. 8 x 3.5 cm or less) and typically distinctly widest in upper half of lamina. 3
   Leaves not both small and widest in upper half. 4
3. Leaves aromatic when crushed. ....**166.** ***Morella kandtiana***
   Leaves not aromatic when crushed. ....**160.** ***Mystroxylon aethiopicum***; **162.** ***Maytenus heterophylla***
4. Leaves small (usually c. 8 x 3.5 cm or less, but sometimes up to 11 x 4.5 cm), without an acumen, base more or less symmetrical, not aromatic when crushed, usually with only a few sharp teeth (some leaves with no teeth). Bark smooth. Slash white or yellow. Mainly a montane forest tree found above 1800 m, with records also from Kibale and Namalala forests. ....**165.** ***Ilex***
   Not as above. 5
5. Leaves small (usually less than 7.5 x 3 cm, but occasionally up to 10 cm long), ovate, base symmetrical, lamina tapering to an acute or acuminate apex. In montane forest above 2000 m. ....**163.** ***Maytenus acuminata***
   Not as above. 6
6. Leaves very hairy below and with a long attenuate base. Mature leaves often c. 20 cm long. A small tree found on Rwenzori and the Bufumbira Volcanoes. ....**171.** ***Vernonia calvoana***
   Not as above. 7
7. Mature leaves c. 20 cm long or longer. 8
   Mature leaves less than 15 cm long. 15
8. Leaves thick and leathery, with numerous very small teeth along margins. ....**153.** ***Campylospermum densiflorum***
   Not as above. 9
9. Leaves very large (c. 60 x 25 cm), very hairy on undersurface. In swamp forest. ....**170.** ***Vernonia conferta***
   Not as above. 10
10. Leaf base conspicuously long attenuate (see Plates 15, 18). Note: the leaves of *Alchornea floribunda* are rounded at the extreme base. 11
    Leaf base various (cuneate to rounded), but not conspicuously long attenuate. 12

11. Branchlets hollow. ....**236. *Barteria***
Branchlets not hollow.
172. *Alchornea floribunda*; 174. *Argomuellera*; 159. *Hugonia* (young plants)

12. Branchlets hollow. ....**236. *Barteria***
Branchlets not hollow. 13

13. Leaves very long (c. 30 x 16 cm). ....**175. *Pseudagrostistachys***
Leaves shorter than above. 14

14. Leaves lanceolate. ....**176. *Discoclaoxylon***
Leaves not lanceolate. ....**152. *Rinorea oblongifolia***

15. Small tree (to 15 m in montane forest, but shorter at lower altitudes). Broken petioles exude brown exudate in small quantities (wait and look very carefully).
....**164. *Maesa***
Broken petioles not exuding brown exudate. 16

16. Leaves with tufts of hairs on margin. Much-branched tree to 20 m, only recorded from Ishasha Gorge. Note: also check *Paropsia* (186). ....**177. *Maesobotrya***
Leaves lacking marginal tufts of hairs. 17

17. Petiole absent. Tree to 10 m, found on the upper slopes of Mt Elgon and possibly also of the Bufumbira Volcanoes. Leaves often c. 12 x 2.5 cm. ....**169. *Conyza***
Petiole present. 18

18. Mature petioles over 5 cm long and leaves with relatively few, large teeth which are regularly arranged and spaced well apart (see Plate 9). **96. *Macaranga monandra***
Petioles either less than 5 cm long or, if longer, then leaves not toothed as above. 19

19. Leaf base long attenuate (see Plate 15). ....**173. *Alchornea hirtella***
Leaf base either not long attenuate of, if so, then markedly less so than above. 20

20. Mature petioles often c. 7 cm long or more. Leaf venation arcuate, with c. 4-7 main lateral veins on each side of the midrib. Teeth on leaf margins normally small and irregularly spaced. ....**132. *Caloncoba***
Petioles less than 7 cm long or, if as long or longer, then either venation not as above or margin regularly toothed or crenate. 21

21. Lamina with more or less regularly arranged teeth or crenations for all or most of its length. 22
Lamina not regularly toothed or crenate as above; often either with irregularly arranged teeth or crenations or, if regularly toothed or crenate, then only toothed or crenate in the upper half of the lamina. 29

22. Forest-edge tree with old leaves turning red. White latex present in branchlets. Slash yellow. ....**179. *Shirakiopsis***
Not as above. 23

23. Large tree to 40 m with rough, dark-coloured, shaggy bark and a reddish slash that turns darker. ....**199. *Prunus***
Not as above. 24

24. Slash smelling of cold cooked chicken. Leaves characteristically distantly toothed (Plate 16). ....**189. *Maesopsis***
Not as above. 25

25. Main lateral veins 12 or more on each side of the midrib. Note: these are not easily distinguished from the secondary lateral veins in some species. 26
Main lateral veins fewer than 12 on each side of the midrib. 27

26. Leaves thick and leathery, with numerous lateral veins. Only recorded from Kigezi at c. 2000 m. ....**188. *Balthasaria***
Not as above. Note: *Hugonia* (159) may key out here.
....**Ochnaceae (see key before 153)**

27. Slash with concentric layers of white to red and yellow. A tree to 30 m. Branchlets simulating pinnate leaves. ....**186. *Paropsia***
Slash and branchlets not as above. 28

28. Leaf base asymmetric. ....**181-185. *Drypetes, Suregada***
Leaf base symmetric or almost so. **Achariaceae, Salicaceae, etc. (see key before 132)**

29. Leaves with translucent dots or lines (seen when held up to the light).
....**207-208. *Casearia***
Not as above. 30

30. Leaves aromatic when crushed. Tree in montane forest above c. 2000 m or in swamps at low as well as high altitudes. ....**166, 258. *Morella***
Leaves not aromatic when crushed. 31

31. Leaf base asymmetric. ....**181-185. *Drypetes, Suregada*; 251. *Microdesmis***
Leaf base symmetric or almost so. 32

32. Leaves yellowish below (at least when young). Leaves up to c. 15 cm long and petiole often c. 1.5-5 cm long. ....**97-98. *Macaranga barteri, M. spinosa***
Leaves not yellowish below. Petiole either less than 2.5 cm long or, if longer, then leaves generally over 15 cm long.
......................................**Achariaceae, Salicaceae, etc. (see key before 132)**

Sub-Key 7

Leaves simple, alternate, margins entire; main lateral veins prominent, easy to count and fewer than 13 on each side of the midrib; no latex, spines, or prominent, upwards extending, lateral veins from the base of the lamina; leaves similar in colour above and below.

Note: *Drypetes bipindensis* (185), *Ilex mitis* (165), *Mystroxylon aethiopicum* (160) and *Rinorea* (148-152a) are not included in the key. Although some leaves of these species are entire, it is common for many to be non-entire.

1. Stipules much divided into thin segments. An understorey tree to 10 m.
....**245. *Antidesma laciniatum***
Stipules, if present, not as above. 2

2. Mature leaves over 18 cm long. 3
Mature leaves less than 18 cm long. 12

3. Stilt roots usually present. Common trees on the Ssese Islands and in lake-shore swamp forest. ....**240-241. *Spondianthus, Uapaca***
Stilt roots absent. 4

4. Slash with some shade of red as the dominant colour. 5
Slash with some other colour, apart from red, as the dominant colour. 7

5. Petiole normally over 5 cm long. ....**241. *Spondianthus***
Petiole less than 5 cm long. 6

6. Leaves hairy below, at least on the midrib. ....**245-246a. *Antidesma***
Leaves glabrous, or almost so, below. ....**123. *Strombosia***

7. Slash very soft, white to yellow, sometimes with brown spots or lines, very rapidly turning darker. ....**114. *Ehretia***
Slash not as above. 8

8. Mature leaves usually over 8 cm broad. 9
Mature leaves usually less than 8 cm broad. 10

9. Petiole purplish. ....**219. *Monodora myristica***
Petiole not purplish. ....**117. *Cola congolana***

10. Mature leaves noticeably hairy beneath. ....**226. *Turraea floribunda***
Mature leaves glabrous or almost so beneath. 11

11. Leaves widest in upper half. Small forest-edge species. **255. *Pittosporum viridiflorum***
Leaves long and thin. In maturer forest types than above. ...**213. *Greenwayodendron***

12. Petiole markedly swollen at base and apex, quite long (1-4 cm). Small trees. ....**231-232. *Baphia capparidifolia, Baphiopsis***
Petiole, if present, not swollen as above. 13

13. Leaves with tufts of hairs on margin. Much-branched tree to 20 m, only known from Ishasha Gorge. ....**177. *Maesobotrya***
Leaves without marginal tufts of hairs. 14

14. Petiole often over 3 cm long. 15
Petiole less than 3 cm long. 17

15. Midrib and main veins red as seen on undersurface. ....**237. *Apodytes***
Midrib and main veins not red. 16

16. Leaves yellowish below, at least when young (due to a covering of scales). ....**97-98. *Macaranga barteri, M. spinosa***
Leaves not yellowish below. ....**73. *Ficus ingens*; 132. *Caloncoba***

17. Leaves with small translucent dots or lines as seen when held up to the light. Note: this does not include species with leaves having only a translucent marginal strips. Tall trees. ....**207-208. *Casearia***
Not as above. 18

18. Scrape (lightly scrape the bark) bright red, under thin, dark-coloured, smooth bark. Scrape not turning darker. A small tree. ....**233. *Maerua***
Not as above. 19

19 Intrapetiolar stipules present. Understorey tree to 7 m. ....**257. *Erythroxylum***
Intrapetiolar stipules absent. 20

20. Lamina widest in upper half. 21
Leaves either widest in centre or in lower half, or more or less parallel-sided, or only very indistinctly widest in upper half. 40

21. At least some leaves markedly sinuate near apex (see Plate 18). Small understorey tree. ....**226. *Turraea vogelioides***
Leaves not as above. 22

22. Leaf apex rounded to obtusely pointed. 23
Leaf apex acute to acuminate. 25

23. Leaves usually densely hairy below. Main lateral veins very conspicuous on lower surface. ....**227. *Turraea robusta***
Leaves not densely hairy below. 24

24. Slash with a shade of red as the dominant colour. ....**263-263a. *Euclea***
Slash not red. ....**255-255a. *Pittosporum***

25. At least some leaves mucronate. 26
Leaves not mucronate (or only very indistinctly so). 28

26. Leaf axils on lower surface with conspicuous tufts of hairs. A tall tree, only known from Bwamba. ....**122. *Nesogordonia***
Not as above. Leaves sometimes hairy below. 27

27. Leaves hairy below. ....**245-246a. *Antidesma***
Leaves glabrous, or almost so, below. ....**247. *Margaritaria*; 249. *Thecacoris***

28. Leaves with conspicuous tufts of hairs in the axils of the main veins below. 29
Not as above. 30

29. Petiole c. 2 cm long. A tall tree, only known from Bwamba. ....**122. *Nesogordonia***
Petiole c. 1 cm long or less. Widespread species. **226. *Turraea vogelii*; 234. *Tapura***

30. Mature leaves densely hairy below, at least on veins. 31
Mature leaves glabrous below or with comparatively few veins. 33

31. Petiole often over 2 cm long. ....**97. *Macaranga spinosa***
Petiole less than 1.5 cm long. 32

32. Main lateral veins conspicuous, parallel with one another. Understorey or forest-edge trees to 10 m. ....**245-246a. *Antidesma***
Venation not as above. ....**216-218. *Xylopia*; 222-225. *Uvaria***

33. Base of lamina gradually tapering to an acute or cuneate base (see illustrations). ...34
Base of lamina rounded to slightly cuneate, not gradually tapering as above. 35

34. Slash white to orange-brown, often with orange lines or spots. Forest tree with a straight trunk. ....**242. *Tetrorchidium***
Not as above. Small trees, often on forest edges or in secondary forest. ....**255-256a. *Pittosporum, Peddiea***

35. Petiole often over 2 cm long. ....**97. *Macaranga spinosa***
Petiole less than 1.5 cm long. 36

36. Leaves thick and leathery. Tall trees with straight trunks. ....**216-218. *Xylopia***
Leaves not thick and leathery. Small trees, usually trunks not especially straight. ..37

37. Young shoots glabrous or almost so. 38
Young shoots hairy. 39

38. Slash with shade of red as the dominant colour. ....**247. *Margaritaria***
Slash not red. ....**220. *Monodora angolensis*; 222-225. *Uvaria***

39. Young branches covered with light-coloured lenticels. ....**222-225. *Uvaria***
Young branches not covered with light-coloured lenticels. ....**226-229a. *Turraea***

40. Petiole indistinct or absent. Lateral veins c. 3-8 on each side of the midrib, conspicuously arcuate. Small tree to 7 m, only known from Ishasha Gorge. ....**126. *Brazzeia***
Petiole distinct. 41

41. Branchlets simulating pinnate leaves (this is obvious). Tree to 10 m. ....**248. *Phyllanthus inflatus***
Not as above. 42

42. Tree occurring above c. 2000 m, with thick, rough bark and leaves less than 3.5 cm wide, aromatic when crushed. ....**258. *Morella salicifolia***
Combination of characters not as above. 43

43. At least some leaves mucronate. Small to medium-sized trees. 44
Leaves not, or only very indistinctly, mucronate. 47

44. Veins conspicuous on undersurface, parallel to one another, hairy. Small trees. ....**245-246a. *Antidesma***
Not as above. 45

45. Leaf axils on lower surface with conspicuous tufts of hairs. A tall tree, only known from Bwamba. ....**122. *Nesogordonia***
Not as above. 46

46. Leaf margin conspicuously translucent when held up to the light. **247. *Margaritaria***
Leaf margin not conspicuously translucent. ....**249-251. *Thecacoris, Cleistanthus, Microdesmis***

47. Midrib and main veins red as seen on lower surface. ....**237. *Apodytes***
Midrib and main veins not red. 48

48. Slash very soft, white or yellow, sometimes with brown spots or lines, very rapidly turning darker. Leaves quite large, c. 15 x 6.5 cm. ......,.............**114. *Ehretia***
Not as above. If slash white to yellow and turning darker, then slash not particularly soft. 49

49. Axils of veins below with conspicuous tufts of hairs. 50
Axils of veins below lacking conspicuous tufts of hairs. 52

50. Hairs in leaf axils brown. Tree to 30 m, only recorded from Bwamba and Kigezi. ....**122. *Nesogordonia***
Hairs in leaf axils white. Small trees. 51

51. Leaves usually very hairy below. Venation prominent (see Plate 18). ....,....**227. *Turraea robusta***
Not as above. ....**226. *Turraea vogelii*; 234. *Tapura***

52. Petiole often over 2 cm long. ....**97. *Macaranga spinosa***
Petiole less than 1.5 cm long. 53

53. Main lateral veins prominent on undersurface of leaf and more or less parallel to one another (see Plates 18, 19). Leaves very to moderately hairy below. ....**226-229a. *Turraea*; 245-246a. *Antidesma***
Not as above. If veins prominent, then undersurface of leaf more or less glabrous. 54

54. Slash a shade of red as the dominant colour. 55
Slash with some other colour, apart from red, dominant. 60

55. Shoots ending in conspicuous claw-shaped scales. A small tree. ....**256. *Peddiea***
Shoots not ending as above. 56

56. Leaves markedly hairy below, at least on main veins. ....**245-246a. *Antidesma***
Leaves glabrous, or almost so, below. 57

57. Leaf margin conspicuously translucent when held up to the light. **247. *Margaritaria***
Leaf margin not as above. 58

58. Montane forest tree, mainly above 2000 m, known from Bwindi Forest and the Imatongs. Leaves often c. 12 x 5 cm. ....**206. *Ocotea kenyensis***
Not as above. Probably always below 2000 m. 59

59. Leaf apex more or less prominently acuminate. ....**250. *Cleistanthus***
Leaf apex obtuse to obscurely acuminate. ....**204. *Beilschmiedia***

60. Shoots growth peculiar. Each node bears a leaf or leaf-scar, an axillary shoot (that grows onwards), and a 'main shoot' that aborts or terminates immediately in an inflorescence (look very carefully). Understorey trees to 10 m. ....**238-239. *Leptaulus***
Shoot growth normal. 61

61. Relatively tall trees with straight trunks. 62
Understorey or forest-edge species to 15 m, generally with crooked trunks. 67

62. Leaves usually over 2.5 times as long as broad. 63
    Leaves usually less than 2.5 times as long as broad. 64

63. Slash scented. ....**213-214. *Greenwayodendron*; *Cleistopholis***
    Slash not scented. ....**73. *Ficus ingens*; 182-185. *Drypetes*; 211-211a. *Diospyros***

64. Leaf base asymmetric. ....**182-185. *Drypetes***
    Leaf base more or less symmetric. 65

65. Slash exuding large quantities of brown sap. ....**73. *Ficus ingens***
    Slash not producing large quantities of brown sap. 66

66. Slash smelling of mangoes. ....**210. *Irvingia***
    Slash not smelling of mangoes. ....**93. *Holoptelea*; 216-218. *Xylopia***

67. Leaf vein reticulum small and very prominent, being of a different colour to the intervening tissue as seen held up to the light. ....**181. *Suregada***
    Vein reticulum not as above. 68

68. Leaf base asymmetric. ....**182-185. *Drypetes***
    Leaf base symmetric. 69

69. Young shoots hairy. ....**222-225. *Uvaria*; 226-229a. *Turraea***
    Young shoots glabrous or almost so. 70

70. In Karamoja. ....**225. *Uvaria schefflera***
    In parts of Uganda other than Karamoja. 71

71. Leaves usually less than 9.5 cm long. ....**180. *Gymnanthes*; 230. *Baphia wollastonii***
    Leaves usually over 9.5 cm long. 72

72. Leaves leathery. ....**210. *Irvingia***
    Leaves not leathery. ....**212. *Uvariopsis*; 220. *Monodora angolensis***

Sub-Key 8

Leaves simple, alternate, margin entire, broadest in upper half; no latex or spines; main lateral veins either not easily distinguishable from secondary lateral veins or more than 12 on each side of the midrib.

1. Tree found above an altitude of c. 2200 m. Note: *Rapanea* is included on this side of the dichotomy; exceptionally, it grows in lake-shore forests in Masaka. 2
   Trees mostly found below c. 2200 m. 4

2. Leaves with translucent lines. ....**262. *Rapanea***
   Not as above. 3

3. Small tree to 5 m, found in the Ericaceous Belt of Elgon and near the summit of Kadam. Leaves thick and leathery. ....**260. *Protea***
   Not as above. ....**255-255a. *Pittosporum*; 256. *Peddiea***

4. Intrapetiolar stipules present. Small understorey tree. ....**257.** ***Eyrthroxylum***
Intrapetiolar stipules absent. 5

5. Leaf apex rounded. 6
Leaf apex usually acute or acuminate. 7

6. Leaves thick and leathery. Slash a shade of red. ....**263.** ***Euclea***
Leaves comparatively thin. Slash not red. ....**255-255a.** ***Pittosporum***

7. Leaves often c. 9 x 2.5 cm, with very indistinct venation. Leaf apex asymmetric. Tree to 40 m with rough bark and fibrous red slash. ....**201.** ***Warburgia***
Not as above. 8

8. Shoots ending in conspicuous claw-shaped scales. Understory tree to 10 m. ....**256.** ***Peddiea***
Shoots not terminating as above. 9

9. Shoots growth peculiar. Each node bears a leaf or leaf-scar, an axillary shoot (that grows onwards), and a 'main shoot' that aborts or terminates immediately in an inflorescence (look very carefully). Understorey trees to 10 m. ....**238-239.** ***Leptaulus***
Leaves not as above. 10

10. Leaves normally mucronate. ....**249.** ***Thecacoris***
Leaves not mucronate. 11

11. Leaf base long cuneate. ....**255-255a.** ***Pittosporum***
Leaf base various, sometimes cuneate, but not as markedly so as above. ....**215.** ***Isolona***; **220.** ***Monodora angolensis***

Sub-Key 9

Leaves simple, alternate, margins entire, not broadest in upper half; no latex or spines; main lateral veins not prominent, being either not easily distinguable from secondary lateral veins or more than 12 on each side of midrib.

Note: *Ilex mitis* is not included here. Although some leaves of *Ilex* are entire, it is common for others to be non-entire.

1. Leaves with gland dots or lines. ....**201.** ***Warburgia***; **207-208.** ***Casearia***
Leaves without gland dots or lines. 2

2. Intrapetiolar stipules present. Small understorey tree. ....**257.** ***Erythroxylum***
Intrapetiolar stipules absent. 3

3. Branchlets simulating pinnate leaves (this is obvious). A small understory tree. ....**248.** ***Phyllanthus inflatus***
Branchlets not, or only a little, simulating pinnate leaves. 4

4. Leaves often mucronate. Understorey to medium-sized trees, often with low foliage. 5
   Leaves not or only very slightly mucronate. 6

5. Leaf margin translucent when held up to the light. ....**247. *Margaritaria***
   Leaf margin not translucent. ....**249. *Thecacoris*; 252. *Chaetachme***

6. High altitude tree (above c. 1500 m) with thick, rough bark. 7
   Not as above. If growing above 1500 m, then bark not thick and rough. 12

7. Leaf venation very indistinct. ....**261. *Agarista***
   Leaf venation clearly visible. 8

8. Leaves aromatic when crushed. ......**206. *Ocotea kenyensis*; 258. *Morella salicifolia***
   Leaves not aromatic when crushed. 9

9. Leaf veins red. ....**259. *Faurea***
   Leaf veins not red. 10

10. Small tree to 5 m, in the Ericaceous Belt of Mt. Elgon and near the summit of Mt. Kadam. Leaves often c. 12 x 2 cm. ....**260. *Protea***
    Not as above. 11

11. Slash a red colour. ....**206. *Ocotea kenyensis***
    Slash yellow to orange. ....**211. *Diospyros***

12. Young leaves very conspicuously covered with orange-brown hairs. Leaves large (often over 18 cm long) with well-defined lateral veins, hairy below. Tree to 35 m with a straight trunk and branches at right angles. ....**202. *Pycnanthus***
    Not as above. 13

13. Petiole c. 1-4 cm long, markedly swollen at both ends. Understorey tree to 15 m. ....**232. *Baphiopsis***
    Petiole not markedly swollen at both ends. 14

14. Dominant colour of slash a shade of red. 15
    Slash not red. Note: there are some understorey trees that normally have red slashes, but the red colour may not be well shown on some specimens. They are included on both sides of the dichotomy. 19

15. Bark rough, splitting into rectangles. Leaf apex asymmetric. Leaf venation indistinct. ....**201. *Warburgia***
    Not as above. 16

16. Shoots ending in conspicuous claw-shaped scales. Understory tree to 10 m. ....**256. *Peddiea***
    Shoots not terminating as above. 17

17. Leaf margin translucent when held up to the light. ....**247. *Margaritaria***
    Not as above. 18

18. Leaf apex usually obtusely pointed, sometimes slightly acuminate. ....**204. *Beilschmiedia***
    Leaf apex acuminate. ....**226-229c. *Turraea***

19. Shoot growth peculiar. Each node bears a leaf or leaf-scar, an axillary shoot (that grows onwards), and a 'main shoot' that aborts or terminates immediately in an inflorescence (look very carefully). Understorey trees to 10 m. ....**238-239. *Leptaulus***
    Shoot growth normal. 20

20. Stipules very long (5-10 cm). Leaves thick and leathery, with narrow lateral veins. A very large buttressed tree. ....**209. *Klainedoxa***
    Stipules less than 5 cm long. 21

21. Young leaves covered by narrow, curved stipular sheaths. Fallen stipules carpet the ground beneath the tree. Crown dense (like a mango). Slash smelling of mangoes. Tree to 20 m, occasionally more. ....**210. *Irvingia***
    Stipules, if present, either not as above or, if so, then crown not dense and tree smaller. Slash not smelling of mangoes. 22

22. Understorey tree to 10 m, with shoots ending in conspicuous claw-shaped scales. Leaves acute at apex and cuneate at base (see illustration). ....**256. *Peddiea***
    Leaves not terminating in conspicuous claw-shaped scales or, if so, leaf shape not as above. 23

23. Leaf base comparatively long cuneate (see illustration). ....**255-255a. *Pittosporum***
    Leaf base various; if cuneate, then not as markedly so as above. 24

24. Trunk straight; bark thick, rough, dark-coloured; slash fibrous, yellow to orange, not or only slightly turning darker, not or only slightly fragrant. ....**211. *Diospyros***
    Not as above; if shape and bark as above, then slash turning darker. 25

25. Main lateral veins prominent on lower surface of leaf and parallel to one another (see Plate 18). Leaves hairy below, often c. 15 x 6 cm. Small tree on forest edges. ....**226. *Turraea floribunda***
    Not as above. 26

26. Understorey tree to 8 m. Leaves often c. 11 x 3.5 cm, apex acute to acuminate, base rounded to obtuse. Bark on young stems not reticulately fibrous and lacking prominent lenticels. ....**226. *Turraea vogelii***
    Not as above. Specimens may display one of more of the above characters, but not all. 27

27. Understorey tree to 8 m, only recorded from Ishasha Gorge and Kalinzu Forest. Leaves comparatively small (often c. 8 x 3 cm) with rather obscure venation. ....**180. *Gymnanthes***
    Not as above. ....**Annonaceae (see key before 212)**

Sub-Key 10

Leaves simple, opposite or whorled; no latex or conspicuously coloured (e.g. red) sap in slash.

Note: *Ocotea usambarensis* is not included here. Its leaves can be (sub-) opposite, but are also alternate and usually of a whitish colour below (so would normally key out to Sub-Key 5).

1. At least some of the leaves in whorls. 2
   Leaves opposite. This side of the dichotomy also includes those species of Rubiaceae and *Cassipourea* that normally have opposite leaves, but occasionally have leaves in 3s; they are all distinguished by having interpetiolar stipules. 4

2. White latex present in young parts. ....**268-270. *Pleiocarpa, Rauvolfia***
   White latex absent from young parts. 3

3. Petiole less than 2.5 cm long. ....**264-265. *Nuxia***
   Petiole usually over 2.5 cm long. ....**266. *Premna***

4. Leaves less than 1 cm wide. Trees to large shrubs found above 2000 m. ....**284-288. *Hypericum***
   Leaves over 1 cm wide. 5

5. Tree with very large leaves, often over 30 cm long (but can be as short as c. 22 cm on crown leaves). Leaves in clusters at ends of branches, resembling cabbages. Interpetiolar stipules absent. ....**289-291. *Anthocleista***
   Leaves not as above. If very large, then interpetiolar stipules present. 6

6. Leaves with only a few (fewer than 8) main lateral veins on each side of the midrib. Veins conspicuously arcuate (see Plate 23). 7
   Veins not arcuate or, if so, not as conspicuously as above. 10

7. Lamina with a pair of relatively inconspicuous lateral veins from the base and pair of more conspicuous veins arising asymmetrically c. 1 cm up the midrib. ....**292. *Strychnos***
   Not as above. 8

8. Leaves ovate, more or less heart-shaped. Petiole to 10 cm long. ....**297. *Mallotus***
   Not as above. 9

9. Lateral veins 4-5 on each side of the midrib. ....**293. *Afrocrania***
   Lateral veins 1-2 on each side of the midrib. ...**294-296. *Dichaetanthera, Lijndenia***

10. Leaves more or less rounded and relatively large (see illustration of *Fleroya rubrostipulata* on Plate 24). ....**Rubiaceae (see key before 303)**
    Leaves not more or less rounded or, if so, then narrower than 10 cm. 11

11. Interpetiolar stipules present (these fall off early in some species, so look carefully at the ends of the shoots). Small to medium-sized trees. 12
    Interpetiolar stipules absent. Trees of all sizes. 14

12. At least some leaves toothed, crenate or with conspicuously wavy margins. Note: *Cassipourea gummiflua* (301), which is rare except in Bwindi Forest, may sometimes have entire leaf margins; it has a straight trunk, small crown and rather stiff leaves; it is included in this part of the dichotomy. 13
Leaf margin entire. ....**Rubiaceae (see key before 303)**

13. Branches at right angles. ....**298-301. *Cassipourea***
Branches not obviously at right angles. ....**302. *Lasiodiscus***

14. Leaf margin with teeth or crenations. 15
Leaf margin entire or wavy. 18

15. Leaves very hairy below. ....**331. *Buddleja***
Leaves not very hairy below. 16

16. Leaves gland-dotted and fragrant when crushed; leaf margin coarsely toothed (see Plate 26). ....**328. *Xymalos***
Leaves not gland-dotted and not fragrant when crushed. 17

17. Slash smelling of cold cooked chicken. ....**189. *Maesopsis***
Slash not smelling of cold cooked chicken.
.....................................**302. *Lasiodiscus*; 329-330. *Catha*, *Elaeodendron***

18. Slash with an unpleasant burnt smell. Young stems 4-angled. Tree found above 1600 m altitude. ....**338. *Olinia***
Slash and young stems not as above. 19

19. White latex present in young parts. ....**Apocynaceae (see key before 267)**
White latex absent from young parts. 20

20. Leaves more or less sessile. A swamp forest species. ....**340. *Syzygium cordatum***
Note as above. 21

21. Leaves c. 7.5 x 2 cm, yellow to brown below (due to a covering of scales).
....**335. *Olea europaea***
Leaves not as above. 22

22. Understorey tree with leaf venation rather inconspicuous on both surfaces.
....**342. *Memecylon***
Leaf venation distinct, at least on lower surface. 23

23. Leaves less than c. 7 cm long. ....**341. *Eugenia***
Leaves over 7 cm long. 24

24. Slash brown to red-brown. Main lateral veins on lamina rather numerous (see Plate 26). ....**339-339a. *Syzygium guineense*, *S. congolense***
Slash white to orange to yellow, sometimes turning darker. Main lateral veins on leaf comparatively few. ....**Oleaceae (see key before 332)**

Sub-Key 11

Leaves bifoliolate, trifoliolate or digitate, or else simple and very deeply palmately lobed; no latex or conspicuously coloured (e.g. red) sap in slash.

1. CHOOSE FROM ONE OF THESE 3 OPTIONS
   Leaves always with 2 leaflets. Large tree with a much fluted trunk and usually spines on branches. ....**343. *Balanites***
   Leaves usually with 3 leaflets. 2
   Leaves digitate, usually with more than 3 leaflets, or very deeply palmately lobed. 8

2. Spines present, sometimes persistent on larger trees as conical woody bosses on trunk. 3
   Spines absent. 5

3. Leaflets gland-dotted. Conical woody bosses absent from trunk. **346. *Balsamocitrus***
   Leaflets not gland-dotted. Conical woody bosses present or absent. 4

4. Conical woody bosses present. Young branches spiny. ....**344-345. *Erythrina***
   Conical woody bosses absent. Young branches not spiny. ....**351-353. *Allophylus***

5. Leaflets gland-dotted (sometimes gland dots few). ....**347-349. *Vepris***
   Leaflets not gland-dotted. 6

6. Margins of at least some leaflets toothed or crenate. ....**351-353. *Allophylus***
   Margins of leaflets entire. 7

7. All leaves normally with 3 leaflets. ....**354. *Euadenia***
   Leaves with from 1 to 5 leaflets on the same tree. ....**355. *Ritchiea***

8. Tree with shape of a pawpaw tree, with only a few thick branches and with large leaves clustered at their ends. ....**356-357. *Cussonia***
   Not as above. 9

9. Leaves deeply palmately lobed, with 11 or more segments. ....**366. *Musanga cecropioides***
   Leaves with 11 or fewer leaflets. 10

10. Leaves opposite. A small tree. ....**368. *Vitex***
    Leaves alternate. 11

11. Tall trees with straight trunks. Crowns not one leaf thick or umbrella-shaped. 12
    Small to medium-sized trees with spreading crowns. Trunks crooked or, if straight, then crowns one leaf thick and umbrella-shaped. 13

12. Conical woody bosses present on trunk. ....**363. *Bombax***
    Conical woody bosses absent from trunk. ....**362. *Ricinodendron***

13. Leaflets with conspicuously large marginal teeth (see illustration). ....**364-365. *Myrianthus***
    Leaflets entire or with comparatively small marginal teeth. 14

14. Leaflets normally 3, but varying from 1 to 5 on the same tree. ....**355. *Ritchiea***
    Leaflets normally more than 3. 15

15. Leaflets distinctly stalked. ....**358-361. *Schefflera***
    Leaflets not stalked. ....**367. *Musanga leo-errerae***

Sub-Key 12

Leaves pinnate; no latex or conspicuously coloured (e.g. red) sap in slash.

Notes: *Paropsia guineensis* (186) and *Phyllanthus inflatus* (248) may appear to have pinnate leaves, but the leaves are actually simple. *Baikiaea insignis* (427) may appear to have simple leaves, but they are actually pinnate.

1. Leaves opposite. 2
   Leaves alternate. 4
2. Leaflets usually 5. Petiole and rachis winged. A montane tree. **379. *Schrebera alata***
   Leaflets usually more than 5. 3
3. Underside of bark bright orange. Slash fragrant. ....**372. *Fagaropsis***
   Not as above. ....**Bignoniaceae (see key before 369)**
4. Spines present (persistent as conical woody bosses on trunks in some cases). 5
   Spines absent. 7
5. Leaf rachis winged. 6
   Leaf rachis not winged. ....***Zanthoxylum* (see key before 373)**
6. Leaflets gland-dotted. ....**378. *Citropsis***
   Leaflets not gland-dotted. ....**382. *Harrisonia***
7. Small tree or shrub with gland-dotted leaflets. ....**377. *Clausena***
   Not as above. 8
8. Leaf rachis with prominent wings or irregular outgrowths. Small to medium-sized trees. 9
   Leaf rachis not prominently winged or with irregular outgrowths. 11
9. Rachis with irregular outgrowths. ....**381. *Hagenia***
   Rachis winged. 10
10. Intrapetiolar stipules present. ....**380. *Bersama***
    Intrapetiolar stipules absent. ....**382. *Harrisonia***
11. Trunk straight (sometimes leaning), ending abruptly and bearing a number of whorled ascending branches without the trunk continuing. The branches continue to branch in the same manner. ....**384. *Polyscias***
    Tree shape not as above. 12
12. Leaflets thick and leathery, venation rather indistinct, particularly on upper surface. Tree to 35 m with a granular slash. Usually in damp places. ....**427. *Baikiaea***
    Leaflets not thick and leathery. 13

13. Leaves imparipinnate. Note: in some cases, the terminal leaflet may have aborted, so look carefully. 14
Leaves paripinnate. Note: in some cases, one of the two terminal leaflets may have aborted, so look carefully. 18

14. Slash with some shade of red as the dominant colour (but the red may be poorly developed on young specimens), sometimes scented. ....**Anacardiaceae, *Canarium*, Meliaceae (see key before 390)**
Slash not red (except sometimes with *Millettia psilopetala*, which may have concentric rings of yellow and red). 15

15. Rare tree (except in Kalinzu Forest), recorded from Ankole, Kigezi and Tooro. Bark light-coloured, fairly smooth, thin and brittle and coming away easily from the slash. ....**383. *Quassia***
Not as above. 16

16. Bark light brown, rather fibrous. Leaflets usually over 5 cm wide. ....**394. *Trichilia prieuriana***
Bark not fibrous or, if so, leaflets narrower than 5 cm. 17

17. Intrapetiolar stipules present. Usually a small tree. ....**380. *Bersama***
Intrapetiolar stipules absent. ....**Connaraceae, Fabaceae (see key before 421)**

18. CHOOSE FROM ONE OF THESE 3 OPTIONS
Leaves mostly with fewer than 7 leaflets. 19
Leaves mostly with 7-24 leaflets. 22
Leaves often with more than 24 leaflets. 27

19. Slash with some shade of red as the dominant colour. 20
Red not dominant slash colour. If red is present, then it is restricted to a fairly thin layer close to the bark. 21

20. Very badly shaped tree recorded from Mabira Forest, Busoga and stream valleys at the base of mountains in north-eastern Uganda. Leaflets small, c. 8 x 2.5 cm. ....**414. *Lecaniodiscus cupanioides***
Not as above. Note: saplings may be badly shaped in some species. ....**Anacardiaceae, *Canarium*, Meliaceae (see key before 390)**

21. Leaflets mostly 4, characteristically shaped (see illustration). Tall tree with large thin buttresses. ....**428. *Cynometra***
Leaflets not shaped as above. Buttresses usually absent. ....**Sapindaceae (see key before 409)**

22. Slash with some shade of red as the dominant colour. 23
Slash not red. If some red is present, then it is a subordinate colour and only in a thin layer near the bark. 25

23. Rare tree recorded from Bwamba, Zoka, Budongo, Kalinzu-Maramagambo and Bugoma forests. Slash dark red with brown fibres. Leaflets 10-24, small (c. 6.5 x 2.5 cm). ....**424. *Cassia***
Not as above. 24

24. Understory tree to 15 m, either with a crooked trunk and found in Mabira and Bwamba forests, Busoga and stream valleys in north-east Uganda or with a weak, leaning trunk. ....**412. *Lecaniodiscus*; 413-414. *Lychnodiscus***
    Taller tree with straight trunk (but saplings may be crooked). ....**Anacardiaceae, *Canarium*, Meliaceae (see key before 390)**

25. Rare tree to 40 m in Bwamba. Trunk straight with a thick crown. Bark usually red-brown, scaling. Slash pale brown. ....**426. *Afzelia***
    Not as above. 26

26. Poorly shaped tree found near water in forests along the Albertine Rift. Leaflets 20-36. Bark grey, scaling in places. Slash yellow. ....**408. *Turraeanthus***
    Not as above. ....**Sapindaceae (see key before 409)**

27. CHOOSE FROM ONE OF THESE 3 OPTIONS
    Leaflets 8-27, often narrow (c. 8 x 2.5 cm). ....**409. *Majidea***
    Leaflets 20-36, often c. 13 x 3.5 cm, with well-marked lateral veins, not gland-dotted. ....**408. *Turraeanthus***
    Leaflets 24-38, c. 5.5 x 2 cm, gland-dotted. ....**423. *Mildbraediodendron***

Special key for tall trees (those over about 20 m tall) and on which the leaves cannot be clearly seen.

Certain trees which sometimes reach a height of over 20 m are excluded. The most important of these fall into two groups: (i) only found or most common above 2000 m altitude, these being *Afrocarpus* spp. (5-6), *Juniperus procera* (7), *Olinia rochetiana* (338), *Podocarpus latifolius* (4), *Rapanea melanophloeos* (262), *Schefflera* spp. (358-361) and *Schrebera alata* (379); (ii) generally rare or of local occurrence, or too poorly known to include in the detailed keys, these being *Balthasaria schliebenii* (188), *Cassia mannii* (424), *Cathormion altissimum* (436), *Croton sylvaticus* (107), *Isolona congolana* (215), *Nesogordonia kabingaensis* (122), *Quassia undulata* (383) and *Xylopia staudtii* (217).

1. Slash with white, off-white, yellow, orange or red latex or with conspicuously coloured (e.g. red) sap. The latex or sap may be produced in very small quantities, so allow time for it to appear. ....**Sub-key 1 (earlier)**
   Slash without latex or conspicuously coloured sap as above. Note: Some species produce colourless, yellow or brown watery sap in large quantities. They are included here. 2

2. Conical woody outgrowth (bosses) present on trunk (look carefully, as they may be inconspicuous on large trees). 3
   Conical woody bosses absent. Narrow sharp spines sometimes present. 4

3. Leaves trifoliolate or digitate. ....**344-345. *Erythrina*; 363. *Bombax***
   Leaves pinnate. ....**373-376. *Zanthoxylum*; 382. *Harrisonia***

4. Stilt roots present. 5
   Stilt roots absent. 7

5. Leaves digitate or deeply palmately lobed. ....**364-367. *Musanga, Myrianthus***
Leaves simple. 6

6. Leaves large (often over 20 cm long) and not 3-lobed.
..........**240-241. *Spondianthus, Uapaca***
Leaves either less than 20 cm long or, if longer, then 3-lobed.
..........**94-99. *Macaranga*; 188. *Balthasaria*; 216-218. *Xylopia*; 243-244b. *Bridelia***

7. Slash with concentric layers of a red shade (or occasionally white) and yellow. Leaves simple, but branchlets strongly simulate pinnate leaves. Bark light-coloured, usually with conspicuous vertical channels. ....**186. *Paropsia***
Slash without concentric layers of red (or white) and yellow. Branchlets not simulating pinnate leaves. 8

8. Slash with some shade of red as the dominant colour (orange is not counted as red). Note: *Maesopsis*, which can have a slash that is both yellow and red in about equal quantities and which smells of cold cooked chicken, is included here. 9
Slash with some other colour, apart from red, dominant (usually white, yellow or brown). There may be a red layer immediately under the bark. 47

9. Buttresses present. 10
Buttresses absent. Note: the basal part of the trunk may flare out in rounded flanges, particularly above the main roots. 16

10. Leaflets very small (usually less than 1 cm wide), giving the crown a feathery appearance as seen from the ground.
..........**435. *Newtonia*; 440. *Parkia*; 442. *Albizia coriaria***
Leaflets or leaves (if simple) over 1 cm wide. Crown not appearing feathery. 11

11. Leaves pinnate (this is usually obvious, the leaves being clustered at the ends of the branches). ....**Meliaceae (see key before 390)**
Leaves simple or bipinnate. 12

12. Leaves relatively large (normally over 6.5 cm wide).
..........**116. *Cola gigantea*; 118. *Sterculia*; 123. *Strombosia***
Leaves or leaflets relatively small (normally less than 6.5 cm wide). 13

13. Leaves covered below with white or yellow-brown hairs, giving the crown a whitish or brownish colour as seen from the ground. ....**200. *Parinari***
Leaves or leaflets not covered below with white or yellow-brown hairs. 14

14. Slash turning distinctly and fairly rapidly darker. ....**199. *Prunus***
Slash not, or only slightly, turning darker. 15

15. Leaves normally opposite. ....**329. *Elaeodendron*; 339-339a. *Syzygium***
Leaves or leaflets alternate. ....**204. *Beilschmiedia*; 433. *Erythrophleum***

16. Slash turning distinctly and fairly rapidly darker. 17
Slash not, or only slightly, turning darker. 20

17. Leaves pinnate (this is usually obvious, the leaves being clustered at the ends of the branches). ....**Meliaceae (see key before 390)**
Leaves simple. Slash not scented (but sometimes smelling of cyanide). 18

18. Leaves normally over 2 times as long as broad. ....**199.** ***Prunus*****; 202.** ***Pycnanthus*****; 207-208.** ***Casearia***
Leaves normally less than 2 times as long as broad. 19

19. Slash with a more or less distinct fibrous reticulum. Leaves alternate. ....**Malvaceae (see key before 115)**
Slash without a fibrous reticulum. Leaves opposite. ....**303-304.** ***Fleroya***

20. Slash smelling of cold cooked chicken. ....**189.** ***Maesopsis***
Slash not smelling of cold cooked chicken. 21

21. Slash granular, at least in places. 22
Slash fibrous, brittle or of even texture. 30

22. Leaves digitate (this is usually obvious). ....**362.** ***Ricinodendron***
Leaves not digitate. 23

23. Mature leaves covered below with white or yellow-brown hairs, giving the crown a whitish or brownish colour as seen from the ground. ....**200.** ***Parinari***
Mature leaves or leaflets not covered below with white or yellow-brown hairs. 24

24. Slash with orange granules, particularly near the bark. ....**202.** ***Pycnanthus*****; 387.** ***Lannea*****; 433.** ***Erythrophleum***
Slash without orange granules. 25

25. Leaflets very small (usually less than 1 cm wide), giving the crown a feathery appearance as seen from the ground. ....**440.** ***Parkia***
Leaflets or leaves over 1 cm wide. Crown not appearing feathery. 26

26. Trunk straight, branches at right angles. ....**202.** ***Pycnanthus*****; 298-301.** ***Cassipourea*****; 427.** ***Baikaea***
Branches not normally clearly at right angles to the trunk. 27

27. Leaves simple. 28
Leaves compound. 29

28. Leaves alternate. ....**94-99.** ***Macaranga*****; 106.** ***Croton macrostachyus***
Leaves normally opposite. ....**329.** ***Elaeodendron*****; 339-339a.** ***Syzygium***

29. Leaves pinnate. ....**390.** ***Ekebergia*****; 427.** ***Baikiaea***
Leaves bipinnate. ....**433.** ***Erythrophleum***

30. Slash strongly scented. 31
Slash not or only slightly scented. 33

31. Leaves whitish below, simple. ....**205.** ***Ocotea usambarensis***
Leaves or leaflets not whitish below. Leaves simple or pinnate. 32

32. Leaf midrib and main veins red as seen from below. Leaves simple. ...**237.** ***Apodytes***
Leaf veins not normally red. Leaves pinnate. ....**Meliaceae and similar-looking trees (see key before 390)**

33. Leaflets very small (usually less than 1 cm wide, giving the crown a feathery appearance as seen from the ground. ....**440. *Parkia*; 442. *Albizia coriaria***
Leaflets or leaves over 1 cm wide. 34

34. Leaves covered below with white or yellow-brown hairs, giving the crown a whitish or brownish colour as seen from the ground. ....**200. *Parinari***
Underside of leaves or leaflets not as above. 35

35. Bark smooth or flaking, sometimes with shallow vertical fissures. 36
Bark markedly fissured, channelled or fibrous. 42

36. Leaves pinnate (this is usually obvious, the leaves being clustered at the ends of the branches). ....**Meliaceae (see key before 390)**
Leaves simple. 37

37. Leaf midrib and main veins red as seen from below. ....**237. *Apodytes***
Leaf veins not normally red. 38

38. Leaves relatively large (often over 20 cm long). ....**123. *Strombosia***
Leaves relatively small (normally under 20 cm long). 39

39. Trunk straight with branches at right angles. ....**298-301. *Cassipourea***
Branches not normally clearly at right angles. 40

40. Leaves alternate. 41
Leaves opposite. ....**329. *Elaeodendron*; 339-339a. *Syzygium***

41. Bark flaking in large pieces. ....**204. *Beilschmiedia***
Bark not flaking in large pieces, sometimes with small vertical fissures. ....**94-99. *Macaranga*; 131. *Grewia mildbraedii*; 207-208. *Casearia***

42. Bark fissuring into rectangles. 43
Bark not fissuring into rectangles. 44

43. Bark tasting of pepper. Leaves simple. ....**201. *Warburgia***
Bark not tasting of pepper. Leaves pinnate. ....**390. *Ekebergia*; 399. *Entandrophragma utile***

44. Leaves pinnate (this is usually obvious, the leaves being clustered at the ends of the branches). ....**Meliaceae (see key before 390)**
Leaves simple. 45

45. Leaves relatively large (often over 20 cm long). ....**241. *Spondianthus***
Leaves relatively small (normally less than 20 cm long). 46

46. Leaves opposite. ....**329. *Elaeodendron***
Leaves alternate. ....**247. *Margaritaria***

47. Slash yellow or white, with brown or black rings or dots. Slash not soft. ....***Celtis* (see keys before 86)**
Slash either not coloured as above or, if so, then slash very soft. Yellow or orange markings may be present. 48

48. Buttresses present. 49
Buttresses absent. Note: the basal part of the trunk may flare out in rounded flanges, particularly above the main roots. 63

49. Leaflets very small (usually less than 1 cm wide), giving the crown a feathery appearance as seen from the ground. 50
Leaflets or leaves over 1 cm wide. Crown not appearing feathery. 51

50. Slash smelling of antiseptic (thymol). ....**409. *Majidea***
Slash not smelling of antiseptic. ....**Fabaceae (see key before 421)**

51. Bark with deep vertical channels or fissures. 52
Bark smooth or flaking, sometimes with shallow vertical fissures. 53

52. Bark fissuring into rectangles. ....**423. *Mildbraediodendron***
Bark not fissuring into rectangles.
..................**93. *Holoptelea*; 210. *Irvingia*; 333. *Olea capensis* subsp. *welwitschii***

53. Leaves large, normally over 12 cm wide. ....**115. *Pterygota***
Leaves or leaflets normally less than 10 cm wide. 54

54. Slash turning conspicuously and fairly rapidly darker, at least in places. 55
Slash not or only slightly turning darker. Note: the slash may turn lighter. 57

55. Buttresses spreading and very thin, plank-like. ....**428. *Cynometra***
Buttresses usually small, not very thin. 56

56. Leaves with prominent lateral veins from base. ....**90. *Celtis philippensis***
Leaves without prominent lateral veins from base.
....**182-185. *Drypetes*; 332. *Schrebera alata***

57. Slash smelling of antiseptic (thymol). ....**409. *Majidea***
Slash not smelling of antiseptic. 58

58. Slash granular, at least in places. 59
Slash fibrous, brittle or of even texture. 60

59. Buttresses usually very large. Stipules very long (5-10 cm). Leaves simple and alternate. ....**209. *Klainedoxa***
Buttresses usually small. If present, stipules less than 5 cm long. Leaves bipinnate or else simple and opposite. ....**339-339a. *Syzygium*; 441-447. *Albizia***

60. Fallen stipules carpet the ground beneath the tree. Slash smelling of mangoes.
....**210. *Irvingia***
Slash not smelling of mangoes. 61

61. Dominant colour of slash a shade of brown. ....**339-339a. *Syzygium***
Dominant colour of slash white to yellow. 62

62. Leaves with prominent lateral veins from base.
..............................**63. *Ficus exasperata*; 73. *Ficus ingens*; 111. *Alangium***
Leaves without prominent lateral veins from base.
....**93. *Holoptelea*; 182-185. *Drypetes***

63. Underside of leaf whitish in colour. Slash with a spicy smell. Old leaves turn conspicuously yellow. ....**109. *Croton megalocarpus***
Underside of leaves or leaflets not whitish in colour. 64

64. Undersurface of bark bright orange. Slash yellow, sometimes with stone cells, aromatic (at least when freshly cut). ....**372. *Fagaropsis***
Undersurface of bark not bright orange. 65

65. Slash turning conspicuously and fairly rapidly darker. 66
Slash not or only slightly turning darker. Sometimes turning lighter. 76

66. Slash very soft, turning greenish (sometimes with brown lines). ....**112-113. *Cordia*; 384. *Polyscias***
Slash not turning greenish. 67

67. Trunk straight (sometimes leaning), ending abruptly and bearing a number of whorled ascending branches without the trunk continuing. The branches continue to branch in the same manner. ....**384. *Polyscias***
Tree shape not as above. 68

68. Leaves relatively large (often over 20 cm long) or, if only 15-20 cm long, then less than twice as long as broad and with large interpetiolar stipules. ...**219. *Monodora myristica*; 289-291. *Anthocleista*; 303-305. *Fleroya, Nauclea***
Leaves relatively small, normally less than 20 cm long; otherwise not as above. ...69

69. Slash turning reddish or reddish-brown. ....**90. *Celtis philippensis*; 207-208. *Casearia*; 332. *Schrebera arborea***
Slash not turning reddish. 70

70. Leaves pinnate. ....**369-371. Bignoniaceae**
Leaves simple or trifoliolate. 71

71. Bark flaking. Leaves opposite. ....**332. *Schrebera arborea***
Bark not, or only indistinctly, flaking; often fissured or challenged. Leaves alternate. 72

72. Slash pleasantly scented. 73
Slash not pleasantly scented. 74

73. Leaves simple. ....**212-214. *Uvariopsis, Greenwayodendron, Cleistopholis***
Leaves trifoliolate. ....**347. *Vepris nobilis***

74. Leaves trifoliolate. ....**347. *Vepris nobilis***
Leaves simple. 75

75. Slash very soft, quite thick, white to yellow-grown, often with brown or yellow markings, rapidly turning brown. ....**112-114. Boraginaceae**
Slash not very soft. ....**165. *Ilex*; 182-185. *Drypetes*; 207-208. *Casearia*; 212. *Uvariopsis*; 234. *Tapura***

76. Bark rough, markedly channelled, fissured or fibrous, or with a raised reticulate pattern. 77
Bark smooth to fairly smooth, or flaking in large pieces, sometimes with shallow vertical fissures. 88

77. Leaflets very small (usually less than 1 cm wide), giving the crown a feathery appearance as seen from the ground. ....**Fabaceae (see key before 421)**
Leaflets or leaves over 1 cm wide. Crown not appearing feathery. 78

78. Slash granular, at least in places. 79
Slash fibrous, brittle or of even texture. 82

79. Leaves whorled, normally over 20 cm long. ....**268. *Rauvolfia caffra***
Leaves opposite or alternate, normally under 20 cm long. 80

80. Leaves with prominent lateral veins from base. ....**106. *Croton macrostachyus***
Leaves without prominent lateral veins from base. 81

81. Leaves alternate. ....**211-211a. *Diospyros***
Leaves opposite. ....**333. *Olea capensis* subsp. *welwitschii*; 335. *O. europaea***

82. Slash soft, whitish, with small golden flecks. Leaves in whorls. ....**266. *Premna***
Slash not as above, but, if so, then leaves small (under 2 cm broad) and leaves not in whorls. 83

83. Leaves relatively large (often over 20 cm long). ....**219. *Monodora myristica***
Leaves relatively small (normally under 20 cm long). 84

84. Slash smelling of mangoes. ....**210. *Irvingia***
Slash not smelling of mangoes. 85

85. Slash strongly and pleasantly scented. ....**214. *Cleistopholis***
Slash not strongly and pleasantly scented. 86

86. Trunk deeply and irregularly fluted for most of its length. Leaves simple and opposite or bifoliolate. ....**335. *Olea europaea*; 343. *Balanites***
Trunk not normally as above. Leaves not as above. 87

87. Slash yellow or orange. ....**93. *Holoptelea*; 179. *Shirakiopsis*; 211-211a. *Diospyros***
Slash not yellow or orange. ....**216-218. *Xylopia*; 426. *Afzelia***

88. Slash smelling of antiseptic (thymol). ....**409. *Majidea*; 416. *Zanha***
Slash not smelling of antiseptic. 89

89. Slash granular, at least in places. 90
Slash fibrous, brittle or of even texture. 97

90. Leaflets relatively small (often less than 2 cm wide on crown leaves). Leaves bipinnate. ....**Fabaceae (see key before 421)**
Leaflets or leaves normally over 2 cm wide. Leaves not bipinnate. 91

91. Trunk straight with branches at right angles. 92
Branches not obviously at right angles. 93

92. Leaves pinnate. ....**427. *Baikiaea***
Leaves simple. ....**242. *Tetrorchidium*; 298-301; *Cassipourea***

93. Leaves less than twice as long as broad. ....**106. *Croton macrostachyus*; 110. *Discoglypremna***
Leaves over twice as long as broad. 94

94. Leaves simple. 95
Leaves pinnate. 96

95. Leaves whorled. ....**268. *Rauvolfia caffra***
Leaves alternate or opposite. ....**242. *Tetrorchidium*; 334. *Olea capensis* subsp. *hochstetteri*; 339-339a. *Syzygium***

96. Trunks of mature trees straight, with branches near apex. ....**427. *Baikiaea***
Trunks normally not straight, with branches from fairly low down. ....**415, 419. *Blighia*; 418. *Glenniea***

97. Leaves bifoliolate. Trunk with deep and irregular fissures. ....**343. *Balanites***
Leaves not bifoliolate. Trunk various. 98

98. Slash coming away easily from the wood. Face of exposed wood smooth and shiny, with conspicuous white lines. Leaves simple, opposite, with arcuate venation. ....**292. *Strychnos***
Slash and wood not as above. If the slash comes easily away from the wood, then face of exposed wood not shiny and without white lines. 99

99. Slash smelling of mangoes. ....**210. *Irvingia***
Slash not smelling of mangoes. 100

100. Slash pleasantly scented. ....**213-214. *Cleistopholis, Greenwayodendron***
Slash not pleasantly scented. 101

101. Trunk straight with branches at right angles. Leaves opposite. **298-301. *Cassipourea***
Lacking combination of branches at right angles and leaves opposite. 102

102. Leaves pinnate. ....**417-418. *Pancovia, Glenniea*; 425-426. *Dialium, Afzelia***
Leaves simple. 103

103. Base of leaf with prominent lateral veins. ....**63. *Ficus exasperata*; 93. *Holoptelea*; 111. *Alangium***
Base of leaf without prominent lateral veins. 104

104. Leaves opposite. Slash a shade of brown. ....**339-339a. *Syzygium***
Leaves alternate. Slash various. 105

105. Leaves distinctly widest in upper half and with a long cuneate base. ....**242. *Tetrorchidium***
Leaves not or only indistinctly widest in upper half. ....**93. *Holoptelea*; 182-185. *Drypetes*; 207-208. *Casearia*; 216-218. *Xylopia*; 230. *Baphia wollastonii***

Part 4

Descriptions of the Species

Order and Numbering of the Species

The species are ordered primarily according to their vegetative features and secondarily by taxonomy. The use of these two different approaches to grouping the species means that there are compromises and exceptions. However, it is interesting to note that the resulting arrangement works reasonably well on both accounts.

The overall arrangement and its logic are depicted in Table 4.1. The vegetative features used to order the species in this table are preceded by letters (A, B, C, etc.), centre-spaced on the page and given in bold, upper case type, as in the example below. The numerical ranges included in these headings (35-342 in this example) show which species possess the character in question using the numbers assigned to species in this book.

B. LEAVES SIMPLE (35-342)

There are two items in the table fronted by each of the letters used (B in this case). These are couplets in a dichotomous key. Note that the method of construction of this key is different from that used for the other keys in this book (which are mostly dichotomous). The two parts of the couplets are placed apart, rather than being placed together, the second of the two being marked with an asterisk (*). The other half of the above couplet is:

B*. LEAVES COMPOUND (343-447)

The items left aligned in the table, given in light, lower case, type and fronted by numbers, are taxonomic categories (267 and 268 in the example below). All taxonomic categories included in Table 4.1 have descriptions (D), keys to lower-ranking taxonomic categories (K) or both descriptions and keys (D, K) embedded in the text. The numbers at the beginning (e.g. 267) are the numbers of the first species that follow the descriptions and/or keys.

D. WHITE LATEX PRESENT IN THE SLASH (267-278)

267. Apocynaceae (D, K)
268. *Rauvolfia* (K)

The numbering sequence used for the species is similar to that in *A Field Guide to Uganda Forest Trees* (UFT – Hamilton 1981). We leave gaps in the sequence when species in UFT are no longer retained. Newly added species are assigned the same numbers used in UFT for the species that they most closely resemble, but distinguished by adding a letter. For example, a new species that most closely resembles the existing species 75 (already in the book) is designated 75a.

Table 4.1. Dichotomous key showing how the species are ordered.

A. UNUSUAL-LOOKING TREES (1-34)

1	Tree ferns (D)
4	Gymnosperms (D)
8	Monocotyledons (D)
8	Palms and screw palms (D)
12	*Dracaena* (D)
17	Bamboos (D)
19	Giant groundsels (D, K)
24	Ericaceous trees (D, K)
31	Succulent *Euphorbia* (D)

A*. NORMAL-LOOKING TREES (35-447)

B. LEAVES SIMPLE (35-342)

C. LEAVES ALTERNATE (35-263)

D. WHITE LATEX PRESENT IN THE SLASH (35-85)

35	Sapotaceae (D, K)
56	Moraceae (D, K)

D*. NO WHITE LATEX IN THE SLASH (86-263a)

E. THREE OR MORE PROMINENT VEINS (INCLUDING THE MIDRIB) FROM THE BASE OF THE LAMINA OR FROM NEARBY (86-131d)

86	Ulmaceae (D, K)
94	Euphorbiaceae and related families (D)
94	*Macaranga* (K)
106	*Croton* (K)
115	Malvaceae (D, K)
119	*Dombeya* (D)
123	Olacaceae (K)
128	*Desplatsia* (K)
130	*Grewia* (D, K)

E*. BASAL LEAF VEINS NOT AS PROMINENT AS ABOVE (132-263a)

F. LEAF MARGIN NON-ENTIRE (132-211A)

132	Achariaceae, Salicaceae and similar-looking species (K)
132	Achariaceae and Salicaceae (D)
144	*Dovyalis* (D)
148	*Rinorea* (K)
153	Ochnaceae (D, K)
160a	*Gymnosporia* and *Maytenus* (D)
169	Asteraceae (D)
172	Euphorbiaceae and related families (D)
172	*Alchornea* (K)
182	*Drypetes* (D, K)

F*. LEAF MARGIN ENTIRE (212-263a)

212	Annonaceae (D, K)
216	*Xylopia* (K)
222	*Uvaria* (D, K)

226 *Turraea* (D, K)
230 *Baphia* and *Baphiopsis* (D)
240 Euphorbiaceae and related families (D)
243 *Bridelia* (K)
245 *Antidesma* (K)

C*. LEAVES OPPOSITE OR WHORLED (264-342)

264 *Nuxia* (K)

G. WHITE LATEX PRESENT IN THE SLASH (267-278)

267 Apocynaceae (D, K)
268 *Rauvolfia* (K)
273 *Tabernaemontana* (K)

G*. NO WHITE LATEX PRESENT IN THE SLASH (279-342

H. YELLOW OR ORANGE LATEX PRESENT IN THE SLASH (279-286)

279 Clusiaceae, Calophyllaceae and Hypericaceae (D)
285 *Hypericum* (D)

H*. NO YELLOW OR ORANGE LATEX IN THE SLASH (289-342)

289 *Anthocleista* (K)

I. INTERPETIOLAR STIPULES PRESENT (298-327)

298 *Cassipourea* and *Lasiodiscus* (D)
298 *Cassipourea* (K)
303 Rubiaceae (D, K)

I*. INTERPETIOLAR STIPULES ABSENT (332-342)

332 Oleaceae (D, K)
339 Myrtaceae (D)
339 *Syzygium* (K)

B*. LEAVES COMPOUND (343-447)

J. LEAVES BIFOLIOLATE, TRIFOLIOLATE OR DIGITATE (343-447)

351 *Allophylus* (D)
358 *Schefflera* (D, K)

J*. LEAVES PINNATE OR BIPINNATE (369-447)

K. LEAVES OPPOSITE (369-372)

369 Bignoniaceae (D, K)
372 Rutaceae (D)

K*. LEAVES ALTERNATE (373-447)

373 *Zanthoxylum* (K)
390 Meliaceae (D, K)
409 Sapindaceae (D, K)
421 Fabaceae and Connaraceae (D, K)

Significance of Font Size in Species Descriptions

Larger font size is used for those species that are more likely to be encountered. Smaller font size is used for species that are generally rare or only locally distributed.

The Names of the Trees

The scientific names of the species (given in bold type in the first lines of the descriptions) consist of three parts, the first two being the names of the genus and species and the last the name or names of those responsible for its scientific description and naming. The botanical family follows each name.

Synonyms are scientific names that have been applied to species in the past, but are now considered to be invalid. Some species have numerous synonyms, too many to be accommodated here. Our main concern has been to provide, as synonyms, the names of species included in *The Indigenous Trees of the Uganda Protectorate* (Eggeling and Dale 1951) and *A Field Guide to Uganda Forest Trees* (Hamilton 1981) and which have since been renamed. These are the out-of-date names that are most likely to be familiar to field workers in Uganda.

Vernacular names are names commonly used by a particular people or in a particular place. We include the English and trade names of the species, as well as names used in indigenous Ugandan languages. The abbreviations used for the various languages are explained in Part 7, which also contains information about the linguistic relationships between the various Ugandan languages and shows the places that are their traditional homelands.

Characters of the Species

Information about some of the characters used in the descriptions is provided in Part 2 and explanations of technical terms in Part 5 (Glossary). The descriptions are based on the features of typical mature plants.

Height

The figures given for the heights of trees refer to normal maximum heights at maturity.

Leaf and Leaflet Size

These are commonly given in the form of length x width (e.g. 16 x 4 cm). The measurements are for typical leaves or leaflets. The measurements of typical leaves are often more useful for field identification purposes than the total ranges of sizes found, as given in many floras.

Geographical Distribution

The areas into which Uganda is traditionally divided for the purpose of recording plant distributions are shown in Fig. 4.1. They are based on the administrative divisions of the country during the 1950s, not those used today. Having a stable geographical reference system is invaluable in biological recording because it facilitates monitoring and recording change. The sites of some of the individual forests mentioned in the descriptions are shown in Fig. 1.4. There has been large-scale reduction in the extent of forest during recent decades and it is certain that some of the species no longer grow at some of their former localities. The altitudes for species occurrences given in the present book refer to their altitudes in Uganda, not those applying to their whole global ranges.

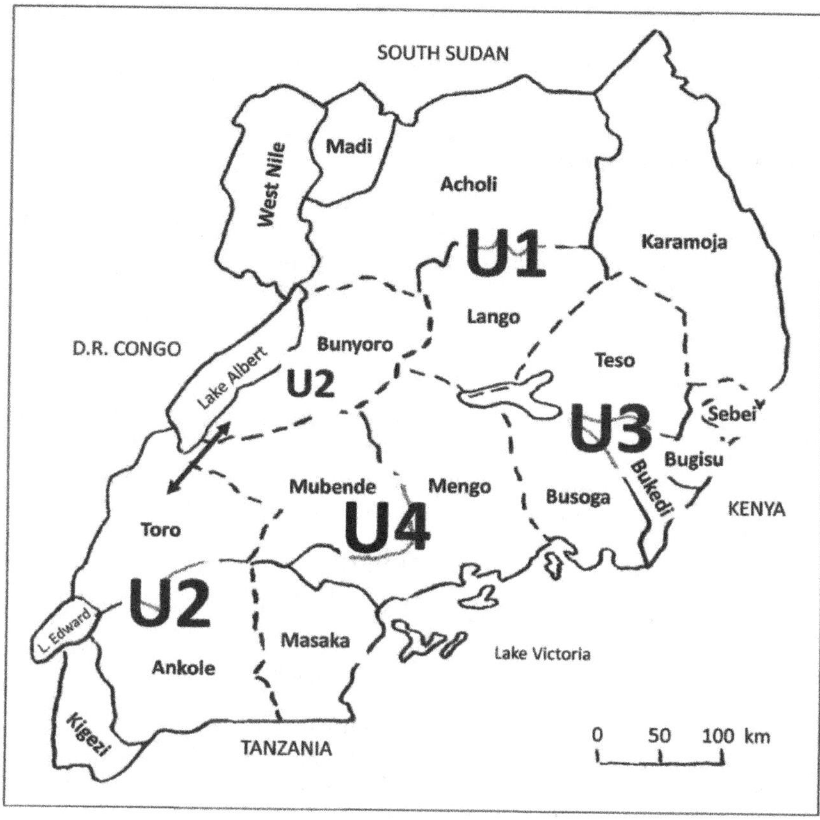

Fig. 4.1. Flora areas and districts used for botanical recording in Uganda. Note: Toro is the spelling of Tooro used by botanists in the 1950s.

CITES Listing and Conservation Status

The Convention on International Trade in Endangered Species of Wild Fauna and Flora (CITES) is a multilateral treaty intended to regulate trade in endangered species. Species may be listed in one of three Appendices (CITES 2017). Species in Appendix I are the most at risk of extinction due to trade. None of the species included in the present book fall into this category. Species in Appendix II are not necessarily now threatened with extinction, but may become so unless trade is closely controlled. Only a few species in the present book are Appendix II listed. One of them is *Encephalartos whitelockii* (7a), all cycads being at least in CITES Appendix II, partly because of fear of unscrupulous collection by specialist collectors. Succulent species of *Euphorbia* (31-33) are listed for the same reason. *Prunus africana* (199) is included in Appendix II because of fear of over-harvesting for the European pharmaceutical sector. Uganda, as a signatory to CITES and potential country of export for CITES Appendix II species, has the option under international law of requiring exporters to obtain export permits to ship abroad.

The information given under Conservation Status refers to the results of three evaluations of the degrees of risk of extinction faced by the species. One (Kalema and Beentje 2012) covers an almost identical list of species to that in the present book. Both TOU and one of the other evaluations (IUCN 2019) are concerned with risk of extinction globally; the other is about risk of extinction at the national level (WCS 2016). All the evaluations refer to a standard list of threat criteria and categories (Table 4.2) (IUCN 2012, 2019). Part 1 of this field guide contains some of the conclusions reached from the results of these surveys.

Cultivation and Propagation

The information given is from Meunier *et al.* (2010). It is hoped that it will be useful for those intending to grow the species to use for their own purposes or contribute towards their conservation. Many of the trees covered here have known uses.

	EXTINCT
EX Extinct	A taxon is extinct when there is no reasonable doubt that the last individual has died.
EW Extinct in the Wild	Known only to survive in cultivation or as a naturalized population (or populations) well outside the past range.
	THREATENED
CR Critically Endangered	Facing an extremely high risk of extinction in the wild.
EN Endangered	Facing a very high risk of extinction in the wild.
VU Vulnerable	Facing a high risk of extinction in the wild.
	NOT THREATENED
NT Near Threatened	Likely to qualify for a threatened category in the near future.
LC Least Concern	Evaluated against the Criteria and found non-threatened
	INADEQUATE INFORMATION
DD Data Deficient	Assessed but inadequate information available to come to a conclusion.
	NOT EVALUATED
NE Not Evaluated	Has not been evaluated against the Criteria.

Table 4.2. IUCN Red List Categories of Threat. For further details (including about the Criteria used when making evaluations), see IUCN (2012, 2019).

The Plates

Leaves are reproduced at 50% of their natural size, except where stated otherwise. Typical leaves of species were selected to draw. The sketches of tree profiles (mainly in the lower left-hand corners of the plates) are reproduced at a scale of 1.25 cm to 10 m (the trees are 800 times bigger than depicted) and those of trunk bases (mainly in the lower right-hand corners of the plates) at a scale of 1.25 cm to 1 m (the trunks are 80 times bigger than depicted). The profiles and trunk bases of typical mature trees were selected to draw.

Accounts of the Species

TREE FERNS

These are ferns that have their foliage elevated above ground level on (usually unbranched) trunks. The growth form is reminiscent of palms. Leaves (fronds) much-divided. Unlike all other species included in this book, tree ferns lack flowers (or cones) and seeds. They reproduce through spores, which are borne, in *Cyathea*, in roundish yellowish to brown structures (sori) on the underside of the leaves.

***Cyathea manniana* Hook. (1)** Cyatheaceae
SYNONYM: *Cyathea deckenii* Kuhn (of ITU)
Tree fern (en); Ekigunju (ki); Segukio (ku); Lisuguku, Lusuguku (ms); Kinyaruba (na); Malere (to).
9 m. Trunk unbranched, bearing a crown of long, tripinnate (thrice-divided) leaves. The lower part of the petiole is very prickly.
OCCURRENCE: U2-4. The most abundant tree fern in Uganda, commonest in valleys, along rivers and near springs, especially at 1500-2500 m, e.g. in Bwindi, Kalinzu and Kayonza forests and on Mt Elgon and Rwenzori. Also on the Ssese Islands.
NOTES: The soft central part of the trunk forms part of the diet of the mountain gorilla.

***Cyathea dregei* Kunze (2)** Cyatheaceae
Tree fern (en); Kinyarabe (na); Malere (to).
A rare species of tree fern, differing from *C. manniana* in lacking prickles on the lower part of the petiole, which is, however, rough.
OCCURRENCE: U2-4. Known from Butiti Hills (Tooro), south Ankole (including near Rwashamaire), Bukasa Island (Ssese Islands) and Mt Elgon, 1135-2100 m. It grows at swamp edges, and along streams in grassland.

***Cyathea camerooniana* Hook. (3)** Cyatheaceae
Tree fern (en).
3 m. Differs from *C. manniana* and *C. dregei* in having bipinnate or bipinnatifid (twice-compound) leaves, rather than tripinnate (thrice-compound).
OCCURRENCE: U2 and 4. Mengo and Kigezi. Swamp and riverine forest, 1300-1500 m.

GYMNOSPERMS (ACROGYMNOSPERMAE)

Gymnosperms are seed-producing plants that lack protective envelopes covering theirs seeds. They are represented in Uganda by four species of indigenous conifers (species 4-7) and four species of cycads in the genus *Encephalartos* (7a). The conifers belong to the families Podocarpaceae (4-6) and Cupressaceae (7), the former having short, very narrow, leaves and the latter with very small leaves (reduced to scales). Species of *Encephalartos* (family Zamiaceae) bear a resemblance to palms, having a single very broad stem (or a few clumped stems) bearing a tuft of very large leathery pinnate leaves at the end. Only one of the cycads (*Encephalartos whitelockii*) can sometimes be a forest plant. Apart from tree ferns (species 1-3), all other species of forest trees in Uganda (species 8-447) belong to the Magnoliophyta (Angiospermae, otherwise Angiosperms or flowering plants), characterized by having their young female reproductive organs (ovules) enclosed within protective ovaries.

The most widely grown of the introduced conifers are the Mexican white cedar *Cupressus lusitanica* Mill. and the pines *Pinus caribaea* Morelet, *P. patula* Schiede *ex* Schltdl. & Cham. and *P. radiata* D. Don. *Cupressus lusitanica* Mill. (cypress) has very small leaves – like *Juniperus*, from which it differs in having 4-angled (rather than round) young stems (as seen in cross-section). Species of *Pinus* (pine) have long, very thin, leaves (needles) borne in bundles. The leaves are longer than 18 cm in *P. caribaea* and *P. patula* and shorter than 18 cm in *P. radiata*. The leaves are borne in 3s in the leaf bundles of *P. patula* and are variable in number in those of *P. caribaea*.

Podocarpus latifolius **(Thunb.) Mirb. (4)** **Podocarpaceae**

SYNONYM: *Podocarpus milanjianus* Rendle
Kiringi (am); Museenene (ga); Omuhurire, Omusenene (ki); Obwipe (ko); Sitetet (ku); Akikache (lo); Gumuhalamwa, Musagali (ms); East African yellow-wood, Podo (tn).
30 m. Trunk straight with a small crown or (in more open places) trunk branched low-down and crown spreading. Bark brown, fairly thin, rough and fibrous, peeling off in long narrow strips. Slash pink. Leaves c. 9 x 1 cm. Seeds blue-green, two together on a red fleshy receptacle.
OCCURRENCE: U1-4. In montane forest, 2000-3000 m. In Kigezi and on Rwenzori, Mt Elgon and the Imatong Mountains. Formerly abundant at 1140 m in Sango Bay swamp forest on the margin of Lake Victoria in Masaka, but large trees have been largely cut out from this locality.
CONSERVATION STATUS: Global LC (IUCN, TOU); National VU (WCS).
CULTIVATION AND PROPAGATION: Collect seeds from mother trees or from the ground beneath. Remove seeds from red fleshy receptacle and dry quickly to remove the seed coat. Soak the seeds in cold water for 24 hours and plant as soon as possible. The seedlings are slow-growing and require at least 6 months in the nursery. Protect young trees against being swamped by weeds.
NOTE: The wood of Podo was formerly much used (when the trees were more abundant).

Afrocarpus gracilior **(Pilg.) C.N. Page (5)** Podocarpaceae

SYNONYM: *Podocarpus gracilior* Pilger
Museenene (ga); Sapta, Saptet (ku); East African yellow-wood, Podo (tn).
35 m. Large tree, usually with a wide trunk and very large spreading branches from low down. Bark brown, fissured. Slash pink. Mature leaves small (c. 5 x 0.3 cm). Fruit blue-green, borne singly without a red receptacle.
OCCURRENCE: U1 and 3. At 2000-3000 m on Mts Elgon, Kadam and Moroto. Grows in climatically drier areas than *Podocarpus latifolius*.
CONSERVATION STATUS: Global LC (IUCN, TOU); National EN (WCS).

Afrocarpus dawei **(Stapf) C.N. Page (6)** Podocarpaceae

SYNONYM: *Podocarpus usambarensis* Pilger var. *dawei* (Stapf) Melville
Museenene (ga).
Leaves of similar size to those of *Afrocarpus gracilior*.
OCCURRENCE: U2 and 4. Grows with *Podocarpus latifolius* in Sango Bay swamp forest on the margin of Lake Victoria in Masaka. There is a record from Kayonza Forest.
CONSERVATION STATUS: Global NT (IUCN), LC (TOU); National CR (WCS).

Juniperus procera **Endl. (7)** Cupressaceae

Ethayoit (ka); Torokio (ku); African pencil cedar (tn).
40 m (usually much less in Uganda). Trunk with low branches. Crown pyramidal, spreading with age. Bark brown, thin, peeling in long narrow strips (similar to *Podocarpus latifolius*). Juvenile leaves c. 2.5 cm long. Leaves on adult shoots very small and scale-like, paired.
OCCURRENCE: U1 and 3. North-east Mt Elgon and mountains in Karamoja, 2000-2750 m. In dry montane forest.
NOTE: The timber has been much used in Kenya.

Encephalartos whitelockii **P.H.J. Hurter (7a)** Zamiaceae
Cycad (en).
SYNONYMS: *Encephalartos hildebrandtii* A. Braun & Bouché (of Melville FTEA 1958: 6 *pro parte*); *Encephalartos laurentianus* De Wild. (*sensu* ITU: 104); *Encephalartos successibus* Vorster
Trunk to 5 m and 1 m diameter, with a terminal 'shuttlecock' of leaves to 4 m long, each with many leathery leaflets with toothed margins. Plants are either male or female. Female reproductive structure a large green ovoid cone.
OCCURRENCE: U2. Only found near Mpanga River Falls, Tooro. Riverine forest, steep rocky slopes of a river gorge and open grassland. A Ugandan endemic.
CONSERVATION STATUS: Global CR (IUCN, TOU); National CR (WCS).
CITES: Appendix II.

MONOCOTYLEDONS

The Monocotyledons (8-18) form one of three divisions into which the flowering plants (Angiosperms) have recently been divided – Magnoliids, Monocotyledons (or simply monocots) and Eudicots (APG IV 2016). The Magnoliids and Eudicots together constitute the group traditionally known as Dicotyledons (or simply dicots), a category no longer used in formal scientific taxonomy. The great majority of species of forest trees in Uganda are dicots (19-447). A typical way in which monocots are distinctive is in the venation of the leaves. The leaves of monocots usually have numerous, more or less parallel, veins of about equal size running along the length of the leaf blade. In contrast, the leaves of dicots typically have a central main vein (midrib), with lateral veins emerging from it on either side, or else with a network (reticulum) of veins.

Bananas and their allies are Monocotyledons, though not trees from a botanical point of view (their 'trunks' are composed of the bases of the leaves, i.e. leaf sheaths). The cultivated banana (*Musa*) was introduced into Uganda from Asia probably during the early centuries AD. *Ensete ventricosum* (Welw.) Cheesman is a banana-like species indigenous to Uganda. Known in Luganda as *ekitembe*, this wild banana grows to a height of 3 m and is especially common in open wet valleys. It lacks edible fruits (bananas), though some other parts of the plant can be eaten. It is a staple food in many parts of Ethiopia.

PALMS AND SCREW PALMS

Only three species of trees in the palm family (Arecaceae, formerly Palmae) are found in forest vegetation in Uganda (8-10). Cycads (7a) and screw palms (11) resemble palms superficially. *Borassus aethiopum* Mart. is the tall unbranched palm with palmately-divided leaves found in savanna. The large prickly climbing palm (rattan) found in Semuliki National Park and Budongo and Mabira forests is *Calamus deerratus* G. Mann & H. Wendl. Isolated trees of the coconut, *Cocos nucifera* L., can be seen near Lake Victoria and at Butiaba and yield good fruit.

Phoenix reclinata **Jacq. (8)** Arecaceae
Otit (ac, la); Tit (al, la); Enkinu (am); Emusogot (at); Wild date palm (en); Lukindu (ga, to); Lukindukindu, Lukomakoma (ga); Mukindu (ga); Ekingol (ka); Itchi, Kia (md); Makendu (ms); Kikindu (na): Muiiti (nl); Omukindu (no); Muyiti (sa); Lusansa (so).
10 m. Stems often clumped. Trunk straight or curved, unbranched, with prominent leaf scars. Leaves clustered at ends of trunk, up to 3 m long, pinnately divided. Leaflets induplicate (V-shaped in cross-section, with the margins higher than the central midrib). Fruit a yellow drupe, c. 2.5 cm long, borne in large clusters.
OCCURRENCE: U1-4. Often abundant on the edges of swamps and other damp places. Normally at lower altitudes.

NOTES: The wood is resistant to termites and often used as poles (*enkoma* in Luganda). The trunk is reported to be tapped for wine in Buganda. The fruits are eaten by people and animals (including vervet monkeys). The tree is closely related to the date palm, *Phoenix dactylifera* L., which is not widely grown in Uganda.

Raphia farinifera (Gaertn.) Hylander (9) Arecaceae

Raphia palm (en); Kibo (ga); Omuswaale (no).

15 m (but usually much less). Leaves very large (up to 8 m long), pinnately divided. Leaflets reduplicate (Λ-shaped in cross-section, with the margins lower than the central midrib). Inflorescence large, terminal. Fruit brown and shiny, cone-like, covered with scales. The stem dies after flowering.

OCCURRENCE: U1-4. In swamp forest. Abundant in lake-belt forests, rare in western Uganda.

NOTES: The genus *Raphia* is distinguished by its huge leaves. *Raphia* fibre (*obuso* in Luganda) is obtained from the young leaflets. The fruits are eaten by bush pigs.

Elaeis guineensis Jacq. (10) Arecaceae

Esa, Mba (am); Oil palm (en).

Trunk to 15 m. Leaves pinnately divided, 3-5 m long. Leaflets reduplicate (Λ-shaped in cross-section, with the margins lower than the central midrib). Fruit a drupe c. 3.5 cm long, not scaly.

OCCURRENCE: U2. Occurs as a wild plant in swamp forest in Bwamba and in Nyamugasani delta forest in Queen Elizabeth National Park. Extensively planted on the Ssese Islands and Buvuma Island in Lake Victoria.

Pandanus chiliocarpus Stapf (11) *Pandanaceae*

SYNONYM: *Pandanus ugandaensis* H. St. John

Kigagara (am); Screw palm (en); Biskere (to).

15 m. Tree resembling a palm, stems procumbent or erect, little branched. Spines present on some stems. Stilt roots present, sometimes forming an intertwined mass. Leaves simple, spiny, very large, up to 3 m long, borne spirally in three ranks, bent over towards their ends.

OCCURRENCE: U1-3. In swamps and stream beds in forest. Gregarious and locally abundant, e.g. along the Dura River, Tooro. Recorded from Tooro, Bunyoro, Madi, Busoga and Kasyoha-Kitomi Forest.

CONSERVATION STATUS: Global NE (IUCN), NT (TOU); National NE.

DRACAENA

Trees (also shrubs and stragglers) with long, strap-like, leaves with parallel veins. *Dracaena laxissima* Engl. (numbered 15 on Plate 1) is an understorey shrub or straggler to 8 m, common in lake-belt forests and western Uganda. Its leaves are elliptic (c. 15 x 5 cm) and wider in proportion to length than the species listed below.

Dracaena steudneri Engl. (12) Dracaenaceae

Kagorogoro (am); Dragon tree (en); Kajjolyanjovu (ga); Gusiompo (ms); Mugorogoro (na); Omukyora (no); Oluwano (sa); Ngorogoro (to).

18 m. Trunk thick. Branches few, ascending. Leaves up to 1 m long and 10 cm wide, clustered at ends of branches.

OCCURRENCE: U1-4. In open forests, sometimes conspicuous in open valleys, 1300-2000 m. Abundant in Central Kibale Forest. Also in gardens, where it is spared or planted for medicinal reasons.

NOTE: The luganda name *kajjolyanjovu* means the 'sugar cane of the elephant'.

Dracaena afromontana **Mildbr. (13)** Dracaenaceae
Omugorogora (ki); Mahati, Muhoti (ko): Muramura (to).
10 m. Spreading understorey tree. Leaves clustered at the ends of the branches, long and thin (c. 23 x 2.25 cm).
OCCURRENCE: U1-3. Between 1800 and 2450 m in Kigezi and on Elgon, Kadam and Rwenzori.

Dracaena fragrans **(L.) Ker-Gawl. (14)** Dracaenaceae
Luwaanyi, Mulamula (ga); Karamura (to).
15 m. Shrub to small tree. Leaves long and thin (c. 50 x 3 cm), not markedly clustered at the ends of the branches.
OCCURRENCE: U2-4. Often abundant in wetter, lower altitude, forests, 650-2200 m. Commonly planted as a live hedge and marker between land holdings.

BAMBOOS

Bamboos are members of the grass family, Poaceae (formerly known as Gramineae), distinguished by being woody perennials. They are the largest members of this family, the larger ones being tree-like and sometimes referred to as 'tree grasses'. Unlike other trees, bamboos have underground stems (rhizomes) and the stems are segmented, usually hollow, and lack bark. They have long and irregular flowering cycles. Bamboos grow worldwide mostly in wet forest types, but one species in Uganda, *Oxytenanthera abyssinica* (A. Rich.) Munro, grows in savanna in northern Uganda. It is solid-stemmed. The widely-grown bamboo with yellow lines on green stems is an introduced species of *Bambusa*.

Sinarundinaria alpina **(K. Schum.) C.S. Chao & Renvoize (17)** Poaceae
SYNONYM: *Arundinaria alpina* K. Schum.
Luma (am); Mountain bamboo (en); Bbanda (ga); Omugano (ki); Tegandet (ku); Madega (ms); Mugano (na); Omuseke (no; Museke (to).
15 m. Hollow-stemmed bamboo. Stems not markedly clumped. Flowering in patches of a few hectares at intervals estimated to lie between 15 and 40 years. The stems die after flowering. However, gregarious flowering of this species has never been recorded from either Rwenzori or Kigezi.
OCCURRENCE: U2-3. Covering large areas and often dominant at 2450-3050 m on Rwenzori, the Bufumbira Volcanoes and Mt Elgon. Also in Bwindi Forest and Echuya Forest (2260-2450 m).
NOTES: The stems are used for making chairs and baskets and for other purposes. The young stems are eaten in Kigezi and Bugisu.

Oreobambos buchwaldii **K. Schum. (18)** Poaceae
Forest bamboo (en).
15 m. Hollow-stemmed bamboo with spreading stems.
OCCURRENCE: U2 and 4. In open swamps in Bunyoro, Busoga, Mengo and Masaka. Uncommon.

GIANT GROUNSELS (DENDROSENECIO)

Giant groundsel, Giant senecio (en); Mukoni (ko).
8 m. Trunk either unbranched or sparsely branched. Bark on older stems very thick. Youngest branches very broad, bearing terminal bunches of very large leaves. Dead leaves sometimes persistent. Flowers yellow, borne in huge terminal inflorescences.
OCCURRENCE: U2 and 3. On mountains above 2750 m. Most abundant above 3750 m, where giant groundsel forest or woodland can be a locally dominant vegetation type.

Key to *Dendrosenecio* (NB: the numbering of species diverges from that in UFT).

1. Plant in eastern Uganda. ....**20. *Dendrosenecio elgonensis***
   Plant in western Uganda. 2
2. On the Virunga Volcanoes. ....**21. *Dendrosenecio erici-rosenii***
   On the Rwenzori Range. 3
3. Primary stem very thick (over 4 cm wide); foliage withers but remains attached to stem, or with retained leaf bases; ray florets often absent or, if present, not prominent. ....**19. *Dendrosenecio adnivalis***
   Primary stem relatively thin (under 4 cm wide); leaf bases retained, but lamina decomposes after leaf death; ray florets prominent. **21. *Dendrosenecio erici-rosenii***

***Dendrosenecio adnivalis* (Stapf) E.B. Knox (19)** Asteraceae
OCCURRENCE: Rwenzori Mts, endemic.
NOTE: There are two subspecies:
(1) subsp. *adnivalis*. The lower surface of the lamina is glabrous to sparsely to densely pubescent. This subspecies has two varieties, var. *adnivalis* (Syn.: *Senecio adnivalis* Stapf; *S. erioneuron* Cotton), whose leaf lamina is not constricted to form a pseudo-petiole but extends along the midvein towards the base; and var. *petiolatus* (Hedberg) E.B. Knox (Syn.: *Senecio petiolatus* Hauman), whose lamina is constricted towards the base to form a pseudo-petiole.
(2) subsp. *friesiorum* (Mildbr.) E.B. Knox. The lower surface of the lamina is densely covered with a felty indumentum.

***Dendrosenecio elgonensis* (T.C.E. Fr.) E.B. Knox (20)** Asteraceae
OCCURRENCE: Mt Elgon.
NOTE: There are two subspecies:
(1) subsp. *barbatipes* (Hedberg) E.B. Knox (Syn.: *Senecio barbatipes* Hedberg; *S. gardneri* Cotton). Lower lamina surface with a dense felty indumentum. Mainly above 3960 m.
(2) subsp. *elgonensis* (Syn.: *Senecio amblyphyllus* Cotton; *S. elgonensis* T.C.E. Fr.) Lower lamina surface without dense felty indumentum but villose along midvein. Mainly below 3960 m. In uppermost montane forest and Ericaceous Belt.

***Dendrosenecio erici-rosenii* (R.E. Fr. & T.C.E. Fr.) E.B. Knox (21)** Asteraceae
NOTE: There are two subspecies:
(1) subsp. *erici-rosenii* (Syn.: *Senecio erici-rosenii* R.E. Fr. & T.C.E. Fr.). Lower surface of lamina glabrous or sparsely pubescent (but may be villose along midvein).
OCCURRENCE: Ruwenzori Mts and Mt Muhavura.
(2) subsp. *alticola* (T.C.E. Fr.) E.B. Knox (Syn.: *Senecio alticola* T.C.E. Fr.). Lower surface of lamina with woolly tufts of hairs or tomentose (with dense felty indumentum).
OCCURRENCE: Mt Muhavura only.

Plate 1. Unusual-looking trees (1-34)

1. *Cyathea manniana* 4. *Podocarpus latifolius* 5. *Afrocarpus gracilior*
8. *Phoenix reclinata* 9. *Raphia farinifera* 11. *Pandanus chiliocarpus*
12. *Dracaena steudneri* 14. *Dracaena fragrans* 15. *Dracaena laxissima*
20. *Dendrosenecio elgonensis* subsp. *elgonensis* 29. *Erica trimera* subsp. *trimera*
31. *Euphorbia teke* 34. *Euphorbia drupifera*

Actual sizes: leaves and fruit x 2; others various.

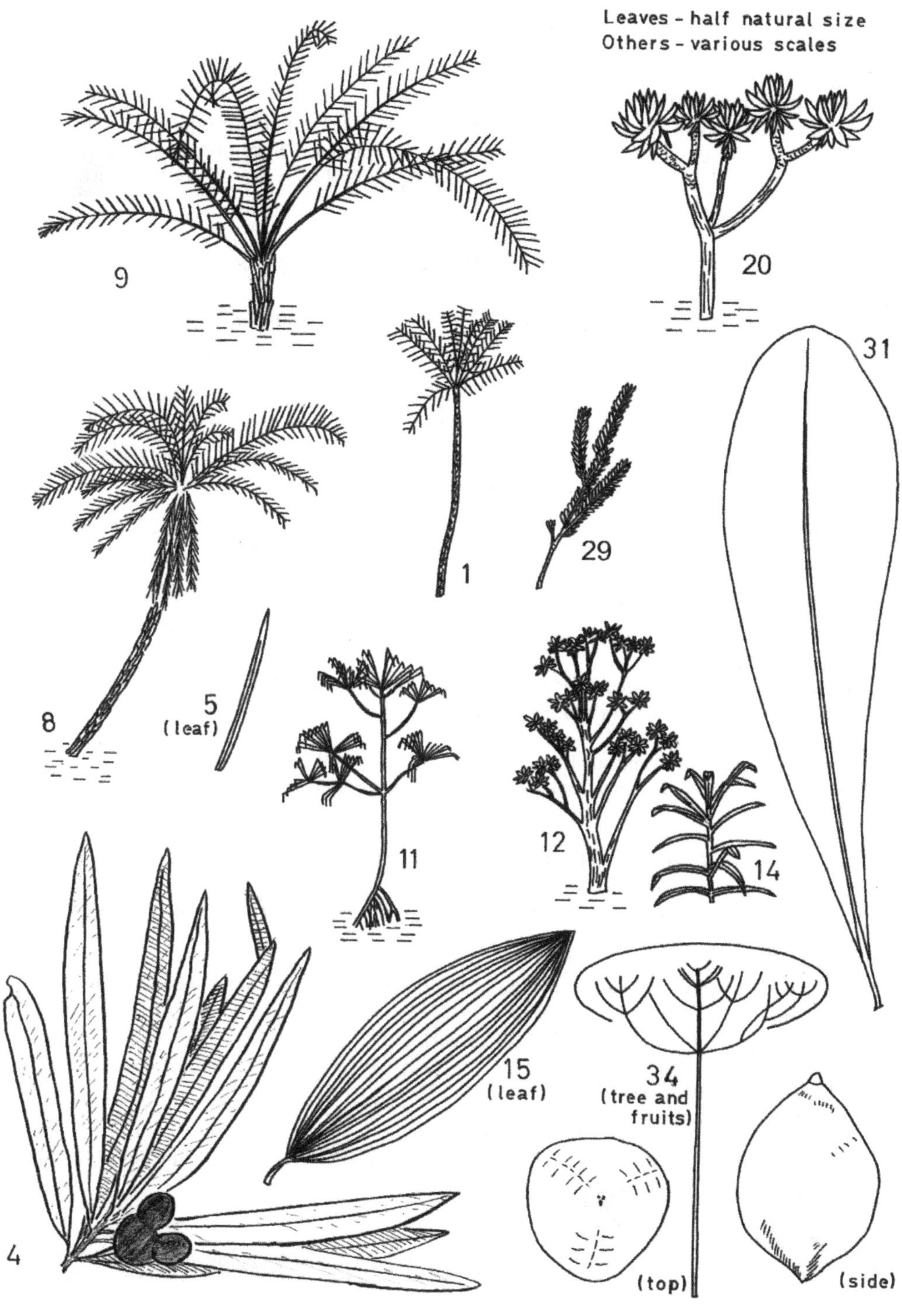
Leaves - half natural size
Others - various scales
9
20
31
1
29
8
5
(leaf)
11
12
14
15
(leaf)
34
(tree and
fruits)
4
(top)
(side)

Plate 1. (1-34)

ERICACEOUS TREES

Much-branched small trees (to 15 m) with very small leaves (less than 1 cm long), found only above an altitude of 2000 m (except for *Erica benguelensis*, which is usually a bush rather than a tree). Often dominant in the Ericaceous Belt (c. 3000-3700 m). The only other indigenous tree with such small leaves is *Juniperus procera* (7). Included here are *Seriphium* (family Asteraceae) and *Erica* (family Ericaceae). Some species of *Erica* are difficult to distinguish from one another in the vegetative state.

Key to Ericaceous trees (NB: the numbering of species of *Erica* diverges from that in UFT).

1. Foliage silvery. Elgon, Moroto, Kadam. ....**24. *Seriphium kilimandscharicum***
   Foliage green. 2
2. Calyx lobes equal in size. Flower stalk glabrous. 3
   Calyx lobes unequal in size. Flower stalk hairy. ....**25. *Erica bengulensis***
3. Young stems noticeably hairy to the naked eye. ....**26. *Erica arborea***
   Young stems not noticeably hairy to the naked eye. 4
4. Corolla lobes in threes. ....**27. *Erica rossii***
   Corolla lobes in fours. 5
5. Stigma capitate (head-shaped at tip). ....**28. *Erica kingaensis***
   Stigma flat, peltate. ....**29. *Erica trimera***

Seriphium kilimandscharicum **(O. Hoffm.) Koekemoer (24)** Asteraceae

SYNONYM: *Stoebe kilimandscharica* O. Hoffm.
Sesindot (ku); Mututu (ms).
8 m. Shrub with silvery leaves and small yellow flowers. Leaves c. 0.3 cm long.
OCCURRENCE: U1 and 3. In Ericaceous forest and burnt grassland on Mts Elgon, Moroto and Kadam, 2450-3300 m.

Erica benguelensis **(Engl.) E.G.H. Oliv. (25)** Ericaceae

SYNONYM: *Philippia benguelensis* (Welw. ex Engl.) Britten
Tree heather (en); Ekihungwe, Omuhungye (ki); Hungi, Kisusuti (ko); Kapsigaga, Kipsigaga (ku); Muhugye (na).
8 m, but usually a bush less than 3 m tall.
OCCURRENCE: U2. 1500-2500 m. Mainly on open stony ground. Very common on open ridges in the Rukiga Highlands.
NOTE: Only the variety *benguelensis* occurs in Uganda.

Erica arborea **L. (26)** Ericaceae

Tree heather (en); Kapsigaga, Kipsigaga (ku).
8 m. Shrub or tree. Flowers white, sometimes abundant.
OCCURRENCE: U1, 2 and 4. Bufumbira Volcanoes, Rwenzori, Mt Elgon and mountains in Karamoja.
NOTE: An important constituent of the Ericaceous Belt.

Erica rossii **Dorr (27)** Ericaceae

SYNONYMS: *Philippia excelsa* Alm & T.C.E. Fr. (of ITU & UFT); *Philippia johnstonii* Engl. (of ITU & UFT)
Tree heather (en); Hungi (ko); Kapsigaga, Kipsigaga (ku).
8 m. Leaves 1-3 mm long, in whorls of three.
OCCURRENCE: U2 and 3. Rwenzori, Mt Muhavura and Mt Elgon.

Erica kingaensis Engl. (28) Ericaceae
SYNONYMS: *Erica bequaertii* De Wild.; *Erica ruwenzoriensis* Alm & T.C.E. Fr.
Tree heather (en).
15 m (5 m on bogs in Kigezi). Flowers white to pale pink.
OCCURRENCE: U2. Rwenzori and Kigezi. On bogs and ridges.
NOTE: Two subspecies occur in Uganda: (1) subsp. *bequaertii* (De Wild) R. Ross, which has leaf margins and flower stalks with no or simple hairs and is a Rwenzori endemic, occurring on both sides of the Uganda/D.R. Congo border; (2) subsp. *rugegensis* (Engl.) Alm & Fries, which has leaf margins and flower stalks with branched hairs and occurs in Kigezi.

Erica trimera (Engl.) Beentje (29) Ericaceae
SYNONYM: *Philippia trimera* Engl.
Tree heather (en); Hungi (ko).
15 m. Leaves in whorls of three. Leaves are relatively large (4-6 mm long) and thick, and lie in well-defined rows.
OCCURRENCE: U2-3. Rwenzori and Mt Elgon. The dominant species on Rwenzori between 3300 and 3660 (-4000) m.
NOTE: Two subspecies occur in Uganda, one on Rwenzori (subsp. *trimera*) and the other on Mt Elgon (subsp. *elgonensis* (Mildbr.) Beentje). The branchlets of subsp. *trimera* have stalked glandular hairs only, while those of subsp. *elgonensis* have minute non-glandular hairs.

SUCCULENT EUPHORBIA

Trees with succulent leaves, bearing spines or prickles, producing copious white latex. Note: all succulent species of *Euphorbia* are listed in Appendix II of CITES (see introduction to Part 4).

Euphorbia teke Pax (31) Euphorbiaceae
Mukoni, Nabanteta (ga).
7 m. Spreading succulent understorey tree with a dense crown. Bark light brown, bearing small paired spines. Slash soft, white to yellow, exuding copious white latex. Branches green and succulent, 4-angled, with paired spines. Leaves fleshy, obovate, c. 18 x 6.5 cm.
OCCURRENCE: U2 and 4. Mengo, Masaka, Bunyoro. In lower to mid-altitude forests. Mainly in damp places.

Euphorbia ampliphylla Pax (32) Euphorbiacae
MISAPPLIED NAME: *Euphorbia obovalifolia* A. Rich.
Large tree to 30 m. Trunk cylindrical. Bark thin, light-coloured. Slash pale green with white latex, shoots green and succulent, 3- to 4-angled. Spines paired, Leaves obovate.
OCCURRENCE: U1 and 3. Mts Elgon and Morongole (Karamoja), 2000-2500 m.

Euphorbia bwambensis S. Carter (33) Euphorbiaceae
Euphorbia sp. of ITU (specimen Eggeling 3368).
7m. Understorey tree. Trunk straight. Branches 3-angled. Spines single and very small.
OCCURRENCE: U2. Bwamba.
CONSERVATION STATUS: Global VU (IUCN), LC (TOU); National VU (WCS).

Euphorbia drupifera Thonn. (34) Euphorbiaceae
SYNONYMS: *Elaeophorbia* sp. nov. of UFT; *Elaeophorbia drupifera* (Thonn.) Stapf
Kididi (am), Nkukuru (to).
22 m. Tree with straight trunk, branches succulent and angular, at first at right angles, then curving up. Leaves borne in clusters at ends of branches. Bark grey, with small rounded ridges, thin and smooth, flaking in places. Phellogen green. Slash very soft, fairly thick, white, exuding copious white latex. Leaves fleshy, obovate, c. 20 x 8 cm.
OCCURRENCE: U2. Maramagambo, South Kibale and Bwamba forests, usually with *Cynometra*.

SAPOTACEAE

Trees with simple, alternate (spirally-arranged) leaves and white latex in the slash. Common small to very large trees, found in many forests in Uganda. Flowers small. Fruits indehiscent. Confusion is only likely with members of the Moraceae. Characters for the separation of the two families are given in Sub-Key 1 in Part 3.

Key to Sapotaceae.

1. Undersurface of leaves densely covered with red to red-brown hairs. 2
   Not as above. Undersurface of leaves sometimes densely covered with white, yellow or brown hairs. 6

2. Leaf base asymmetric. Lateral veins 8-11 on each side of the midrib, widely spaced. Only known from Bwamba Forest. .... **44. *Gambeya beguei***
   Leaf base symmetric or almost so. 3

3. Tree found above 1300 m (e.g. in Central Kibale, Kalinzu, Kayonza and Bwindi forests and on Rwenzori and Elgon). 4
   Tree found below 1300 m (e.g. in lake-belt forests, Budongo and Maramagambo forests). 5

4. Leaves relatively broad in proportion to length (often c. 10 x 4.5 cm). Margins of exposed leaves inrolled. ....**36. *Pouteria adolfi-friedericii***
   Leaves relatively narrow in proportion to length (often c. 16 x 5 cm), margins not inrolled. ....**42. *Gambeya gorungosana***

5. Leaves bright red-brown below, relatively broad (often c. 20 x 7.5 cm). ....**41. *Gambeya perpulchra***
   Leaves not as strikingly red-brown as above, relatively narrow (often c. 15 x 4.5 cm). ....**43. *Gambeya gorungosana***

6. Undersurface of leaves with a dense covering of white, yellow or brown hairs and therefore conspicuously different in colour from upper surface. 7
   Undersurface of leaves not as above, of about the same colour on both surfaces or of different shades of green. 13

7. Shoots growing by repeated subapical branching. Leaves in clusters separated by bare stems. ....**37-38. *Englerophytum***
   Method of branching not as above. 8

8. Lateral veins narrow and numerous or difficult to see. ....**50-51. *Manilkara***
   Lateral veins not narrow and numerous, prominent on leaf undersurface. 9

9. Leaves large (over 20 x 7 cm). 10
   Leaves small (less than 20 cm long and/or less than 7 cm wide). 12

10. Petiole c. 1 cm long or less. ....**54. *Synsepalum msolo***
    Petiole usually over 1.25 cm long. 11

11. Lamina usually distinctly widest in upper half. Main lateral veins c. 18-22 on each side of the midrib. ....**52. *Manilkara dawei***
    Lamina either not or only slightly widest in upper half. Main lateral veins c. 9-16 on each side of the midrib. ....**40. *Gambeya albida***

12. Leaves whitish below. Secondary lateral veins present. ......**39. *Gambeya muerensis***
Leaves brownish below. Secondary lateral veins absent. **42. *Gambeya gorungosana***

13. Lateral veins narrow and numerous, or difficult to see. 14
Lateral veins not narrow and numerous, being comparatively wide and prominent. 16

14. Long, narrow acumen present (see illustration). Lateral veins very numerous, closely parallel, at right angles to the midrib or slightly ascending. Recorded from Itwara and Kayonza forests only. ....**45-46. *Donella ubangiensis, D. pruniformis***
Acumen relatively short or absent. 15

15. Leaves clustered at ends of branches. Leaves often noticeably widest in upper half of lamina, apex various. ....**50-52. *Manilkara***
Leaves not particularly clustered at ends of branches. Leaves often, but not always, widest in about the centre of the lamina, apex usually acuminate. ....**47-48. *Mimusops***

16. Leaves large (often c. 35 x 14 cm). ....**54. *Synsepalum msolo***
Leaves smaller than above. If over 25 cm long, then relatively narrow in proportion to length. 17

17. Base of leaf rounded or only slightly cuneate; leaves comparatively broad in proportion to length (often c. 13 x 4.5 cm). ....**35. *Pouteria altissima***
Base of leaf cuneate or, if not cuneate, then leaves narrower in proportion to length than above. 18

18. Leaves glabrous or nearly so below. .. **53. *Synsepalum brevipes*; 55. *S. cerasiferum***
Leaves markedly hairy below. ....**42. *Gambeya gorungosana***

***Pouteria altissima* (A. Chev.) Baehni (35)** Sapotaceae
SYNONYM: *Aningeria altissima* (A. Chev.) Aubrév. & Pellegr.
Nkalati (ga); Mutokye (na); Osan (tn); Mutoke (to).
50 m. Large tree with a fairly spreading crown. Trunk straight, usually fluted for some height, the flanges sometimes spreading out at the base to form buttresses. Bark brown, with vertical fissures, but general effect smooth. Slash fibrous, most often red-brown with white lines, but varying from white to pink or yellow-brown, exuding white latex. Leaves simple, alternate, entire, c. 13 x 4.5 cm (but variable in size), with 13-24 main lateral veins on each side of the midrib. Mature leaves glabrous except (sometimes) for the midrib below. Petiole c. 1 cm long. OCCURRENCE: U1-4. Widespread below 1500 m. Abundant in lake-belt forests and in Tooro and Ankole.

***Pouteria adolfi-friedericii* (Engl.) A. Meeuse (36)** Sapotaceae
SYNONYM: *Aningeria adolfi-friedericii* (Engl.) Robyns & Gilbert
Sosi (ko); Lulyo (ku); Mwiruni (ms).
50 m. Trunk straight with a medium-sized crown which appears brown from below. Trunk usually fluted, the flanges sometimes spreading out at the base to form buttresses, which may be large. Bark smooth, light-coloured. Slash white, often with yellow stone cells, exuding white latex. Young shoots and undersurface of mature leaves covered with yellow hairs. Leaves simple, alternate, c. 8.5 x 4 cm to c. 12 x 5 cm (but much larger on young plants), with c. 15-20 main lateral veins on each side of the midrib, margin inrolled on crown leaves. Petiole c. 1.5 cm long.

OCCURRENCE: U1-3. A montane tree found between 1800 and 2450 m in Kigezi and on Elgon, the Imatongs and Rwenzori. Abundant in 1971 on the western slopes of Elgon.
NOTE: *Gambeya gorungosana* (42) is rather similar and grows at similar altitudes, but is absent from Mt Elgon.

Englerophytum oblanceolatum (S. Moore) T.D. Penn. (37) Sapotaceae
SYNONYM: *Bequaertiodendron oblanceolatum* (S. Moore) Heine & J.H. Hemsl.
Munyamata (to).
15 m (occasionally to 25 m). Trunk straight or crooked, fluted on larger trees. Branches at all heights. Bark brown, moderately thick, flaking. Slash fibrous, white, pink to light brown, exuding white latex. Shoot growth is by repeated subapical branching, clusters of leaves being separated by bare stems. Leaves simple, alternate, c. 14 x 4.5 cm, with c. 12-25 main lateral veins on each side of the midrib, these veins sometimes not readily distinguishable from the secondary lateral veins, whitish below. Petiole c. 1 cm long.
OCCURRENCE: U2 and 4. Mengo, Tooro, Bunyoro. An abundant understorey tree in Central Kibale, Budongo and Mabira forests.

Englerophytum natalense (Sond.) T.D. Penn. (38) Sapotaceae
SYNONYM: *Bequaertiodendron natalense* (Sond.) Heine & J.H. Hemsl.
25 m. Differs from *E. oblanceolatum* in lacking stipules. Leaves c. 11 x 2.75 cm.
OCCURRENCE: U2. Only recorded from a swamp forest on the edge of Lutoto Crater, Ankole.

Gambeya muerensis (Engl.) Liben (39) Sapotaceae
SYNONYM: *Chrysophyllum muerense* Engl.
Omunyamata (no).
40 m. Understorey or canopy tree with a straight cylindrical trunk. Branches at right angles on young trees. Buttresses present on old trees. Bark brown, thin and smooth, with vertical fissures. Slash fibrous, red on old trees, varying from white to pink to yellow on young trees, exuding white latex. Leaves simple, alternate, regularly and alternately arranged on the branchlets, thin and narrow, c. 16 x 4 cm, with 13-20 main lateral veins on each side of the midrib, glabrous above, hairy and whitish below. Petiole c. 1.5 cm long.
OCCURRENCE: U1, 2 and 4. Widely distributed. Abundant in Mabira Forest.
CONSERVATION STATUS: Global NE (IUCN), LC (TOU); National VU (WCS).
NOTE: The leaves differ from those of other species of *Gambeya* in having secondary lateral veins. They are quite similar to those of *Englerophytum*, which is, however, easily recognized by its peculiar method of shoot growth.

Gambeya albida (G.Don) Aubrév. & Pellegr. (40) Sapotaceae
SYNONYM: *Chrysophyllum albidum* G. Don
Badongulo (am); White star apple (en); Mululu (ga, so, tn); Nkalati (ga); Omushayu (ki); Muhinguba, Mulyanyoni (na); Omubakampungu, Omululu, Omunyamata (no); Muhubu (sa); Muhambulya (to).

Plate 2. Sapotaceae (35-46)

35. *Pouteria altissima* 36. *Pouteria adolfi-friedericii* 37. *Englerodendron oblanceolatum* 39. *Gambeya muerensis* 40. *Gambeya albida* 41. *Gambeya perpulchra* 42-43. *Gambeya gorungosana* 46. *Donella pruniformis*

Actual sizes: leaves and fruit x 2; trunk bases x 80; tree profiles x 800.

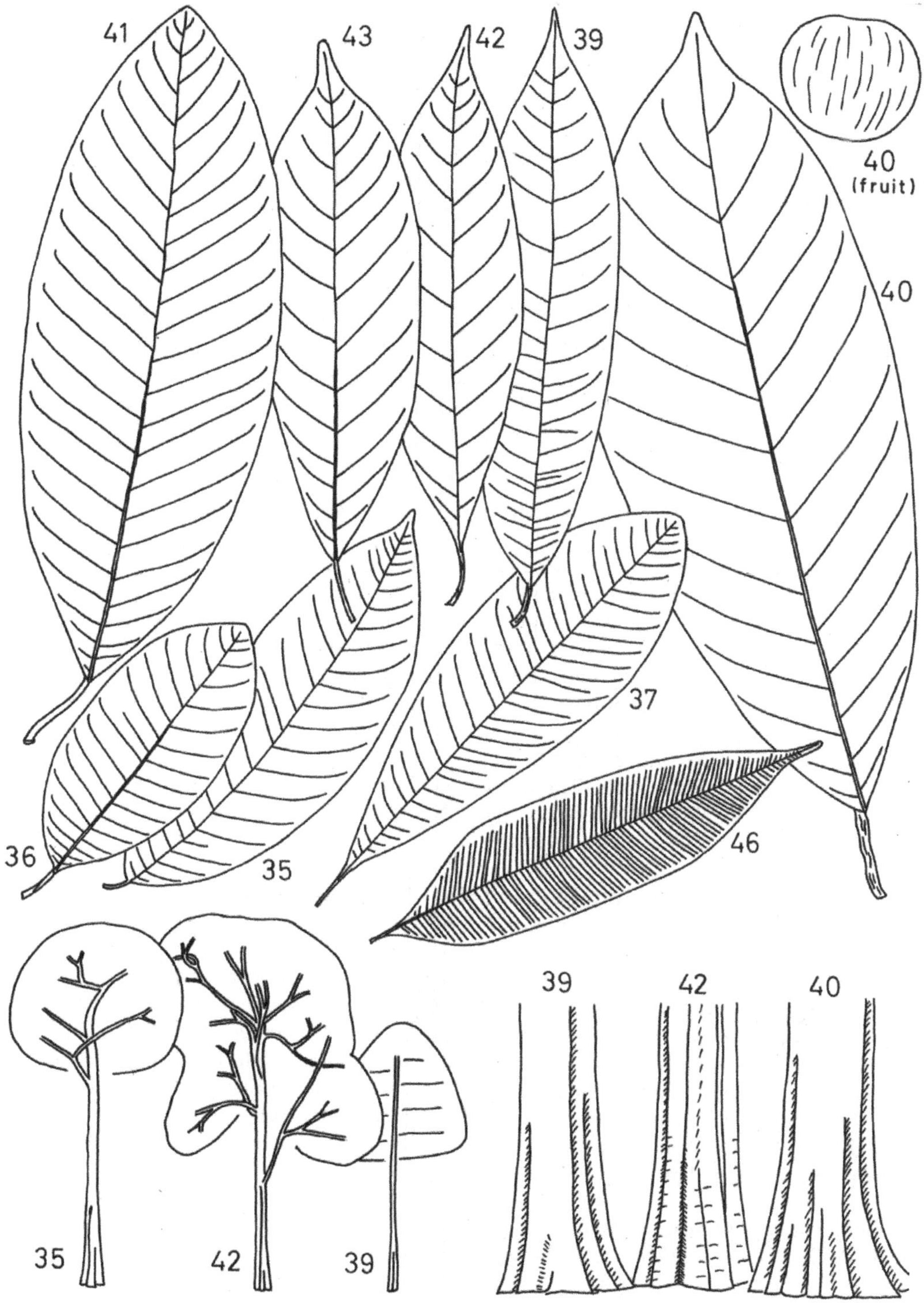

Plate 2. (35-46)

45 m. Tall tree with a straight trunk and dense, rounded, crown. Underside of leaves white as seen from the ground. Trunk fluted for some distance, the flanges often spreading out at the base to form buttresses. Bark quite thin, brown, vertically fissured, but general effect smooth. Slash fibrous, most often white with brown lines, sometimes red-brown, exuding white latex mainly from near the wood. The slash turns slowly darker. Leaves simple, alternate, c. 25 x 9 cm, upper surface glabrous, lower surface hairy and light brown to whitish (young leaves may be green below), with c. 9-16 main lateral veins on each side of the midrib. Petiole c. 2.5 cm long. Fruit depressed-spherical, c. 3 cm in diameter, yellow to yellow-brown when ripe, 5-ribbed.
OCCURRENCE: U1-4. A widespread lower altitude species, abundant in Budongo, Mabira and other forests.
CONSERVATION STATUS: Global: NE (IUCN), LC (TOU); National VU (WCS).
CULTIVATION AND PROPAGATION: Moderately fast-growing. Seedlings require shade during first year of growth. Prefers a mixed stand. Collect fruits from mother trees or from the ground beneath and remove the seeds from the fleshy pulp. Germination rate likely to be high.
NOTES: The tree may be distinguished from *Gambeya muerensis* by the broader leaves. The timber is strong. The fruit is edible.

Gambeya perpulchra (Mildbr. ex Hutch. & Dalziel) Aubrév. & Pellegr. (41)
Sapotaceae
SYNONYMS: *Chrysophyllum perpulchrum* Hutch. & Dalziel
Monkey star apple (en); Omubakampungu (no); Mululu (tn).
35 m. Trunk thin and straight. Crown rounded. Undersurface of leaves very conspicuously red or red-brown as seen from the ground. Trunk fluted. Bark smooth, pale brown. Slash brown with paler streaks, exuding white latex. Leaves simple, alternate, c. 20 x 7.5 cm, with c. 14-24 main lateral veins on each side of the midrib, upper surface glabrous, lower surface with a very dense and characteristic covering of red-brown hairs. Petiole c. 2.5 cm long. Fruit spherical, c. 2.5 cm diameter, covered with red hairs.
OCCURRENCE: U2 and 4. Abundant in Budongo Forest. Occasional in West Mabira.
CONSERVATION STATUS: Global NE (IUCN), LC (TOU); National VU (WCS).
NOTE: The fruit is edible.

Gambeya gorungosana (Engl.) Liben (42) Sapotaceae
SYNONYMS: *Chrysophyllum gorungosanum* Engl.; *Chrysophyllum fulvum* S. Moore of ITU; *Chrysophyllum delevoyi* De Wild.
40 m. Trunk long and straight. Crown dense, small to spreading. Trunk characteristically fluted for much of its length. Buttresses sometimes present. Bark light brown, fairly thin, with vertical fissures (the bark is rougher than *Pouteria altissima*, but much smoother than *Mimusops*). Slash fibrous, brown, red-brown or red, with white streaks, exuding white latex. The slash turns slowly darker. Leaves simple, alternate, c. 16 x 5 cm, with c. 13-18 main lateral veins on each side of the midrib, undersurface with red to silvery-brown hairs. Petiole c. 1.5 cm long.
OCCURRENCE: U1, 2 and 4. Altitudinal range 1300-2300 m. Abundant in Bwindi, Kayonza and Kalinzu forests and parts of Kibale.

Gambeya beguei Aubrév. & Pellegr. (44) Sapotaceae
SYNONYM: *Chrysophyllum beguei* Aubrév. & Pellegr.
30 m. Easily distinguished from other species of *Gambeya* by its asymmetric leaf base and the relatively few and widely spaced main lateral veins (8-11 on each side of the midrib).
OCCURRENCE: U2. Only recorded from Bwamba at 750 m.

Donella ubangiensis **(De Wild.) Aubrév. (45)** Sapotaceae

SYNONYMS: *Chrysophyllum ubangiense* (De Wild.) D. J. Harris; *Chrysophyllum pentagonocarpum* Engl. & K. Krause

25 m. Leaves simple, alternate, c. 14 x 5 cm, acuminate, glabrous on both surfaces. Main lateral veins numerous, more or less straight, running almost perpendicular to the midrib. Petiole c. 0.5 cm long. This species and *Donella pruniformis* differ from *Manilkara* in having hairy young parts.

OCCURRENCE: U2. Only recorded from Itwara Forest (Tooro) at 1500 m.

Donella pruniformis **(Engl.) Pierre ex Engl. (46)** Sapotaceae

SYNONYM: *Chrysophyllum pruniforme* Engl.

30 m. Leaves similar in shape and venation to those of *Donella ubangiensis*, but smaller (c. 12 x 3.5 cm), acuminate, glabrous above and below (except for the midrib). Petiole c. 0.8 cm long.

OCCURRENCE: U2. Only recorded from Ishasha Gorge (Bwindi Forest), 1500 m.

Mimusops bagshawei **S. Moore (47)** Sapotaceae

Mbande (am); Musaali, Musandasanda (ga).

40 m. Trunk straight and cylindrical, sometimes slightly fluted at base. Crown large and spreading. Buttresses absent. Bark brown, layered, very thick, with deep vertical fissures and sometimes also horizontal fissures (cutting the bark into rectangles). Slash fibrous, red, sometimes with paler lines, exuding white latex. Leaves simple, alternate, c. 10 x 3.5 cm, glabrous, with many rather obscure veins of all sizes joining a single vein parallel with and just inside the leaf margin. Midrib not sunk into a groove above. Petiole c. 1.75 cm long. Fruit yellow, c. 2.5 cm long.

OCCURRENCE: U2-4. A widespread and common lower altitude species.

NOTES: Differs from *Manilkara* in not having the leaves markedly clustered towards the ends of the branchlets.

Mimusops kummel **A. DC. (48)** Sapotaceae

Njenjeka (am); Elepolepo (at); Aitareng (ka).

A similar-looking tree to *Mimusops bagshawei*, but smaller (25 m). The flower stalks are over 2 cm long (longer than those of *M. bagshawei*).

OCCURRENCE: U1-4. Drier areas, especially in riparian forest (for instance at the base of Mt Moroto). Possibly absent from true rainforest.

Manilkara butugi **Chiov. (50)** Sapotaceae

35 m. Trunk long and straight, cylindrical to slightly fluted. Crown spreading. Bark grey-brown, rough and fissured. Leaves simple, alternate, c. 14 x 5.5 cm, widest in upper half, apex acuminate, base cuneate, lateral veins numerous and rather inconspicuous. Petiole c. 3 cm long.

OCCURRENCE: U1-3. A higher altitude species, 1500-2300 m.

NOTES: Very similar to *Manilkara obovata*. The altitudinal ranges of the two species appear not to overlap.

Manilkara obovata **(Sabine & G. Don) J.H. Hemsl. (51)** Sapotaceae

SYNONYM: *Manilkara multinervis* (Baker) Dubard

Nkunya (ga).

40 m. Trunk straight, slightly fluted, buttresses sometimes present. Crown spreading extensively. Bark brown, layered, very thick and rough, with deep vertical fissures and a tendency to split into rectangles. Slash fibrous, bright red, sometimes streaked with a paler shade, turning darker, exuding white latex. Leaves simple, alternate, clustered at the ends of the branches, c. 10 x 4 cm (to c. 13 x 7 cm on some trees), distinctly widest in upper half, glabrous, midrib sunk in a groove on the upper surface. Lateral veins numerous, straight, many small and hair-like (similar to *Manilkara butugi*). Petiole c. 2.5 cm long.

OCCURRENCE: U1 and 4. In rainforest, riverine forest, swamp forest (in Buddu and on the Ssese Islands) and woodland, 900-1300 m. On upper slopes in Mabira Forest.

NOTE: Has a strong resemblance to *Mimusops bagshawei*. However, the leaves are clustered at the ends of the branches, the leaf is usually more distinctly obovate and the midrib is impressed into the upper surface of the lamina (rather than being slightly raised).

Manilkara dawei (Stapf) Chiov. (52) Sapotaceae
25 m. Trunk fluted at base. Bark brown, fissured. Slash fibrous, pink to white, turning darker, exuding white latex. Leaves clustered at ends of branches, leathery, c. 25 x 10 cm, usually widest in upper half, with c. 18-22 main and very prominent lateral veins on each side of the midrib, glabrous above, hairy and whitish below. Petiole c. 4 cm long.
OCCURRENCE: U1-4. Widespread, but not common.

Synsepalum brevipes (Baker) T.D. Penn. (53) Sapotaceae
SYNONYM: *Pachystela brevipes* (Baker) Engl.
Nkalati (ga).
30m. Trunk straight (in forest specimens). Crown fairly spreading and quite deep. Trunk fluted, the flanges sometimes spreading at the base and becoming buttresses. Bark light brown, vertically fissured (like *Gambeya gorungosana*). Slash red to almost white, turning darker, exuding white latex. Leaves simple, alternate, clustered at the ends of the branches, c. 14 x 5.5 cm, with c. 7-14 main lateral veins on each side of the midrib, widest in upper half, usually acuminate, base long cuneate, glabrous or slightly hairy beneath.
OCCURRENCE: U2 and 4. Widespread, often near lakes and rivers. Abundant in lake-shore forests.

Synsepalum msolo (Engl.) T.D. Penn. (54) Sapotaceae
SYNONYM: *Pachystela msolo* (Engl.) Engl.
50 m. Trunk deeply fluted. Crown much-branched, spreading. Leaves simple, alternate, large, c. 35 x 14 cm, with c. 10-21 main lateral veins on each side of the midrib, widest in upper half, apex often acuminate, base long cuneate, abruptly obtuse or sub-auriculate right at the end, upper surface glabrous, lower surface hairy and whitish. Petiole c. 1 cm long.
OCCURRENCE: U2 and 3. Recorded from Busoga and also Bugoma and Bwamba forests, rare.
NOTE: Distinguished by its large leaves.

Synsepalum cerasiferum (Welw.) T.D. Penn. (55) Sapotaceae
SYNONYM: *Afrosersalisia cerasifera* (Welw.) Aubrév.
Nkalati (ga); Nkoba (na).
30 m. Trunk straight. Crown often thick. Trunk fluted. Bark brown, thick and rough, vertically fissured, with occasional horizontal fissures. Slash fibrous, red, yellowish red to almost white, exuding white latex. Leaves simple, alternate, tending to be clustered towards the ends of the branches, c. 13 x 5 cm (but sometimes much longer, c. 25 x 7 cm), with c. 8-14 main lateral veins on each side of the midrib, dark green with yellowish-green veins, apex rounded to acute, but not acuminate, glabrous on both surfaces. Petiole c. 0.5 cm long.
OCCURRENCE: U1-4. Widespread. Most abundant between 1300 and 1600 m.
NOTE: The leaves can usually be told from those of *Synsepalum brevipes* by the absence of an acumen.

Plate 3. Sapotaceae (47-55)

47. *Mimusops bagshawei* 48. *Mimusops kummel* 49, 51. *Manilkara obovata*
52. *Manilkara dawei* 53. *Synsepalum brevipes* 55. *Synsepalum cerasiferum*

Actual sizes: leaves and fruits x 2; trunk bases x 80; tree profiles x 800.

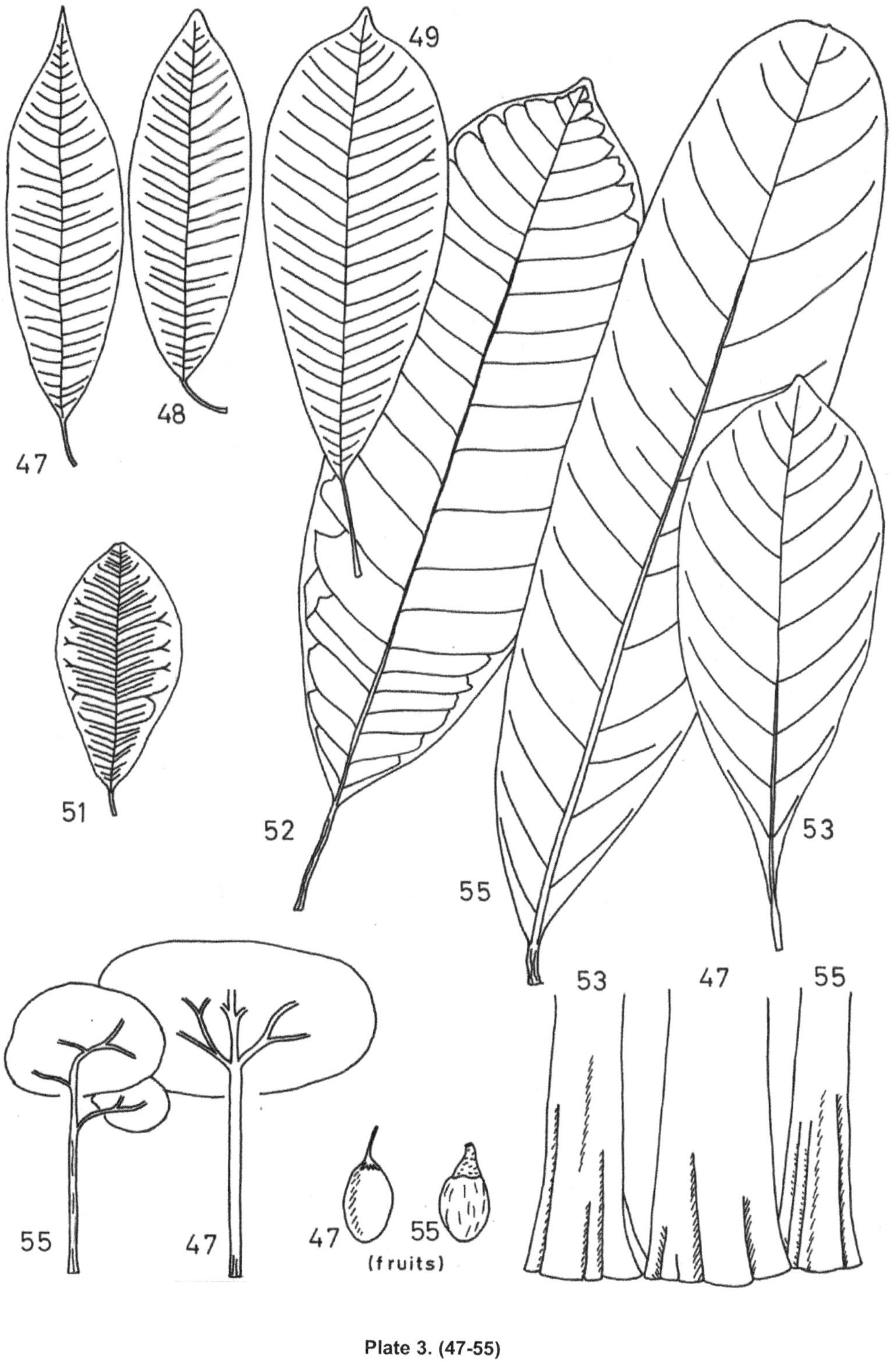

Plate 3. (47-55)

MORACEAE

Trees with alternate (spirally arranged) simple leaves and usually white or off-white latex in the slash, which sometimes turns darker. Bark thin, smooth (except in *Ficus vallis-choudae* and old *Milicia excelsa*), often with prominent lenticels and a green phellogen. Stipules often large, enfolding the young leaves. Members of this family are common forest trees. They may be distinguished from Sapotaceae by the characters given in Sub-Key 1 in Part 3. The genera *Myrianthus* and *Musanga* (364-366) were included in Moraceae in ITU and UFT, but are now placed in Urticaceae.

Fig trees (*Ficus*) belong to the Moraceae. They can generally be recognized by the following characteristics: crowns often rather open, light-coloured; buttresses or aerial roots generally present; bark thin and light-coloured, with a green phellogen; slash usually with abundant white or off-white latex. The flowers of figs are borne on the inside of a hollow receptacle (fig) that has a small opening (ostiole) near its apex. Fig trees are very common in many forests, as well as in cultivated land and savanna. They are not found above about 2400 m and are uncommon towards the upper part of their altitudinal range. Some figs are free-living, but others start life as epiphytes, send down roots to the ground and finally strangle their hosts. After assuming independence, they use the space previously occupied by their hosts and grow as normal trees. Fig trees that were epiphytic earlier in their lives often have trunks with hollow interiors and an abundance of interwoven aerial roots.

Key to Moraceae (excluding *Ficus katendei* – see 75a).

1. Leaves characteristically shaped, with a few large teeth near the apex and a long acumen. Understorey tree to 6 m, known from Mengo, Bunyoro, Tooro and Bwamba. Rare. ....**62. *Dorstenia***
   Leaf shape not as above. 2

2. White or off-white latex absent from the slash. 3
   White or off-white latex present in the slash. 4

3. Leaves very rough on both surfaces. Note: *Ficus sycomorus* (64) is another species of *Ficus* with rough leaves. It is mostly found in savanna. ....**63. *Ficus exasperata*; 68. *F. asperifolia***
   Leaves not very rough on both surfaces. ....**68. *F. asperifolia*; 73. *Ficus ingens***

4. Leaves often c. 7 x 5 cm, 3-veined from the base, margin toothed or crenate (see Plate 4). Can be a tall tree. ....**58. *Morus***
   Leaf shape not as above. 5

5. Leaves large (often c. 33 x 14 cm) with entire margins, red or yellow when young. Latex and slash discolouring. ....**60. *Treculia***
   Leaves not red or yellow when young and (except for *Ficus saussureana* and *F. vogeliana*) smaller than above. The latex in *F. saussureana* does not discolour and the leaf margin in *F. vogeliana* is toothed or wavy. 6

6. Slash granular, at least in part. Bark dark-coloured, often rough, with very large lenticels. Buttresses absent. A tall tree. ....**59. *Milicia***
   Combination of characters not as above. 7

7. Leaf shape characteristic (elliptic to oblong or obovate - see Plate 4). Slash white or yellow, with a yellow, orange or red layer near the bark (very rarely the slash is entirely red), exuding white latex, both the slash and latex discolouring. An understorey tree to 30 m. ....**61. *Trilepisium***
   Leaf shape not as above. 8

8. Trunk straight and cylindrical, with large spreading branches near top. Buttresses usually present. Bark thin and smooth. Slash not red. Latex not or only slightly discolouring. Leaves with c. 7-15 main lateral veins on each side of the midrib (see Plate 4). ....**56. *Antiaris***
   Not as above. 9

9. Leaves with 2 or more well-marked lateral veins from the base of the lamina, at least one of these veins extending about a quarter or farther up the lamina. 10
   Basal veins either not prominent or not extending a quarter of the way (or more) up the lamina. 14

10. Leaves often c. 20 x 20 cm, margin wavy, dark green. A tree of damp places. ....**65. *Ficus vallis-choudae***
    Leaves commonly less than 15 cm broad (at least when mature). Leaf shape not as above. 11

11. Leaf margin conspicuously toothed or lobed. ....**68. *Ficus asperifolia*; 69. *F. sur***
    Leaf margin entire, at least on mature leaves. 12

12. Leaves rather rounded in shape (at least less than 1.75 times as long as broad) and normally over 8 cm broad. ....**66. *Ficus mucuso*; 67. *F. trichopoda***
    Leaves normally over 1.75 times as long as broad (commonly about twice as long as broad) or, if less than 1.75 times as long as broad, then under 8 cm broad. 13

13. Leaves comparatively small (often c. 9 x 4 cm) and thin, with the basal pair of veins ascending at a considerably steeper angle than the others (see Plates 5 & 6). ....**70. *Ficus ottoniifolia*; 72. *F. sansibarica***
    Not as above. ....**69. *Ficus sur*; 70. *F. ottoniifolia***

14. Leaves large, often over 25 cm long. ....**75. *F. saussureana*; 76. *F. vogeliana***
    Leaves normally shorter than 25 cm long. 15

15. Leaves comparatively small (often c. 9 x 4 cm) and thin, with the basal pair of veins ascending at a considerably steeper angle than the others (see Plate 6). ....**72. *Ficus sansibarica***
    Combination of characters not as above. 16

16. Leaf base cordate. 17
    Leaf base not cordate. 18

17. Leaves more or less heart-shaped. ....**73. *Ficus ingens*; 74. *F. polita***
    Leaves not more or less heart-shaped. ....**69. *Ficus sur* (69); 71. *F. ovata***

18. Leaf base rounded and with some prominent veins from near, or at, the base. ....**68. *Ficus asperifolia*; 69. *F. sur*; 71. *F. ovata***
    Leaf base not rounded or, if so, then basal veins not prominent. 19

19. Leaves truncate at apex. ....**81. *Ficus craterostoma***
    Leaves often not truncate at apex. 20

20. Leaf margin toothed. ....**68. *Ficus asperifolia***
    Leaf margin entire or almost so. 21

21. Leaves comparatively large (often c. 17 x 5-7.5 cm).
    ....**77-78. *Ficus cyathistipula, F. densistipulata, F. artocarpoides***
    Leaves smaller than above. 22
22. Leaves long and narrow, often c. 9-15 x 2-3 cm (see illustrations).
    ....**79-80. *Ficus barteri, F. verruculosa***
    Leaves not as narrow in proportion to length as above. 23

23. Figs sessile or sub-sessile. Leaf apex rounded to acute. 24
    Figs stalked. Leaf apex various. 25

24 Figs over 0.5 cm in diameter when dry. ....**83. *Ficus thonningii***
   Figs less than 0.5 cm in diameter when dry. ....**83a. *Ficus lingua***

25. Bracts at base of fig persistent. Leaves more or less acuminate at apex. 26
    Bracts at base of fig falling early. Leaves rounded to obtuse at apex.
    ....**82. *Ficus natalensis***

26. Bracts at base of fig united to one another. ....**85. *Ficus pseudomangifera***
    Bracts at base of fig not united to one another. ....**83. *Ficus thonningii***

Antiaris toxicaria Lesch. (56) Moraceae

Olwaa (ac); Kesuba, Kisuba (am); Eloa (at); False iroko, False muvule, Upas tree (en); Kirundu (ga); Elwa (la); Ripi (md); Lulundu (ms); Mumaka (na); Musende (nl); Mulundulundu (sa); Antiaris (tn); Mbondo, Muhehere (to).

45 m. Big tree, with a straight, thick, cylindrical trunk. Branches high on the trunk, large and spreading. Buttresses usually present on larger trees, of medium size, not extending far up the trunk. Bark light brown, grey or green, thin and smooth, with prominent lenticels and usually very shallow vertical fissures, ring marks often present. Slash fibrous to rather brittle, yellow or white or a combination of these colours, exuding white latex, which turns slightly darker. Leaves simple, alternate, c. 8 x 5.5 cm on tall trees (but considerably larger on small trees), with c. 7-15 main lateral veins on each side of the midrib, entire or with small teeth. Petiole c. 0.6 cm long. Fruit a hairy red drupe, c. 1.5-2 cm long, with a single seed.

OCCURRENCE: 1-4. A widespread and often abundant tree in lower altitude forests. Particularly common in lake-belt forests. The tree is found in farmland in Buganda and Busoga.

CULTIVATION AND PROPAGATION: Fast-growing. Can be grown on a range of soils. Plant alone, as a shade tree or in mixed stands. Collect seeds from the ground and plant as soon as possible. Wildings may be available under mother trees in quantity. Germination time in nursery beds irregular (2-12 weeks).

NOTES: The wood, which is fairly soft and light, has been used for making tea chests and beer canoes in Buganda. The bark is used for making a light-coloured type of bark cloth. The fruit is a favourite food of frugivorous monkeys and hornbills and other birds. The latex is very poisonous.

Plate 4. Moraceae (56-62)

56. *Antiaris toxicaria* 58. *Morus mesozygia* 59. *Milicia excelsa*
61. *Trilepisium madagascariense* 62. *Dorstenia kameruniana*

Actual sizes: leaves and fruit x 2; trunk bases x 80; tree profiles x 800.

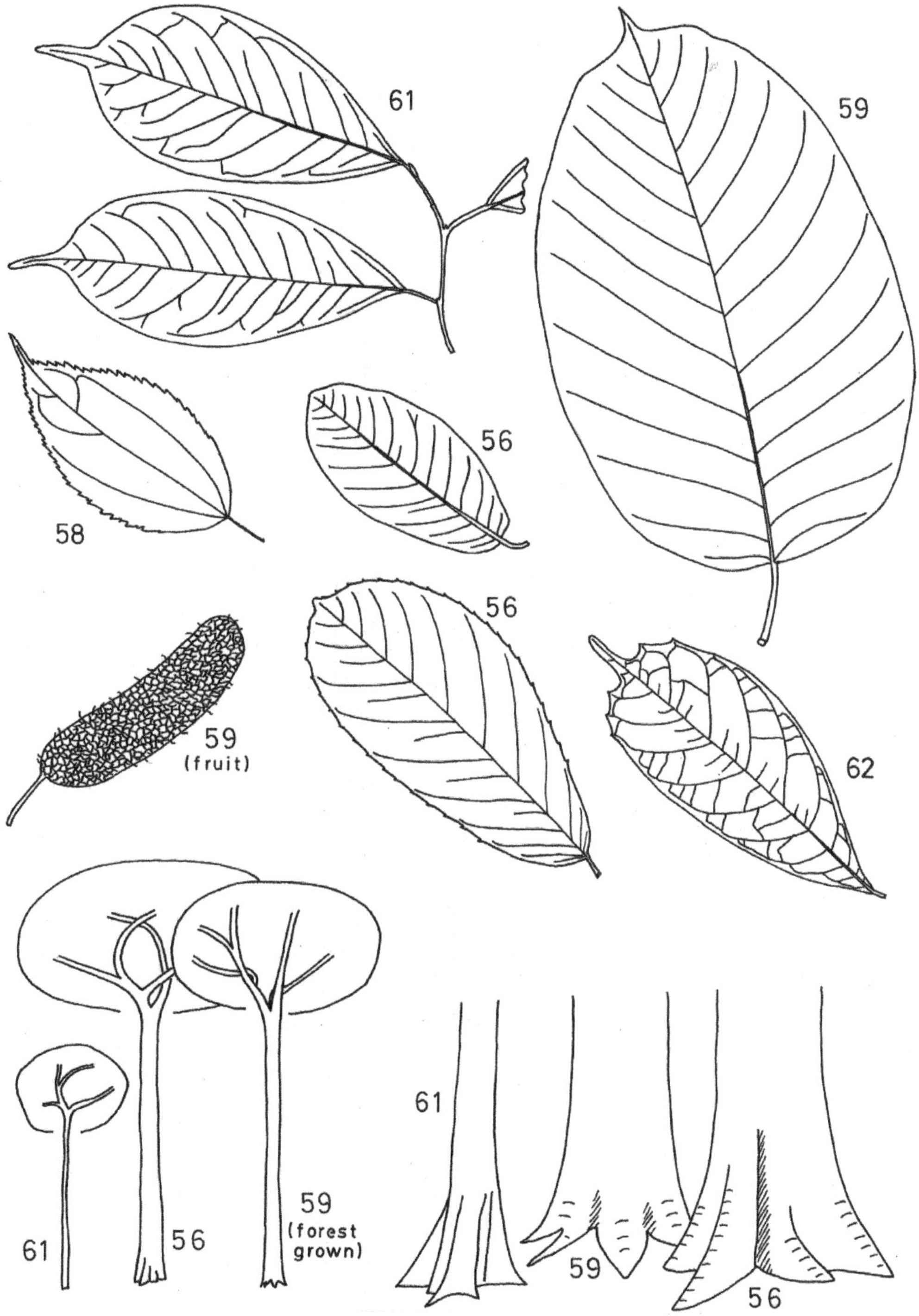

Plate 4. (56-62)

Morus mesozygia **Stapf (58)** Moraceae

SYNONYM: *Morus lactea* (Sim) Mildbr.

Mukooge (ga); Enyakatoma (no); Mufullo (sa); East African mulberry, Mecodze, Uganda mulberry (tn).

45 m. Tree shape similar to *Antiaris toxicaria* (57). Trunk straight and cylindrical. Crown with spreading large branches on older trees. Deciduous. Buttresses absent. Bark brown or green, thin and smooth, with prominent lenticels which are usually arranged in vertical columns. Slash fibrous, white or yellow or a combination of these colours, sometimes red near the bark, exuding white latex which sometimes turns slightly darker. Leaves c. 7 x 5 cm, 3-veined from the base, asymmetric at base, toothed or crenate, acuminate.

OCCURRENCE: U1-4. A widely distributed upperstorey to second storey tree in lower altitude forests. Sometimes common, as in Mengo and parts of Budongo and Bugoma forests.

CULTIVATION AND PROPAGATION: Fast-growing under good conditions. Suitable for single planting or for growing in pure or mixed stands. Can be pruned and pollarded. Collect the fruits on the ground near mother trees and lightly rub between the fingers to separate the tiny seeds. Sow directly or after soaking for half a day in water.

NOTES: The wood is very strong, beautiful and easily worked. The tree somewhat resembles *Antiaris* in general shape and look, but the leaves are very differently shaped.

Milicia excelsa **(Welw.) C.C. Berg (59)** Moraceae

SYNONYM: *Chlorophora excelsa* (Welw.) Benth. & Hook. f.

Olia (al); Mbara (am); Eluwa (at); Muvule (ga, tn); Elwa (la); Olwaa (lo); Vundi (md); Boru, Gutumba, Kimurumba (ms); Omutumba (no); Mutumba (sa); Iroko (tn).

45 m. Deciduous upperstorey tree with large branches. Crown varying from small and rounded (particularly in dense forest) to large and spreading. Trunk straight and cylindrical, flared slightly at base. Buttresses absent. Surface roots often prominent. Bark moderately thick and rough, dark brown to almost black, but occasionally lighter in colour, with prominent large lenticels, flaking raggedly in pieces c. 2-15 cm long. Slash orange and white, granular, sometimes also fibrous, exuding white latex. Leaves simple, alternate, c. 14 x 8 cm, with c. 10-18 main lateral veins on each side of the midrib, shortly acuminate, entire or with small teeth (young leaves are toothed and much larger). The spikes of the male inflorescences are much longer (8-20 cm) than those of the female inflorescences (2-3 cm).

OCCURRENCE: U1-4. Occasional in forests; much commoner in farmland and grassland in the medium to well-watered parts of Uganda, particularly in Buganda and Busoga. Absent from Tooro (except Bwamba) and Kigezi (except Maramagambo Forest).

CONSERVATION STATUS: Global NT (IUCN), LC (TOU); National EN (WCS).

CULTIVATION AND PROPAGATION: Moderately fast-growing on better sites. Best grown in mixed stands, as this may reduce the incidence of insect damage (it is liable to insect attack if grown in pure stands). Collect fruits on the ground near mother trees and crush to remove the tiny seeds from the flesh. Viable seeds can be isolated by submersion in water – those that sink are healthy, those that float should be discarded. Sow the seeds as soon as possible. Transplant seedlings after 4-6 months.

NOTES: One of the most valuable timber trees in East Africa. Trees are normally unisexual. The shapes of male and female trees are reported to be different.

Treculia africana **Decne. (60)** Moraceae

African breadfruit (en); Muzinda (ga).

25 m. Crown spreading. Trunk fluted. Bark smooth, grey. Slash white to light pink, sometimes with brown streaks, exuding copious off-white latex which, together with the slash, turns red-brown. Young leaves red or yellow. Leaves simple, alternate, large, c. 33 x 14 cm, glabrous,

entire. Petiole c. 1 cm long. Fruit large, spherical, c. 30 cm diameter, containing numerous seeds (resembling a breadfruit).
OCCURRENCE: U2 and 4. An uncommon tree, usually (but not always) found near swamps and streams.
CULTIVATION AND PROPAGATION: Moderately fast-growing. Grows best in damp places. Collect the fruits when ripe from mother trees or from the ground. Allow the fruits to rot for 2 days and then pick out the seeds manually. Plant the seeds directly
NOTES: The seeds are dried, fried and eaten by children in Buganda. The breadfruit tree (*Artocarpus altilis* (Parkinson) Fosberg) and the jackfruit tree (*A. heterophyllus* Lam.) are cultivated in Uganda, the latter being by far the commoner. They are both native to south-east Asia.

***Trilepisium madagascariense* DC. (61)** Moraceae
SYNONYM: *Bosqueia phoberos* Baill.
Mugwi (ga, tn); Omukumbwe (ki); Lubelagaiyi (ms); Munyabweya (na); Akatomatoma (no); Nasabi (to).
30 m. Understorey tree with a thin, straight, cylindrical trunk and a small, rounded, dark-coloured crown. Small buttresses sometimes present. Bark brown or greenish, thin and smooth, with prominent lenticels. Phellogen green. Slash fibrous, white or yellow, with a red, orange or yellow layer near the bark (slash very rarely entirely red), turning darker, exuding white latex which also turns darker. Leaves simple, alternate, c. 9 x 3.5 cm, widest in upper half, apex long acuminate.
OCCURRENCE: U1-4. Widespread and often abundant understorey tree. Ascending to 2200 m.
NOTES: The tree can be confused with *Ficus*, but the slash and leaves are distinctive once known. The leaves have been used to feed goats in Buganda.

***Dorstenia kameruniana* Engl. (62)** Moraceae
SYNONYM: *Craterogyne kameruniana* (Engl.) Lanjouw
Understorey shrub or tree to 6 m. The leaves are characteristically shaped, c. 11 x 4.25 cm, with a few large teeth near the apex and a long acumen (see Plate 4).
OCCURRENCE: U2 and 4. Recorded from Mengo, Bunyoro and Tooro. Rare.

***Ficus exasperata* Vahl (63)** Moraceae
Luwawu (ga, so); Muwawu (ga); Musano (sa); Kiseno, Museno (so); Musomoru (to).
25 m. Tree with crooked trunk, branches at all heights and a spreading and deep crown. Buttresses present on larger trees. Flutes sometimes present. Bark very thin and smooth, often with ring marks, greenish to light brown. Phellogen green. Slash white to light yellow, sometimes with yellow lines, exuding copious to little colourless sap, no white latex. Leaves simple, alternate, 3-lobed to unlobed, c. 10 x 4.5 cm, with 2 main lateral veins from the base, these extending more than half way up the lamina, and 2-5 other main lateral veins on each side of the midrib, very rough on both surfaces (like sandpaper). The shape of young leaves is completely different (see Plate 5). Petiole c. 1 cm long. Fig c. 0.9 cm diameter, red when ripe.
OCCURRENCE: U2-4. Usually an understorey tree.
NOTES: A species easily distinguished from all other figs (except *Ficus sycomorus* and sometimes *Ficus asperifolia*) by the rough leaves. The luganda name *luwawu* refers to the rough texture of the leaves. The leaves are used for cleaning dishes.

***Ficus sycomorus* L. (64) Moraceae**
SYNONYM: *Ficus gnaphalocarpa* (Miq.) A. Rich.
Mukunyu (ga)

Leaf shape and roughness as for *Ficus exasperata*, but leaves larger (c. 17 x 8.5 cm). Figs 3-5 cm long.
OCCURRENCE: 1-3. Savanna tree, very rarely found in forest.

Ficus vallis-choudae Delile (65) Moraceae
Kokoowe (ga); Mulabo, Obulo (gb); Murebe (na); Kidondwe, Widwe (so).
20 m. Spreading tree. Bark often rough, light brown. Slash light pink, light brown or red, with much off-white latex. Both the latex and slash turn darker. Leaves simple, alternate, c. 20 x 20 cm, with 2 main lateral veins from the base, these ascending over half way up the lamina, and with c. 3-5 other main lateral veins on each side of the midrib, dark green, usually glabrous, margin undulate.
OCCURRENCE: U1-4. Found in damp places and open vegetation. This species is particularly common in Mengo.
NOTE: Easily distinguished by the broad leaves with wavy margins.

Ficus mucuso Ficalho (66) Moraceae
Kiloko (am); Mukunyu (ga); Omukunyu (no); Muhuyu (sa).
40 m. Very big deciduous tree with a straight, cylindrical, trunk and very large, spreading branches. Large buttresses present. Bark smooth and very thin, brown, with prominent ring marks and lenticels. Phellogen green. Slash normally pink to pink-brown, but off-white on young trees, exuding copious white latex. Both the slash and latex turn darker. Leaves simple, alternate, c. 14 x 12 cm, with 2 main lateral veins from the base, these extending over half way up the lamina, and c. 4-6 other main lateral veins on each side of the midrib, noticeably hairy on undersurface, margin entire. The leaves of young plants have non-entire margins. Figs large, 3-5 cm diameter, orange.
OCCURRENCE: U1-4. Common forest tree. The fruits are much appreciated by birds and monkeys.
NOTE: Easily distinguished from all other fig species (except *Ficus trichopoda*) by the leaf shape. The leaves have fewer lateral veins than *Ficus trichopoda* and the slash and latex differ.

Ficus trichopoda Baker (67) Moraceae
SYNONYM: *Ficus congensis* Engl.
Mutembo (ko); Oduri (lo).
40 m. Probably normally an epiphyte when young, but becoming free-living and reaching a large size. Branches large. Crown thin and spreading. Trunk usually an interwoven, irregular mass of aerial roots below. Buttresses sometimes present. Bark thin and smooth, light brown to grey, sometimes with prominent ring marks and lenticels. Phellogen green. Slash red (sometimes yellow on aerial roots), fibrous, exuding white latex. Usually neither the latex nor the slash discolours, but occasionally the latex discolours slightly. Leaves simple, alternate, c. 14 x 9.5 cm, with 2 main lateral veins from the base, these not reaching (or only just reaching) half way up the lamina, and 6-9 other main lateral veins on each side of the midrib, margin entire, glabrous or hairy below, turning yellow when old. Petiole c. 6 cm long. Figs reddish, c. 2.5 cm diameter.
OCCURRENCE: U1, 2 and 4. Widely distributed, on both wet and raised sites.

Plate 5. Moraceae (63-70); see also Plates 6 and 7

63. *Ficus exasperata* 64. *Ficus sycomorus* 65. *Ficus vallis-choudae* 66. *Ficus mucuso* 67. *Ficus trichopoda* 68. *Ficus asperifolia* 70. *Ficus ottoniifolia*

Actual sizes: leaves and fruit x 2.

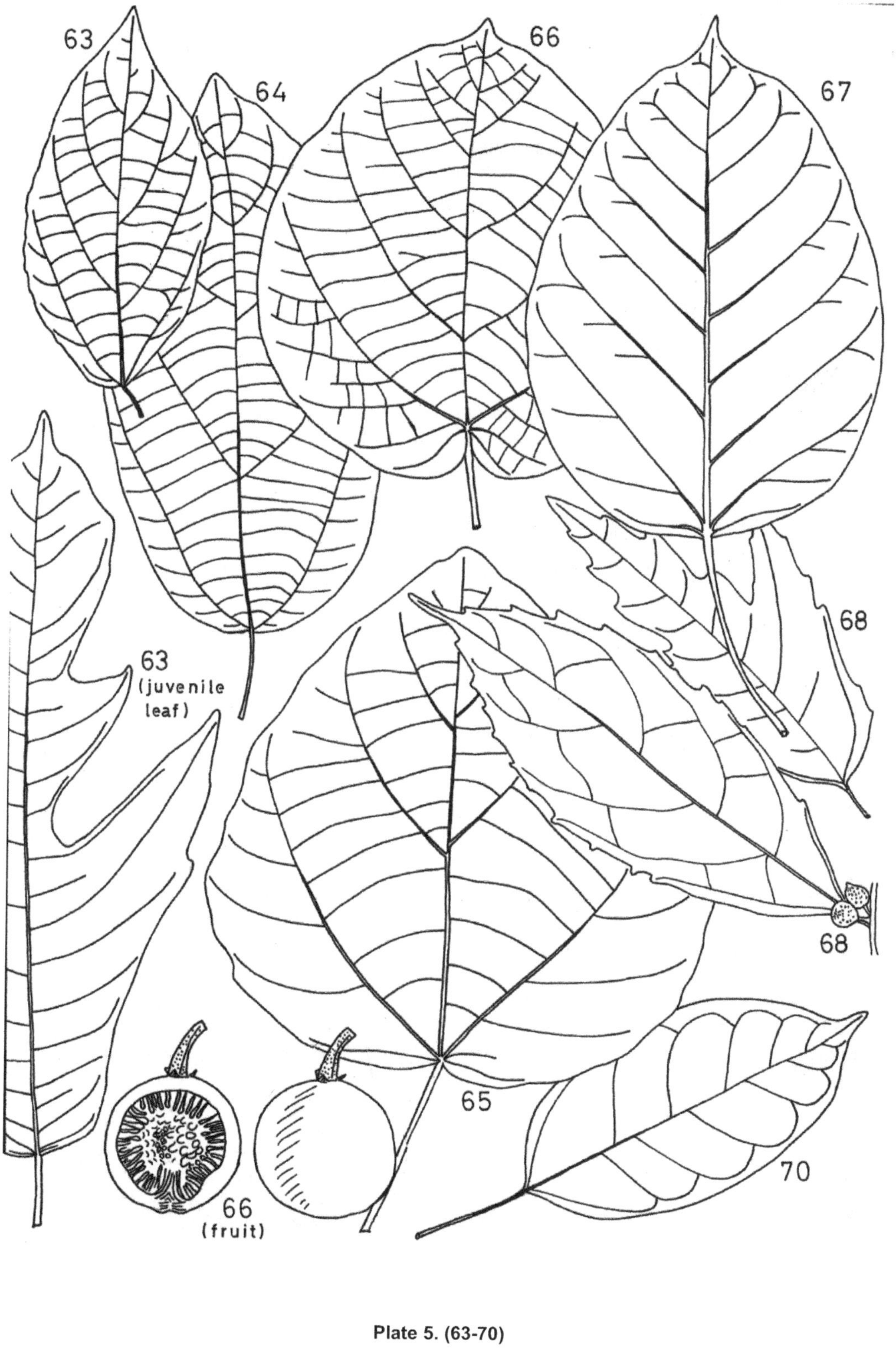

Plate 5. (63-70)

Ficus asperifolia **Miq. (68)** Moraceae

SYNONYM: *Ficus urceolaris* Hiern

Kitonto, Ntonto (ga); Omulyangabe (no).

5 m. Understorey tree or shrub with a weak stem. White latex present or absent (even from young parts). Leaves simple, alternate, c. 15 x 5.5 cm, with 2 main lateral veins from the base, these reaching less than half way up the lamina, and 4-7 other main lateral veins on each side of the midrib, sometimes very rough on both surfaces, margin with widely spaced teeth, acuminate. The leaves on some plants are trilobed or pinnatifid. Figs orange or red, c. 0.7 cm across, usually 2 together in a leaf axil.

OCCURRENCE: U1, 2 and 4. Widely distributed, usually in fairly open forest.

Ficus sur **Forssk. (69)** Moraceae

SYNONYM: *Ficus capensis* Thunb.

Eduro, Edurokoi (at); Kabalira, Mukunyu (ga); Idio (gb); Omurehe (ki); Ebuu (la); Elo (md); Mukunyu (so).

25 m. Crown thin. Buttresses normally absent. Bark thin and smooth, green to light brown, with prominent lenticels. Phellogen green. Slash off-white, pink to red-brown, exuding white latex. The latex and/or slash sometimes turn darker. Leaves simple, alternate, c. 12 x 6.5 cm, but variable in size and sometimes considerably larger (c. 20 x 10 cm), margin entire to non-entire (with widely spaced teeth). Basal lateral veins prominent, reaching quarter to half way up the lamina, and 3-9 other main lateral veins on each side of the midrib. Petiole c. 4 cm or more long. Figs borne on large, woody, much-branched, outgrowths from the trunk and larger branches, c. 2.5 cm across, red.

OCCURRENCE: U1-4. Common in forest and savanna.

Ficus ottoniifolia **(Miq.) Miq. (70)** Moraceae

SYNONYM: *Ficus lucanda* Ficalho

15 m. Understorey tree with a spreading crown, branching from near base. Small buttresses sometimes present. Bark thin, brown. Slash fibrous, yellow, producing off-white latex which does not turn darker. Leaves simple, alternate, c. 12 x 6 cm, with 2 main lateral veins from the base, these veins extending from a quarter to over a half way up the lamina, and with c. 4-6 other main lateral veins on each side of the midrib, margin entire. Petiole c. 4 cm long. Figs yellow, produced in groups of 1-3 on older stems.

OCCURRENCE: U2 and 4. Recorded from Mengo, Masaka, Tooro, Ankole and Kigezi.

NOTE: The leaves are very similar to those of *Ficus sur*, but tend to be more markedly acuminate and are probably darker green.

Ficus ovata **Vahl (71)** Moraceae

SYNONYM: *Ficus brachypoda* Hutch.

Ebule, Ebwolibwol (at); Kokoowe, Mukookoowe (ga); Mukoko (gw); Kobakoba, Odulindri (md); Kukowe (so).

Plate 6. Moraceae (69-83); see also Plates 5 and 7

69. *Ficus sur* 71. *Ficus ovata* 72. *Ficus sansibarica* 73. *Ficus ingens*
75. *Ficus saussureana* 77. *Ficus cyathistipula* 78. *Ficus artocarpoides*
83. *Ficus thonningii*

Actual sizes: leaves x 2.

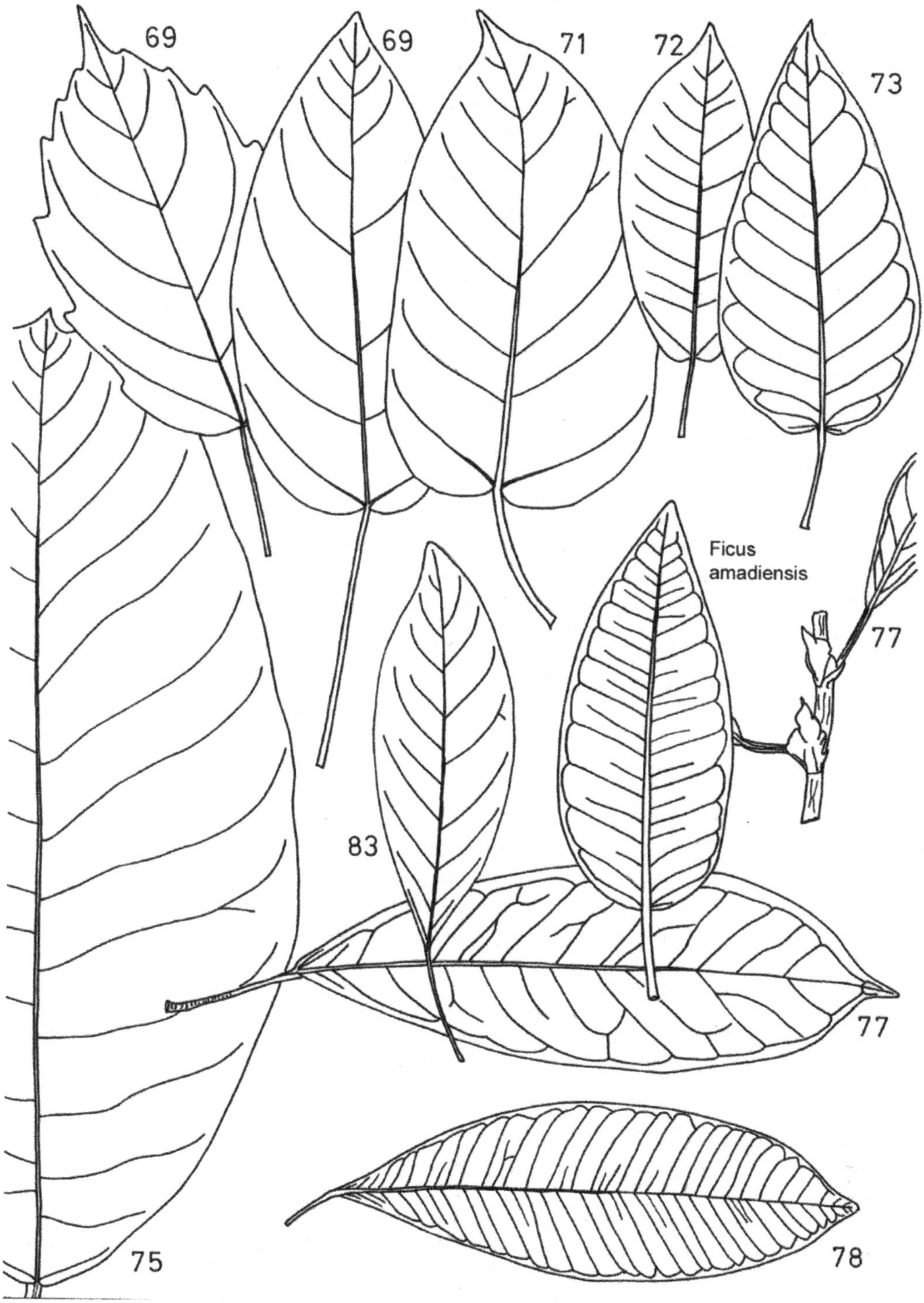

Plate 6. (69-83)

20 m, usually less. Tree with a spreading and rather open crown with large leaves. Deciduous. Bark thin and smooth. Phellogen green. Slash with vertical red and off-white lines, exuding white latex which does not discolour rapidly. Leaves simple, alternate, c. 22 x 12 cm, entire, the basal pair of lateral veins either not noticeably ascending or ascending less than half way up the lamina, with c. 6-13 other main lateral veins on each side of the midrib. Petiole c. 6 cm long.
OCCURRENCE: U1-4. A widely distributed species, most commonly found in farmland, in which situation it is abundant in Mengo, Ankole and Masaka.
NOTE: The leaves of this species are distinguished from those of *Ficus ottoniifolia* by the more markedly pinnate and more numerous lateral veins. They are usually larger than the leaves of *F. ottoniifolia* and *F. sur*.

***Ficus sansibarica* Warb. (72)** Moraceae
SYNONYM: *Ficus brachylepis* Hiern.
40 m. Very large tree that starts life as an epiphyte. Crown large and spreading. Base of trunk a mass of interwoven aerial roots. Bark light brown, smooth, with prominent ring marks and lenticels. Slash fibrous, pink to red, sometimes yellow on outside, exuding off-white latex. The latex (but not the slash) discolours. Leaves simple, alternate, c. 9 x 4 cm, rather thin, entire, with 2 main lateral veins from the base, these ascending at a steeper angle than the other lateral veins, but not reaching more than a quarter of the way up the lamina. There are c. 6-11 other main lateral veins on each side of the midrib. Petiole c. 2.5 cm long.
OCCURRENCE: U2 and 4. The most abundant large species of fig tree in Central Kibale Forest.
NOTES: The leaf venation is characteristic. The leaves are smaller and thinner than those of *Ficus ovata, F. sur* and *F. ottoniifolia.*

***Ficus ingens* (Miq.) Miq. (73)** Moraceae
Eereere (at).
40 m. Trunk straight and cylindrical, with a very large, spreading, crown. Large buttresses present. Bark thin and smooth, grey, appearing yellow from a distance. Phellogen green. Slash fibrous, yellow with white lines, exuding copious brown sap, but no white latex. Leaves simple, alternate, c. 12 x 6 cm, entire and glabrous, heart-shaped, with a cordate base and c. 9 main lateral vein on each side of the midrib. Petiole c. 3 cm long.
OCCURRENCE: U1, 3 and 4. Usually a savanna tree, but also found in disturbed forest.

***Ficus polita* Vahl (74)** Moraceae
35 m. Generally starting life as an epiphyte. Slash yellow, exuding copious white latex, which does not discolour. Leaves c. 12 x 9 cm, glabrous, with a cordate or rounded base and c. 7 main lateral veins on each side of the midrib. The petiole is longer (c. 12 cm) than that of *Ficus ingens*.
OCCURRENCE: U2 and 4. Recorded from Mengo. Tooro and Bunyoro. Probably uncommon.

***Ficus saussureana* DC. (75)** Moraceae
SYNONYM: *Ficus eriobotryoides* Kunth & Bouché
Muwo (ga).
35 m. Large tree, probably always epiphytic at first. Base of trunk a mass of fused aerial roots. Crown deciduous, flat and spreading. Bark thin, light brown, with prominent lenticels. Slash fibrous, pink to red (white on young trees), producing copious white latex. The slash discolours, but the latex does not. Leaves simple, alternate, clustered at the ends of the branches, very large (c. 28 x 14 cm), hairy on undersurface, with c. 7-15 main lateral veins on each side of the midrib, margin entire. Petiole c. 7 cm long.
OCCURRENCE: U1, 2 and 4. Sometimes common.
NOTE: Easily distinguished by the large leaves.

***Ficus katendei* Verdc. (75a) Moraceae**
20 m. Starts as an epiphyte, later free. Bark light brown-grey, smooth. Slash brown-pink to purplish-pink, exuding white latex. Leaves simple, brown above when dry, red-brown beneath, oblong-elliptic or slightly obovate, 12-29 x 7-13 cm, apex shortly acuminate, base slightly sub-cordate, margin entire; glabrous above, minutely puberulous on veins beneath, 3-veined from base and with 4 more prominent veins, secondary venation brochidodromous, other venation reticulate; petiole 2-7 cm, stipules linear-lanceolate, 2-3.5 x 2.5-3.5 mm, glabrous, deciduous. Figs obovoid, glabrous, densely red and white dotted, dark brown when dry, cuneate-obovoid, base slightly narrowed, rugulose, pubescent, in short groups on corrugated spurs 9 x 2.3 cm, forming fascicles of up to 20, stipules small and triangular, 1 x 1 mm, deciduous and forming groups at apex of spurs; peduncles 0.7-1.3 mm.
OCCURRENCE: U2. Only known from Bwindi Impenetrable National Park and Kasyoha-Kitomi Central Forest Reserve. Lower montane rainforest, 1400 m.
CONSERVATION STATUS: Global NE (IUCN), CR (TOU); National NE.

***Ficus vogeliana* (Miq.) Miq. (76)** Moraceae
Large tree. Slash reddish, exuding copious white latex which turns red after a few minutes. Leaves simple, alternate, very large, c. 30 x 12 cm, margin with widely spaced teeth or wavy. Petiole to 30 cm long.
OCCURRENCE: U2. In swamp forest in Bwamba.

***Ficus cyathistipula* Warb. (77)** Moraceae
Munyanyoni (to).
10 m. Small tree, usually epiphytic at first. Crown spreading. Bark thin and fairly smooth, with vertical fissures and prominent lenticels. Slash fibrous, red, with white latex. Both the slash and latex turn slowly bright red. Leaves simple, alternate, c. 18 x 7.5 cm, glabrous and entire, with a short acumen and a long cuneate base. Main lateral veins c. 5-9 on each side of the midrib as seen on the upper surface. Reticulate venation very prominent on undersurface. Petiole c. 4 cm long.
OCCURRENCE: U1-4. Often in damp places.

***Ficus densistipulata* De Wild. (77a)** Moraceae
SYNONYM: *Ficus namalalensis* Hutch.
10 m. Small tree or shrub, epiphytic at first. Slash red with creamy latex rapidly turning rusty red. Leaves simple, c. 14 x 6 cm, glabrous and entire, acuminate, base mostly acute but may be rounded. Main lateral veins often 5 on each side of the midrib, tertiary venation reticulate, more prominent on undersurface. Petiole to 4.8 cm long.
OCCURRENCE: U2 and 4. Moist and secondary forest, 1134-1220 m.
NOTE: Differs from *Ficus cyathistipula* in having a peduncle less than 0.5 cm long, otherwise very similar.

***Ficus artocarpoides* Warb. (78)** Moraceae
25 m. Bark grey. Slash red. Leaves simple, alternate, shiny, c. 17 x 5 cm, with a short acumen. Petiole 1.25-2.5 cm long. Distinguished from *Ficus cyathistipula* by the more numerous main lateral veins (see Plate 6).
OCCURRENCE: U4. Only recorded from Mengo.

***Ficus barteri* Sprague (79)** Moraceae
SYNONYM: *Ficus stipulifera* Hutch.
35 m. Probably always epiphytic when young. Crown spreading and thin. Basal part of trunk either supported on stilt roots or a mass of interwoven aerial roots. Bark thin and smooth, brown. Phellogen green. Slash fibrous, pink to red, sometimes with white lines, exuding white latex. The slash and latex discolour. Leaves simple, alternate, long and thin, often between c. 9 x 2.25 and 14 x 2.5 cm in size, with a long acuminate apex and often a long cuneate base. Petiole c. 1.25 cm long.
OCCURRENCE: U2 and 4. Probably common in parts of Mabira Forest.

***Ficus verruculosa* Warb. (80)** Moraceae

Small tree. Leaves c. 11 x 2.5 cm, differing from those of *Ficus barteri* in being obtuse to acute (but not acuminate) at the apex.
OCCURRENCE: U2 and 4. Usually in swamps.

***Ficus craterostoma* Mildbr. & Burrett (81)** Moraceae

SYNONYM: *Ficus pilosula* De Wild.
15 m. Usually epiphytic when young. Bark grey and thick. Slash red, fibrous, exuding white latex. Leaves simple, alternate, c. 5.5 x 2.5 cm, with a truncate apex.
OCCURRENCE: U2 and 4. Usually near water. One of the commonest figs on Nkose Island.
NOTE: Easily distinguished by the leaf shape.

***Ficus natalensis* Hochst. (82)** Moraceae

Kolawingo (ac); Bark cloth tree (en); Mutuba (ga, sa); Tera (gw); Ituba (la); Kitoma, Mutoma (na); Omutoma (no); Kirianyonyi, Mugaire (so).
15 m. Epiphytic when young. Base of trunk a mass of interwoven aerial roots (in forest-grown specimens). Hanging aerial roots sometimes present. Bark thin and smooth, whitish. Phellogen green. Slash white or yellow, exuding white latex or, more rarely, yellowish sap. Leaves simple, alternate, glabrous and entire, variable in size, often c. 6 x 2.5 cm in forest-grown plants, apex rounded or obtusely pointed (not acute or acuminate), with 6-14 main lateral veins on each side of the midrib. Petiole c. 1.5 cm long. Figs stalked, c. 0.5-1 cm diameter.
OCCURRENCE: U1-4. Usually in more open types of forest. Often the most common *Ficus* on agricultural land, where it is propagated by cuttings.
NOTES: This species is the main source of bark cloth. It can be easily confused with several other species of *Ficus* with similar types of leaf, namely *F. thonningii*, *F. pseudomangifera* and *F. amadiensis* De Wild. (see Plate 6; note: this is not a typical forest species).

***Ficus thonningii* Blume (83)** Moraceae

SYNONYM: *Ficus persicifolia* Welw. ex Warb.
Ekuboi, Emidit (at); Laro (gb); Ananga (la).
15 m. Tree branching from near base, with a dense crown and drooping branches, with aerial roots dangling from the upper part of the trunk and from the branches. Usually epiphytic at first. Bark smooth, pale grey. Leaves simple, alternate, c. 10 x 3-4 cm, entire and dark green, acuminate to rounded at apex, with c. 6-15 main lateral veins on each side of the midrib. Petiole c. 2.5 cm long. Figs sessile, not stalked (as they are with *F. natalensis*).
OCCURRENCE: U1-4. A widely distributed tree.
NOTES: This species is sometimes planted as an avenue shade and is also used for making bark cloth. Cultivation has resulted in a number of varieties, some of which are very similar to varieties of *Ficus natalensis*.

***Ficus lingua* De Wild. & Th. Dur. (83a)** Moraceae

15 m. Epiphytic. Bark light grey, with prominent lenticels. Slash yellow, exuding copious white latex. Leaves simple, oblanceolate to obovate, c. 0.8-3.4 x 0.5-1.1 cm, apex mostly rounded, rarely truncate. Figs small, yellow-green.
OCCURRENCE: U2 and 4. Moist forest, including regenerating forest near water. Most specimens are from Budongo Forest; also at Sango Bay.
NOTE: Only the subsp. *lingua* occurs in Uganda.

Plate 7. Moraceae (66-85); see also Plates 5 and 6

66. *Ficus mucuso* 69. *Ficus sur* 73. *Ficus ingens* 75. *Ficus saussureana*
79. *Ficus barteri* 80. *Ficus verruculosa* 81. *Ficus craterostoma* 82. *Ficus natalensis*
83. *Ficus thonningii* 85. *Ficus pseudomangifera*

Actual sizes: leaves and fruits x 2; trunk bases x 80; tree profiles x 800.

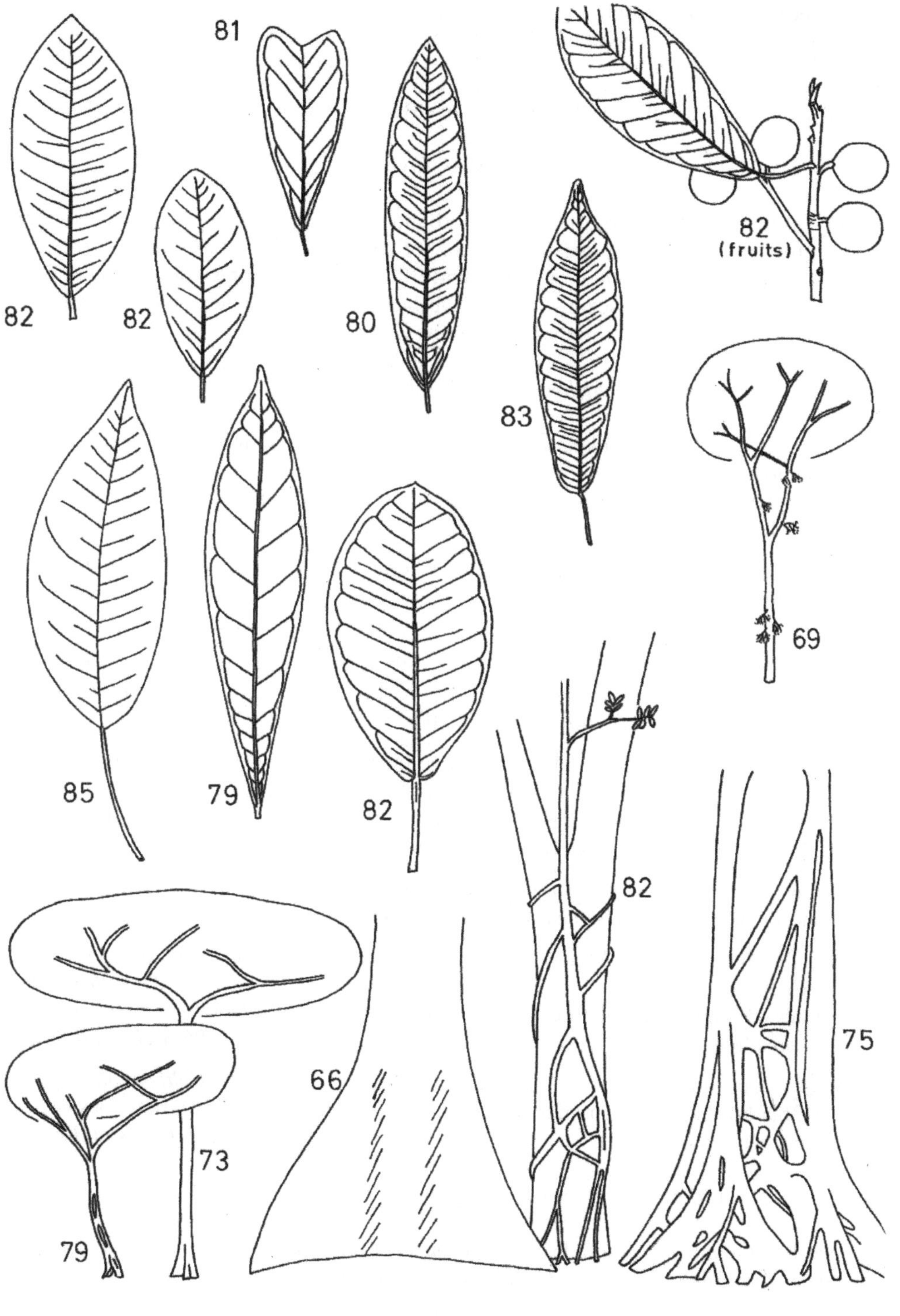

Plate 7. (66-85)

Ficus pseudomangifera Hutch. (85) Moraceae
40 m. Branches spreading. Probably normally epiphytic when young. Bark thin, greenish to light brown, with large lenticels. Slash red with white lines, fibrous, producing white latex. Leaves simple, alternate, c. 10 x 3.5 cm, apex acuminate. Petiole c. 1.25-2.5 cm long.
OCCURRENCE: U2 and 4. Recorded from Kibale and Budongo forests. Said to be widely planted.

ULMACEAE

A family of small to very large trees, often abundant in Ugandan forests. Leaves simple, alternate, stipulate. The bark is usually smooth, but may flake off in fairly large pieces. The slash varies from white to yellow or brown and often has brown rings, dots or other brown markings. The fruit is indehiscent. *Chaetachme* has rather stiff leaves, with inconspicuous venation. It is placed elsewhere in the descriptions (252).

Key to Ulmaceae (except *Chaetachme*), based on leaves.

1. Leaves with c. 5-10 main lateral veins on each side of the midrib, margins entire. Petiole bright green, contrasting with the grey branches. ....**93. _Holoptelea_**
   Main lateral veins usually fewer than 8 on each side of the midrib. Leaves either toothed or entire, but not shaped as above. 2

2. Leaves closely and regularly toothed for all (or at least three-quarters) of length (see Plate 8). ....**92. _Trema_**
   Not as above. 3

3. Lamina coarsely toothed in upper about two-thirds only. Main lateral veins 2 or 3 on each side of the midrib (see Plate 8). ....**89. _Celtis africana_**
   Lamina entire or toothed only in upper half. 4

4. Basal lateral veins extending less than three-quarters of the way up the lamina. Main lateral veins 3-7 on each side of the midrib. 5
   Basal lateral veins extending about three-quarters or farther of the way up the lamina. Main lateral veins 1-4 on each side of the midrib. 7

5. Widest part of leaf usually below its centre (see Plate 8) ....**88. _Celtis gomphophylla_**
   Widest part of leaf usually near to, or above, the centre. 6

6. Stipules 7-10 mm long. Mature leaves c. 14 x 6.5 cm, with some hairs on undersurface. Crown spreading. ....**87. _Celtis zenkeri_**
   Stipules 4-7 mm long. Mature leaves c. 6 x 2.5 cm, glabrous below. Crown not spreading. ....**86. _Celtis mildbraedii_**

7. Leaf blade entire. ....**91. _Celtis adolfi-fridericii_**
   Leaf blade either entire (particularly in Mengo, where *C. adolfi-fridericii* does not occur) or toothed. ....**90. _Celtis philippensis_**

Key to Ulmaceae (except *Chaetachme*), based on the slash.

1. Slash without dark-coloured rings, dots or other markings. 2
   Slash with dark-coloured rings, dots or other markings. 4

2. Slash not turning darker, yellow and hard, smelling of urea. ....**93. _Holoptelea_**
   Slash turning fairly rapidly darker. 3

3. Slash turning reddish after some time. ....**90. *Celtis philippensis***
   Slash not turning reddish. ....**92. *Trema***

4. Slash white to light brown, with brown dots or lines, but without dark-coloured rings. ....**88-91. *Celtis gomphophlyla, C. africana, C. adolfi-fridericii***, occasionally ***C. philippensis***
   Slash with brown or black rings. ....**86. *Celtis mildbraedii*; 87. *C. zenkeri*; 91. *C. adolfi-fridericii***

Celtis mildbraedii **Engl. (86)** Ulmaceae

Bolwe (am); Omukomakoma (no); African celtis (tn).

45 m. Tall tree with a long, straight, rather thin, trunk and a small, rounded, deciduous or evergreen, crown (young trees sometimes with long narrow crowns). Buttresses present, large and thin. Bark thin, light brown, flaking in large pieces. Slash yellow, with conspicuous brown rings, turning darker. Leaves simple, alternate, c. 6 x 2.5 cm on larger trees (but much bigger on young specimens), asymmetric at base, wavy to coarsely toothed in upper half, occasionally entire, with 3-7 main lateral veins on each side of the midrib, glabrous below when mature. Stipules 4-7 mm long. Fruit a small drupe, c. 1 cm long.

OCCURRENCE: U2-4. Abundant in Budongo Forest and Mengo, sometimes co-dominant. Either absent from, or very rare in, lake-shore forests, such as Jubiya Forest.

CULTIVATION AND PROPAGATION: Fast-growing. Probably best grown in mixed stands. Collect fruits from the ground beneath mother trees. Remove flesh manually or dry in the sun to extract the seeds. Sow seeds as soon as possible.

NOTE: The wood produces a strong, general purpose, timber.

Celtis zenkeri **Engl. (87)** Ulmaceae

Kasisa (ga); Omukomakoma (no).

40 m. Tall deciduous tree (but shorter than *Celtis mildbraedii*) with a wide trunk (wider than *C. mildbraedii*). Crown wide, spreading, usually light-coloured. Buttresses large and thin (larger than *C. mildbraedii*). Bark thin, brown, smooth, flaking in large pieces, giving the trunk a mottled appearance with shades of different colours. Slash yellow (possibly sometimes white), with brown rings, turning darker. Leaves simple, alternate, c. 14 x 6.5 cm (usually larger than *C. mildbraedii*), acuminate, asymmetric at base, entire or toothed in upper half, with 3-5 main lateral veins on each side of the midrib, with some hairs on undersurface, even when mature. Stipules 7-10 mm long.

OCCURRENCE: U2, 4 and possibly 3. Usually less abundant than *Celtis mildbraedii*, but locally common.

Celtis gomphophylla **Baker (88)** Ulmaceae

SYNONYM: *Celtis durandii* Engl.

Murundu (am); Ejeeje, Runuuka (ki); Bukemi (ko); Munyamazi (na); Musisa (sa); Mukyemogola (so); Namunuka, Stinkwood (tn); Busiri, Mujunju, Mujunu, Nyamunuka, Nyabununka (to).

25 m. Understorey to (occasionally) small canopy tree. Trunk straight or crooked, with a spreading deciduous crown. Buttresses small or absent. Bark thin and smooth, light brown to whitish. Phellogen green. Slash white, with numerous brown dots and sometimes streaks, occasionally with rather obscure yellow fibrous lines, rather granular. Leaves simple, alternate, c. 9 x 2.5 cm (but much larger on young trees), acuminate, entire or (rarely) with a few coarse teeth, with 4-7 main lateral veins on each side of the midrib.

OCCURRENCE: U2-4. Abundant in many forests, up to an altitude of 2000 m (on Mt Elgon), often on drier sites. Common under *Cynometra*.

NOTE: The wood has a foul smell.

Celtis africana Burm. f. (89) Ulmaceae

Kasisa (ga); Ejeeje, Runuuka (ki); Mastet, Mastitet (ku); Gusotono, Lusa (ms); Muzhunzhu (na); Musisa (sa); Mukyemogola (so); Akasinsa, Camdeboo stinkwood (tn); Nyamanuka (to).
30 m. Deciduous tree with a straight trunk and much-branched, spreading, crown with (often) rather few leaves. Unbuttressed. Bark very thin and smooth, light brown to whitish. Slash white to off-white, with brown dots and markings, turning darker. Leaves simple, alternate, c. 5 x 2.25 cm (but much larger on young plants), acuminate, asymmetric at base, toothed in upper two-thirds, with 2-3 main lateral veins on each side of the midrib.
OCCURRENCE: U1-4. A widely distributed and often abundant species, found up to 2300 m. It usually grows in colonizing forest or on relatively dry sites, for instance on the upper parts of slopes. It is particularly common in Tooro.
CULTIVATION AND PROPAGATION: Fast-growing on well-watered sites. Can grow on poor soils and can be used as a pioneer in afforestation projects. Collect fruits from mother trees or from the ground beneath. Sun-dry to extract the seeds. Germination rate high if fresh seeds used.
NOTE: The leaves are somewhat similar to those of *Morus*, but differ in shape and toothing.

Celtis philippensis Blanco (90) Ulmaceae

SYNONYM: *Celtis wightii* Planch.
Mulundu, Njabutulu (am).
20 m, rarely to 30 m. Evergreen understorey tree with a crooked (rarely straight) trunk and fairly spreading crown. Small buttresses present on larger trees. Bark smooth and thin, green to light brown, usually with prominent lenticels, flaking on older trees. Slash fibrous (rarely granular), yellow, off-white or light brown, occasionally with small brown dots, turning slowly and characteristically reddish or red-brown. Leaves simple, alternate, c. 15 x 6.5 cm, entire (particularly in Mengo) or coarsely toothed (particularly in western Uganda), glabrous, with 1-4 main lateral veins on each side of the midrib, the lowermost extending nearly to the apex.
OCCURRENCE: U2 and 4. Recorded from Budongo, Mengo and Bwamba. Often abundant.
NOTE: There is considerable variation in the texture, size and toothing of the leaves.

Celtis adolfi-fridericii Engl. (91) Ulmaceae

Ekembebakaswa (am).
Large tree to 35 m. Trunk very straight, with a rounded crown. Buttresses present. Bark brown, fairly rough. Slash whitish to light brown, very granular, with numerous dark-coloured rings or spots. Leaves simple, alternate, c. 11 x 5 cm, with 3 main veins from the base (similar in venation to *C. philippensis*). In that part of Uganda in which they both occur, *C. philippensis* has at least some leaves with coarse teeth.
OCCURRENCE: U2. Only recorded from Bugoma Forest and Bwamba, common in the former.

Trema orientalis (L.) Blume (92) Ulmaceae

Opobo-bunga (ac); Yakiyaki (al); Bukingi (am); Ereer (at); Kasisa (ga, to); Omubengabakwe, Omutangiri (ki); Muhera (ko); Mugiryanjole (nl); Lusabusubi (sa); Nkulidho (so); Mutete (to).

Plate 8. Ulmaceae (86-93)

86. *Celtis mildbraedii* 87. *Celtis zenkeri* 88. *Celtis gomphophylla* 89. *Celtis africana* 90. *Celtis philippensis* 92. *Trema orientalis* 93. *Holoptelea grandis*

Actual sizes: leaves and fruit x 2; trunk bases x 80; tree profiles x 800.

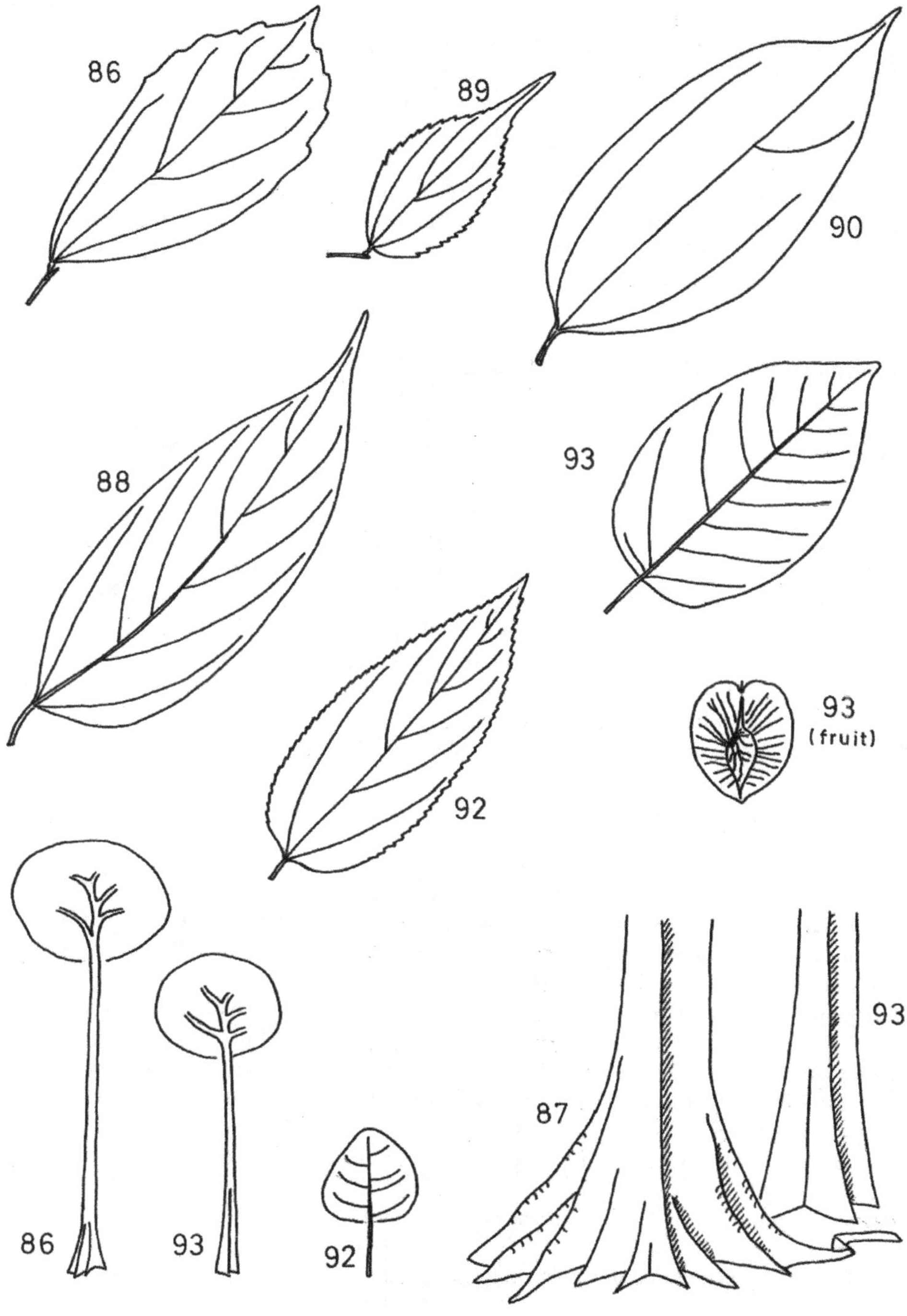

Plate 8. (86-93)

15 m. Trunk straight, with branches at right angles and curving up (similar to *Maesopsis*). Bark very thin, smooth, light-coloured. Lenticels numerous, prominent, brown, more or less arranged in vertical columns. Slash very soft, brown to almost white or pale pink, sometimes with brown markings, turning darker. Leaves simple, alternate, c. 11 x 4 cm (smaller on some high altitude trees and larger on young plants), hairy to glabrous above, hairy below, with many closely-spaced teeth and 3-6 main lateral veins on each side of the midrib.
OCCURRENCE: U1-4. A light-demanding, fast-growing species, abundant on forest edges and in larger forest gaps. Sometimes forms pure stands in recently exploited forests.
CULTIVATION AND PROPAGATION: Fast-growing. Tolerant of a wide range of soil conditions and habitats. Suitable for land reclamation. Plant in pure or mixed stands. Withstands pruning, coppices readily and can provide abundant firewood. Collect the fruits from the ground and remove the seeds manually from the pulp. Sow the seeds as soon as possible.

Holoptelea grandis **(Hutch.) Mildbr. (93)** Ulmaceae
Butungu (am); Mumuli (ga, tn); Mutaa (md); Omumuli (no); Mumuli (tn); Mutawale (to).
40 m. Deciduous tree with a straight or wavy, cylindrical, and often very long trunk. Crown spreading. Buttresses present, small and not spreading, but sometimes extending for some distance up the trunk. The lower part of the trunk may be slightly fluted. Bark brown, moderately thick, moderately rough (at least in places), with vertical fissures, flaking in places, with prominent lenticels. Phellogen green, at least in places. Slash hard, of even texture or fibrous, yellow (sometimes of different shades of yellow), turning very slowly lighter in colour, with a distinctive smell (said by some to be of urea or iodine). Leaves simple, alternate, c. 9 x 5.5 cm, with c. 5-10 main lateral veins on each side of the midrib. Petiole bright green, contrasting in colour with the grey branches, drying to black. Fruit winged.
OCCURRENCE: U1-4. A widely distributed tree. Abundant in Mabira Forest and common in Budongo. Said to be a light-demanding species.
NOTE: It produces a good general-purpose timber which is little used in Uganda.

EUPHORBIACEAE AND RELATED FAMILIES

The Euphorbiaceae, as formerly understood (including in ITU and UFT), has been taxonomically revised, with some genera removed to the related families Pandaceae, Phyllanthaceae and Putranjiviaceae. The species included in this field guide that belong to these four families range from small to very large trees. Some are common in Ugandan forests. The fruit often has three locules (cavities), each with one or two seeds.

The genus *Acalypha* (Euphorbiaceae) includes some large shrubs that can be common in Ugandan forests. One of the most frequent is *A. neptunica* Müll. Arg., a weak-stemmed spreading shrub that can reach a height of 6 m. Its more or less leafless main branches bear leafy shoots with leaves that are thin, simple, alternate and regularly toothed. It is very abundant in lake-belt forests and in Budongo Forest.

A wide variety of vegetative characters is found in the species of the four families included here. Consequently, the species can be found in six different parts of the field guide. Most species have simple alternate leaves (groups 1-4 below).

1) Species 31-34. Succulents. In Euphorbiaceae: *Euphorbia*.

2) Species 94-110. Leaves normally with either more than 3 main veins (including the midrib) from the base of the lamina or with basal lateral veins extending more than a quarter of the way up the lamina. Small to large trees. In Euphorbiaceae: *Acalypha, Alchornea cordifolia, A. laxiflora, Croton, Discoglypremna, Macaranga, Neoboutonia.* Note: *Croton megalocarpus*

and some species of *Macaranga* are placed in this section, even though they lack the characteristic venation.

3) Species 172-185. Venation not as in 2. Leaf margin toothed or crenate. Small to medium-sized trees. In Euphorbiaceae: *Alchornea floribunda*, *A. hirtella*, *Argomuellera*, *Discoclaoxylon*, *Gymnanthes*, *Pseudagrostistachys*, *Shirakiopsis*, *Suregada*. In Phyllanthaceae: *Flueggea*, *Maesobotrya*. In Putranjivaceae: *Drypetes*.

4) Species 240-251. Venation not as in 2. Leaf margin entire. Small to medium-sized trees. In Euphorbiaceae: *Tetrorchidium*. In Phyllanthaceae: *Antidesma*, *Bridelia*, *Cleistanthus*, *Margaritaria*, *Phyllanthus*, *Spondianthus*, *Thecacoris*, *Uapaca*. In Pandaceae: *Microdesmis*.

5) Species 297. Leaves opposite. Small tree. In Euphorbiaceae: *Mallotus*.

6) Species 362. Leaves digitate. Large tree. In Euphorbiaceae: *Ricinodendron*.

Key to *Macaranga*.

1. Leaves distinctly trilobate, often as long as wide. 2
   Leaves not lobate, often longer than wide. 3
2. Leaves (20-)25-50 cm long and wide. ....**94. *M. schweinfurthii***
   Leaves (8-)10-18(-21) cm long and wide. ....**95. *M. angolensis***
3. Leaves palmately 5-11 nerved from base, +/- peltate. ....**99. *M. capensis***
   Leaves 3(-5) nerved from base or nerves pinnately branched. 4
4. Leaves distinctly toothed, 3(-5) nerved. ....**96. *M. monandra***
   Leaves entire or shallowly sinuate-wavy. Nerves may or may not be pinnately branched. 5
5. Branchlets often spiny; petioles tomentellous to pubescent. ....**97. *M. spinosa***
   Branchlets not spiny; petioles turn hairless at maturity. ....**98. *M. barteri***

Macaranga schweinfurthii **Pax (94)** Euphorbiaceae

Mweganza (ga); Gumudoadoa (ms); Mukokoma (na).

15 m, occasionally 25 m. Trunk straight or crooked, sometimes multi-stemmed, with a spreading crown. Stilt roots sometimes present. Spines often present on trunk. Bark very thin, light brown, fairly smooth, sometimes slightly vertically fissured. Slash soft and thick, brown to pink, fibrous to rather granular, sometimes exuding small drops of red exudate. Leaves simple, alternate, very large, c. 35 x 30 cm, normally 3-lobed, wavy-dentate, usually with 5-7 main veins from the base. Petiole c. 30 cm long.

OCCURRENCE: U2-4. Abundant in swamps and swamp forest. Sometimes in secondary forest.

NOTE: The fruit is eaten by the grey parrot.

Macaranga angolensis **(Müll. Arg.) Müll. Arg. (95)** Euphorbiaceae

Luzibaziba (ga); Mukoko (to).

10 m. Shrub, small tree or small climber. Stems spiny. Leaves c. 17 x 17 cm, 3-lobed, smaller than those of *Macaranga schweinfurthii*. Petiole c. 9 cm long.

OCCURRENCE: U2 and 4. Uncommon. Recorded from Kibale Forest and Mengo.

NOTE: The spiny character may cause its confusion with *M. spinosa* and *M. monandra*. However, the trunk and branch spines in *M. angolensis* are blunt, while those in *M. spinosa* are sharp, downward facing and sometimes forked. Also, the leaves of *M. spinosa* are not lobed. *M. monandra* may be distinguished by its distinctly toothed leaves.

Macaranga monandra Müll. Arg. (96) Euphorbiaceae
Omufurafura, Omurara (ki).
20 m. Trunk thin, cylindrical, straight or crooked, sometimes multi-stemmed with a light crown. Stilt roots sometimes present. Long, down-curving, branched spines often present on trunk. Bark very thin and smooth, light brown, slightly vertically fissured. Phellogen green. Slash usually red-brown, sometimes white or light brown, sometimes with brown streaks. Leaves simple, alternate, c. 12 x 6.5 cm (but variable in size), either with widely spaced teeth or with an uneven margin having shallow indentations (see Plate 9). Petiole c. 7 cm long.
OCCURRENCE: U2 and 4. Mengo, Masaka, Ankole, Kigezi. In swamp and secondary forests. Common in lake-shore forests.

Macaranga spinosa Müll. Arg. (97) Euphorbiaceae
SYNONYM: *Macaranga pynaertii* De Wild.
Nabaluka (ga).
20 m. Spreading tree, branched from near base. Stilt roots sometimes present. Long, downward-curving, branched spines usually present on trunk. Trunk sometimes fluted at base. Bark thin and smooth, whitish. Slash fibrous, red-brown. Leaves simple, alternate, c. 10 x 4 cm, with c. 7-11 main lateral veins on each side of the midrib, margin usually entire and often wavy, base rounded to somewhat cordate. Petiole c. 2.5 cm long.
OCCURRENCE: U2 and 4. Often in swamp forest, but also in secondary forest and sometimes on ridges. Abundant in lake-shore forests.
NOTE: Similar to *Macaranga monandra*. Some plants appear to be intermediate between the two species.

Macaranga barteri Müll. Arg. (98) Euphorbiaceae
SYNONYM: *Macaranga lancifolia* Pax
Omuraha (ki).
20 m on the sandy, lake-shore soils of Masaka, but larger (to 35 m) in lshasha Gorge. Trunk straight, branching fairly low down, bearing a dark-coloured, moderately large crown. Buttresses absent. Trunk sometimes slightly fluted. Bark grey, fairly thin and smooth, with lenticels arranged in horizontal lines. Phellogen green. Slash soft and quite thick, brittle, red-brown, with a fibrous yellow layer near the wood. Leaves simple, alternate, c. 12 x 4.5 cm, similar to those of *Macaranga spinosa*, but with a cuneate to attenuate base (subcordate-auriculate at very base). Petiole c. 4 cm long.
OCCURRENCE: U2 and 4. Kayonza, Masaka. Common in Ishasha Gorge.
NOTE: Similar to *Macaranga spinosa*, from which it is distinguished by the shape of the leaf base; also, the petiole turns hairless with age, while in *M. spinosa* it is tomentellous to pubescent.

Plate 9. Euphorbiaceae (95-109)

95. *Macaranga angolensis* 96. *Macaranga monandra* 97. *Macaranga spinosa* 98. *Macaranga barteri* 99, *Macaranga capensis* 100. *Neoboutonia macrocalyx* 102. *Alchornea cordifolia* 106. *Croton macrostachyus* 107. *Croton sylvaticus* 109. *Croton megalocarpus*

Actual sizes: leaves and fruits x 2.

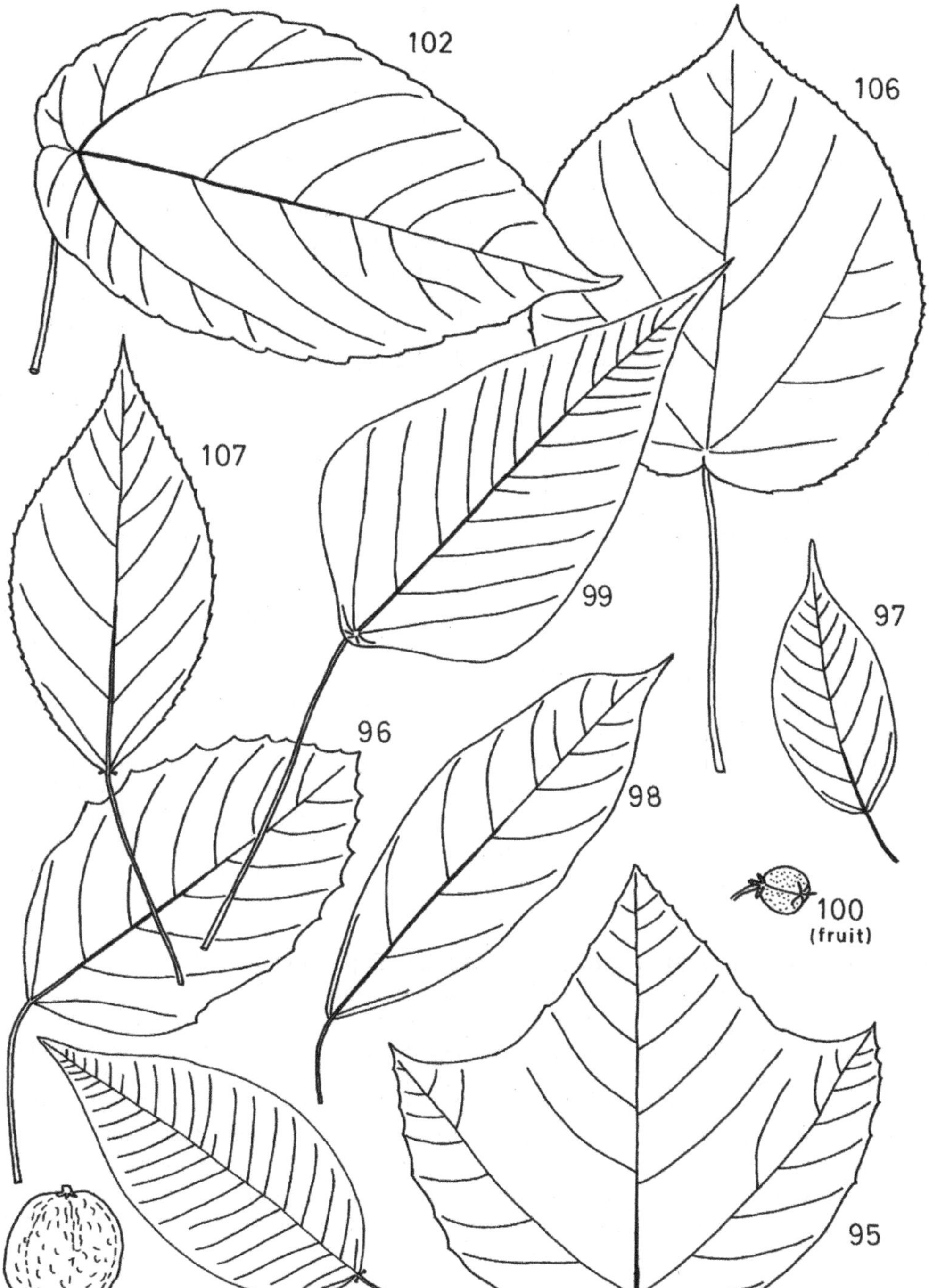

Plate 9. (95-109)

Macaranga capensis (Baill.) Sim (99) Euphorbiaceae

SYNONYM: *Macaranga kilimandscharica* Pax

Omurara (ki); Muhunga (ko); Kaptebema (ku); Kiararwe, Ludesi, Luwessu, Mudwess (ms); Muhoti (to).

40 m. Trunk straight, with a dense crown which is shiny when viewed from above. Sometimes multi-stemmed. Trunk fluted on older trees. Bark thin and smooth, light-coloured. Slash pink to red-brown, sometimes with white lines. Leaves simple, alternate, c. 13 x 8 cm, usually peltate, heart-shaped, with a long-pointed tip. Petiole c. 9 cm long.

OCCURRENCE: U1-3. Abundant in wetter montane forests at 1400-2500 m, e.g. in Kigezi and on Mt Elgon and Rwenzori. Abundant in secondary forest in Kalinzu Forest.

CULTIVATION AND PROPAGATION: Fast-growing pioneer species, which grows best in mixed stands at higher altitudes in high rainfall areas. Suitable for rapid production of fuelwood and poles. Can be used as a cover to shelter slower-growing species. Collect seeds from ground and plant directly.

Neoboutonia macrocalyx Pax (100) Euphorbiaceae

Ekishembabwoki, Ekyanya (ki); Kiona (ko); Chebakwa (ku); Kidoadoa, Mudoadoa (ms); Mukole (na).

20 m. Tree with a short trunk and a spreading crown of large, often insect-damaged, leaves. Bark very thin and smooth, light-coloured. Phellogen green. Slash off-white to pale brown, sometimes mottled. Leaves simple, alternate, large, c. 30 x 30 cm, with 5-9 main veins from the base. Young parts covered with light brown stellate hairs which also form a regular pattern of small dots on the upper surface of the lamina. Main veins with or without spreading hairs on lower surface. Petiole c. 25 cm long. Fruit a 3-lobed capsule at least 10 mm long.

OCCURRENCE: U2 and 3. In Ankole, Kigezi, and on Mt Elgon and Rwenzori. Common in wetter montane forest at 1400-2500 m.

CULTIVATION AND PROPAGATION: Fast-growing tree that can be planted in pure stands. Useful as a cover for slower-growing species. Can be pollarded and coppiced. Suitable for a quick supply of fuelwood. Collect fruits from mother trees before they open, then keep the capsules until they split open and release the seeds. Sow the seeds as soon as possible.

NOTE: The tree is sometimes confused in the vegetative state with *Dombeya torrida* (120), which differs in having red leaf veins.

Neoboutonia melleri (Müll. Arg.) Prain (101) Euphorbiaceae

Kiwumumu (am); Kafunkula (ga); Kiona (ko); Mukoko (to).

15 m. Similar in general appearance to *Neoboutonia macrocalyx*. Trunk straight and cylindrical, with a moderately spreading crown of usually insect-damaged leaves. Bark thin, light brown to brown, slightly vertically fissured though general impression fairly smooth. Slash soft, white with orange granules, turning darker. Leaves simple, alternate, c. 20 x 20 cm, with 5-9 main veins from the base, undersurface either light-coloured or with a uniform covering of long hairs. Petiole c. 13 cm long. Fruit usually less than 10 mm long.

OCCURRENCE: U2-4. A lower altitude species than *Neoboutonia macrocalyx*. Often abundant in valley swamp forest, e.g. in Mengo and in Kibale Forest. Also, on forest edges in Kibale Forest.

Alchornea cordifolia (Schumach. & Thonn.) Müll. Arg. (102) Euphorbiaceae

Luzibaziba (ga).

8 m (sometimes scrambling higher). Semi-recumbent small tree, shrub or straggler, often with spreading leafless branches that bear leafy shoots (like *Acalypha*). Branches hollow. Thorns sometimes present on old stems. Bark quite thin, light brown, with vertical fissures. Slash granular, red-brown (sometimes yellow near wood). Leaves simple, alternate, c. 16 x 10 cm,

ovate to elliptic, with 7-10 main lateral veins on each side of the midrib (3 main veins including the midrib from the base), apex obtusely to acutely acuminate, base distinctly cordate, margin crenate-serrate. Petiole 7-10 cm long, ascending. Lamina drooping over the petiole. Male inflorescence axillary; female inflorescence axillary, often on leafless branches. Fruits red, 2-3 lobed, 1.3 cm long, 1.5 cm across, borne in clusters hanging from the trunk and larger branches.
OCCURRENCE: U2-4. Common on edges of riverine and swamp forest. Abundant in lake-shore forests.
NOTE: According to ITU, the plant is a favourite food of the sitatunga.

***Alchornea laxiflora* (Benth.) Pax & K. Hoffm. (103)** Euphorbiaceae
Shrub or tree to 10 m, with the growth habit of *Polyscias* (384). Green and brown shoots contrast strikingly in colour. Twigs with lenticels. Leaves simple, alternate, c. 12 x 5 cm, ovate to oblanceolate, apex obtusely to acutely acuminate, base usually caudate or rounded, margin crenate to subentire, with 3-7(-9) main veins on each side of the midrib. There is a pair of very small, thread-like, processes (stipels) on the lamina at its junction with the petiole. Petiole 4 cm long. Male inflorescence axillary; female inflorescence terminal. Fruits 3(-4)-lobed, 5-7 mm long, 7-8 mm across.
OCCURRENCE: U1, 2 and 4. Recorded from Bwamba, Mengo, Bunyoro, West Nile, Karamoja and Zoka Forest. Said to be common in Bwamba and Budongo Forest.
NOTE: The leaf differs from that of *Alchornea cordifolia* in usually having a rounded to only slightly cordate base.

Key to *Croton*.

1. Leaves with tiny scales beneath. ....**109. *C. megalocarpus***
   Leaves without scales beneath. 2
2. Leaf blade becoming nearly glabrous above. ....**107. *C. sylvaticus***
   Leaf blade stellate pubescent above. ....**106. *C. macrostachyus***

***Croton macrostachyus* Delile (106)** Euphorbiaceae
Ekwanga (ac, la); Moti, Muhuta (am); Musogasoga (ga); Mwiyo (gw); Omurangara (ki); Toboswa (ku); Ekwango (la); Guyi, Gwihihi, Lwihihi (ms); Mulangara (na); Nahingunya (nl); Muchwi-chwi (sa); Muyemba, Myemberera (so); Muhoti (to).
25 m. Trunk straight or wavy, with a thin crown. Bark dark brown to grey, hard, quite thick, generally smooth (but becoming fissured and granular on old trees). Slash red, light yellow to white (if yellow or white, sometimes becoming deep red near the bark), granular, quite thick. Leaves simple, alternate, c. 12 x 9 cm (but variable in shape and occasionally 3-lobed), 3-7 veined from base, entire or with small teeth, with a pair of stalked glands at the top of the petiole, with stellate hairs. Old leaves turn conspicuously yellow/orange. Petiole c. 10 cm long.
OCCURRENCE: U1-4. Widespread and common in secondary forest and on forest edges. Found up to an altitude of 2300 m; in valleys at higher altitudes.
CULTIVATION AND PROPAGATION: Moderately fast-growing. Tolerant of a wide range of soils. Young trees grow best under some shade. Collect fruits just before maturity (when yellow to brown in colour) either from the tree or just after falling to the ground. Rub the pulp to obtain the stones, which should then be sun-dried. Sow immediately. Seeds lose viability rapidly after storage for a year.

***Croton sylvaticus* Krauss (107)** Euphorbiaceae
SYNONYM: *Croton bukobensis* Pax
25 m. Shrub or small spreading tree, with a rather weak trunk and thin crown. Bark fairly thin, dark brown, longitudinally fissured, moderately rough. Slash light yellow, fibrous, with a strong peppery smell. Leaves simple, alternate, c. 11 x 6 cm, toothed, base obtuse to rounded,

apex acuminate. Petiole c. 6 cm long. Lamina with c. 4-9 main lateral veins on each side of the midrib. Young parts covered with light brown stellate hairs.
OCCURRENCE: U2 and 4. In secondary forest and on forest edges. Abundant in Budongo Forest.

Croton megalocarpus Hutch. (109) Euphorbiaceae
Nkulumire (ga); Omuvune (ki); Mutugunda (na); Musine (tn); Munyabakakuru, Mwenyabakikulu (to).
40 m. Upperstorey tree with a narrow, cylindrical, straight trunk and thin, spreading (to rounded and symmetrical?) crown. Buttresses absent. Trunk sometimes flaring out at base. Bark light brown, fairly thin, fissuring vertically and horizontally to give a pattern of small squares or rectangles c. 5 cm wide. Slash yellow to off-white, layered, brittle, with a spicy smell. Leaves simple, alternate, c. 12 x 5 cm, entire, with a pair of stalked glands at base of lamina and c. 10-25 main lateral veins on each side of the midrib, whitish below. Petiole c. 5 cm long. Old leaves turn conspicuously yellow. Fruit grey-brown, c. 3 cm long, 3-valved.
OCCURRENCE: U2-4. Widely distributed, but generally uncommon. Exceptionally, it is dominant in some of the Tooro forests (e.g. parts of Kibale Forest) and on montmorillonite soils in Mabira Forest.
CULTIVATION AND PROPAGATION: Moderately fast-growing. Can be grown as single specimens or in mixed or pure stands. Collect fruits and gently splinter to release the seeds. Sow seeds as soon as possible.

Discoglypremna caloneura (Pax) Prain (110) Euphorbiaceae
45 m. Deciduous upperstorey tree with a long straight trunk and small crown. Bark smooth, thin, light-coloured. Slash finely and evenly granular, yellow-brown, not turning darker. Leaves simple, alternate, c. 7 x 4.5 cm, toothed to crenate, 3-veined from the base, with only 2-3 other main lateral veins on each side of the midrib. Petiole c. 2.5 cm long.
OCCURRENCE: U2. Only recorded from Budongo Forest, where it is very rare.

Alangium chinense (Lour.) Harms (111) Alangiaceae
Omukofe (ki); Mukoko (ko); Kimuaitit (ku); Guronono, Kistono, Lusontono, Lusotono (ms); Mukerenge (na).
25 m (usually smaller at lower altitudes). Trunk straight and cylindrical (occasionally weak and leaning), with branches at right angles and a layered crown. Small buttresses sometimes present. Bark thin and smooth, grey to green, sometimes slightly vertically fissured. Phellogen green. Slash yellow, with white or yellow lines (white with a yellow border on young trees). Leaves simple, alternate, c. 12 x 6.5 cm, with c. 5-7 main veins from the base and 2-4 other main lateral veins on each side of the midrib, margin entire, base asymmetric (*Begonia*-shaped). Petiole c. 2 cm long. Flowers conspicuous, yellow or white.
OCCURRENCE: U1-4. A widespread, light-requiring, species, reaching up to 2400 m altitude. In open types of forest and secondary forest, and on forest edges. Abundant in Bwindi Forest, where it grows in gullies, and on the western side of Mt Elgon.

Plate 10. Malvaceae, Boraginaceae and others (111-121); see also Plate 11

111. *Alangium chinense* 112. *Cordia millenii* 114. *Ehretia cymosa*
115. *Pterygota mildbraedii* 116. *Cola gigantea* 118. *Sterculia dawei*
119. *Dombeya kirkii* 121. *Leptonychia mildbraedii*

Actual sizes: leaves, flowers and fruits x 2; tree profiles x 800.

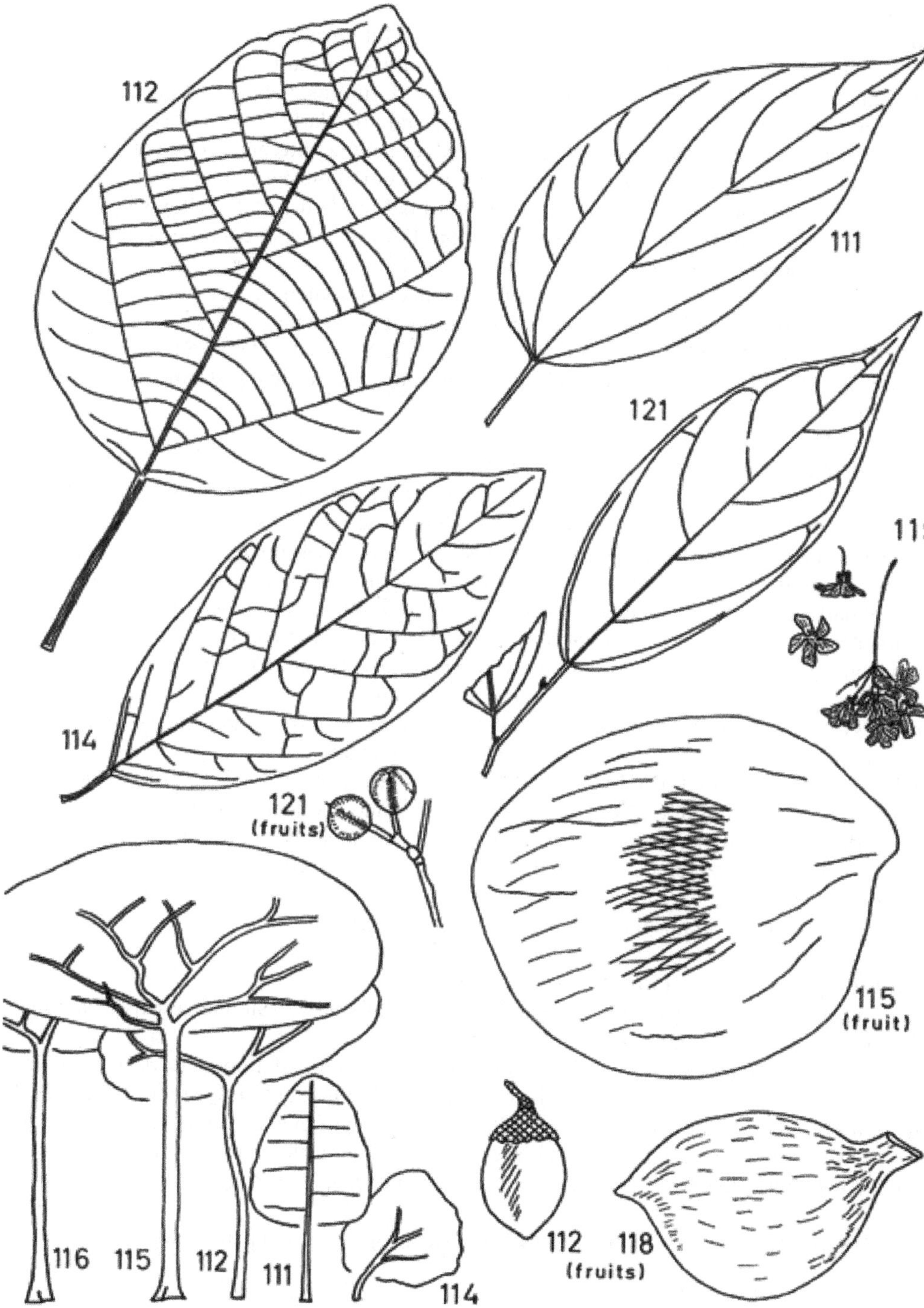

Plate 10. (111-121)

Cordia millenii Baker (112) Boraginaceae

Ketumba (am); Mukebu (ga, tn); Omujugangoma (no); Mutumba (to).

35 m. Trunk wavy, with a thin, spreading crown. Buttresses usually absent. Bark thick and rough, light brown, with deep vertical fissures. Slash soft, fibrous, yellow to white, sometimes with granular orange bands under the bark fissures, rapidly darkening to dark green (with vertical brown lines), eventually turning brown. Leaves simple, alternate, rounded, c. 16 x 13 cm, with 3-7 main veins from base (or arising within 1.5 cm of the base), margin entire or very slightly toothed. Petiole c. 12 cm long. Calyx not strongly ribbed. Petals yellowish, united to form a bell-shaped corolla, which is under 2 cm long. Fruit ovoid, c. 4 cm long, green, cupped by the enlarged calyx.

OCCURRENCE: U1, 2 and 4. Mainly a forest species; sometimes in nearby grassland. Abundant in Budongo and Kibale forests, but less common elsewhere.

CONSERVATION STATUS: Global LC (IUCN, TOU); National EN (WCS).

NOTES: The wood is much used for making drums and other musical instruments, as well as canoes. Bees appreciate the flowers.

Cordia africana Lam. (113) Boraginaceae

Akoiyi (al); Mutumba (am, ko, to); Mukebu (ga, tn); Omujugangoma (ki, no); Mugengere (ku); Chichikiri (ms); Muzugangoma (na); Hinghobe, Khinghobe (nl); Mukumari (tn).

15 m. Shrub or small tree with a curved or crooked trunk. Bark thick, light to dark brown, vertically fissured. Slash soft, white to yellow-brown, sometimes with darker lines, darkening rapidly to dark green-brown. Leaves quite similar to those of *Cordia millenii*, but usually ovate and rather longer in comparison to width (c. 13 x 9 cm) and the main basal lateral veins not reaching into the upper half of the lamina. Calyx strongly ribbed. Corolla white, over 2 cm long. Flowers very conspicuous. Fruit less than 1.5 cm long.

OCCURRENCE: U1-4. A savanna and forest-edge species, not found in primary forest. This is the common species on Mt Elgon and in Ankole and Kigezi.

CULTIVATION AND PROPAGATION: Moderately fast-growing. Tolerant of a wide range of soil conditions. Young trees grow best under some shade. Collect fruit just before maturity (when yellow to brown in colour) either on the tree or just after falling. Rub the pulp to obtain the stones, which should then be sun-dried. Sow immediately. Seeds lose viability rapidly after storage for a year.

NOTE: The tree never attains the dimensions of *Cordia millenii*.

Ehretia cymosa Thonn. (114) Boraginaceae

Musuga (ga); Omukobakoba (ki); Mondarariet (ku); Chibondwe, Sibondwe (ms); Nkabwa, Mukobokobo (na, to); Ikobokobo (so).

20 m. Shrub or small tree with a crooked trunk. Sometimes multi-stemmed. Bark light brown, moderately thick, with prominent lenticels, fissuring and flaking, but fairly smooth. Slash soft, white, with brown or yellow markings, rapidly turning brown. Leaves simple, alternate, c. 15 x 6.5 cm, with c. 6-11 main lateral veins on each side of the midrib, margin entire, often much damaged by insects. Petiole c. 1.5 cm long. Flowers white, c. 0.75 cm long. Fruit a small yellow or red berry.

OCCURRENCE: U1-4. Common small tree on forest edges and in open forests.

MALVACEAE

The family Malvaceae, as defined today, is represented by three groups of tree species in Ugandan forests. Group 1 (115-122), with species formerly placed in the family Sterculiaceae, contains small to large trees characterized by having simple, alternate and often large leaves, stellate hairs and often more than 3 main veins (including the midrib) from the base of the lamina. The slash often shows reticulate markings. Members of Group 2 (127-131d), with species formerly placed in the family Tiliaceae, are small trees (except for *Grewia mildbraedii*, which only occurs in Ishasha Gorge). The leaves are simple and alternate, and tend to have a pair of prominent lateral veins originating from (or from close to) the base of the lamina and extending more than a quarter of the way up the lamina. Group 3 has only one species, *Bombax buonopozense* (363), formerly in the family Bombacaceae. It is distinguished by having alternate digitate leaves.

Cocoa (*Theobroma cacao* L.), a member of this family, is native to Central and South America. It has been cultivated on a small scale in Uganda, especially in Busoga, Bwamba and Kyaggwe.

Key to Malvaceae (species 115-122 only).

1. Main lateral veins reaching from base of lamina to over half way up its length. 2
   Not as above. 4
2. Slash yellow to white. Bark fairly smooth. ....**115. *Pterygota***
   Slash with some red colour or, if white, then bark vertically fissured. 3
3. Leaf margin entire or almost so. ....**116. *Cola gigantea***
   Leaf margin not entire. ....**119-120. *Dombeya***
4. Leaves with more than 3 main veins from the base of lamina and leaves over 6 cm wide. ....**118. *Sterculia***
   Leaves lacking more than 3 main veins from the base of the lamina or, if present, then leaves less than 6 cm wide. 5
5. Leaves usually over 18 cm long. ....**117. *Cola congolana***
   Leaves usually less than 18 cm long. 6
6. Leaves comparatively long and thin, often c. 11 x 3.5 cm. ....**121. *Leptonychia***
   Leaves comparatively broad, often c. 10 x 4.5 cm. ....**122. *Nesogordonia***

Pterygota mildbraedii Engl. (115) Malvaceae

Mwira (am); Buteri (gb); Endawula, Omukoko (no); Mukoko (tn, to).

45 m. Emergent or canopy tree, with a wide, straight, cylindrical trunk, bearing large spreading branches high up. Buttresses often large. Bark light brown, thin to fairly thick in places, often with prominent lenticels, with small vertical fissures, but general effect smooth. Slash hard, yellow or occasionally white, with a reticulum of fibres (best seen if cut shallowly), not or only slightly turning darker. Leaves simple, alternate, c. 16 x 14 cm, sometimes 3-lobed. Petiole c. 9 cm long. Fruit a woody fibrous follicle, more or less rounded, 10-15 cm long, containing large winged seeds.

OCCURRENCE: U1, 2 and 4 (only in Mubende in 4). Said to be mainly on forest edges. Abundant in South and Central Kibale Forest and parts of Maramagambo, occasional in Bunyoro, rare elsewhere. Often gregarious.

CULTIVATION AND PROPAGATION: Fast-growing. Suitable for rapid production of wood fuel. Withstands pruning. Collect fruit from the ground under mother trees and extract seeds by hand. Sow the seeds as soon as possible.

Cola gigantea A. Chev. (116) Malvaceae

Kitoko (am); Mutumbwe (ga); Awe-awe (gb); Omurehe (ki); Kikura (ko); Mujugangoma (na); Omujugangoma (no).

35 m. Trunk straight or wavy, with a fairly spreading crown of large, dark-coloured, leaves. Small buttresses present. Bark brown, quite thick and rough, with deep vertical fissures. Slash fibrous, pink to red, with a reticulum of fibres, turning darker. Leaves simple, alternate, large, c. 30 x 18 cm, with more than 3 main veins from the base of the lamina, one pair reaching half-way or more up the lamina. Petiole c. 8 cm long. Fruit a follicle, red at first, later brown, containing arillate seeds.

OCCURRENCE: U1, 2 and 4. A widespread species, abundant in Budongo Forest.

CULTIVATION AND PROPAGATION: Moderately fast-growing. Probably best grown in mixed stands. Collect seeds from fallen fruits or on the tree when ripening (becoming brown).

NOTE: Chimpanzees eat the fruit walls and arils.

Cola congolana De Wild. & T. Durand (117) Malvaceae

SYNONYM: *Cola bracteata* De Wild.

Understorey tree to 10 m. Trunk short. Bark brown to light-coloured, with small vertical fissures. Slash off-white, fibrous. Leaves simple, alternate, large, c. 25 x 10 cm, with c. 8-11 main lateral veins on each side of the midrib. Petiole c. 5 cm long.

OCCURRENCE: U2. Recorded from Ankole, Kigezi and Tooro. Uncommon, except in Itwara Forest.

CONSERVATION STATUS: Global LC (IUCN, TOU); National VU (WCS).

Sterculia dawei Sprague (118) Malvaceae

Kitoko, Kitokwe (am); Mutumbwe (ga); Muhanga (sa); Musandasanda (so).

25 m. Trunk straight and cylindrical, with a fairly small crown. Buttresses generally absent. Bark quite thin, moderately rough, with vertical and horizontal fissures, flaking in small pieces. Slash fibrous, pink to red on a lighter coloured background, with a fibrous reticulum, turning rapidly brown. Leaves simple, alternate, c. 15 x 12 cm (but much larger on young plants), with more than 3 main veins from the base, the laterals not extending more than half way up the lamina, shortly acuminate. Petiole c. 3.5 cm long. Fruit a follicle, red, turning brown. Seeds black when ripe, lacking arils.

OCCURRENCE: U2-4. A widespread species, but absent from Kigezi and Ankole (except Kasyoha-Kitomi Forest). Common in lake-belt forests and in Bwamba. Rarer than *Cola gigantea* in Budongo Forest.

Plate 11. Malvaceae (115-120); see also Plate 10

115. *Pterygota mildbraedii* 116. *Cola gigantea* 118. *Sterculia dawei*
119. *Dombeya kirkii* 120. *Dombeya torrida*

Actual sizes: leaves x 2.

Plate 11. (115-120)

DOMBEYA

A genus of shrubs and small to medium-sized trees with heart-shaped or palmately-lobed, hairy leaves, found in forest and savanna. The flowers are showy, white to pink in colour, massed in conspicuous inflorescences. *Dombeya burgessiae* Harv. (Syn.: *D. nairobensis* Engl.) is a shrub, sometimes found growing together with *D. torrida* in montane forest. Its inflorescence is simple umbellate, rather than compound umbellate (as is that of *D. torrida*). *Dombeya rotundifolia* Harv. is a small tree to 9 m found in grassland and woodland up to an altitude of 2200 m on Mts Elgon and Moroto. It has pink flowers.

***Dombeya kirkii* Mast. (119)** Malvaceae

SYNONYM: *Dombeya mukole* Sprague

Mukole (am, ko); Omukole (no); Mufudufu (sa).

15 m, occasionally more. Trunk crooked, bearing an open spreading crown. Trunk sometimes fluted. Bark brown, sometimes very dark-coloured on outside, thick and rough, deeply vertically fissured, sometimes flaking. Slash soft, fibrous, red to pink (to white on young trees), usually with white lines, reticulate, turning darker. Leaves simple, alternate, c. 12 x 10 cm (much smaller than those of *Dombeya torrida*), variable in shape, sometimes 3-lobed, margin toothed to crenate. Petiole c. 4 cm long. Flowers white.

OCCURRENCE: U2-4. A widespread lower altitude species, in secondary forest, open forest and on upper slopes.

Dombeya torrida* (J.F. Gmel.) Bamps (120) subsp. *torrida Malvaceae

SYNONYM: *Dombeya goetzenii* K. Schum.

Omukore (ki); Borowa, Borowetamoi (ku); Chikole, Gabaluwa (ms).

20 m. Trunk fairly straight, but branched low down, bearing a spreading crown. Bark grey, smooth to rough. Slash fibrous, pale pink to red, turning darker. Leaves simple, alternate, c. 20 x 15 cm, with red veins, margin toothed. Petiole c. 10 cm long. Flowers white, sometimes tinged with pink.

OCCURRENCE: U1-3. In montane forest, 1800-3200 m. In Kigezi and Acholi and on Mt Elgon and Rwenzori. Often abundant.

CULTIVATION AND PROPAGATION: Fast-growing in mixed stands. It can be coppiced and used for the fast production of fuelwood. Collect fruit from mother trees or from the ground beneath. Dry for 1-2 days. Remove seeds by slightly rubbing the dry fruits, taking care to avoid the fine hairs of the fruit since these may cause irritation to the eyes. Seeds can be sown directly or stored.

NOTE: The tree is distinguished from *Dombeya kirkii* by the difference in habitat and in having a compound (rather than simple) umbellate inflorescence. It is unlikely that the altitudinal ranges of the two species overlap.

***Leptonychia mildbraedii* Engl. (121)** Malvaceae

Nkomakoma (na).

10 m. Understorey tree, trunk sometimes leaning, crown spreading. Bark thin and smooth, dark brown to almost black, sometimes with small vertical fissures. Phellogen sometimes red. Slash white to light yellow, with a reticulate pattern of darker fibres, turning rapidly red-brown. Leaves simple, alternate, c. 11 x 3.5 cm, with c. 5 main lateral veins on each side of the midrib, apex acuminate. Petiole c. 0.5 cm long.

OCCURRENCE: U2 and 4. Abundant in Kibale Forest.

Nesogordonia kabingaensis **(K. Schum.) R. Germ. (122)** Malvaceae

SYNONYM: *Cistanthera kabingaensis* K. Schum.
Mabaka (am).
Tree to 30 m with fibrous, brown bark and a pink to red slash. Leaves simple, alternate, c. 10 x 4.5 cm, with c. 6-8 main lateral veins on each side of the midrib, apex more or less acuminate, base rounded, with conspicuous tufts of brown hairs in the axils of the main veins below.
OCCURRENCE: U2. Bwamba (where it is fairly common) and Kigezi. Mainly on the edge of riparian forest.

Key to Olacaceae.

1. Stamens more numerous than the petals. ....**125. *Heisteria parvifolia***
   Stamens the same number as the petals. 2
2. Flowers 5-merous (parts in 5s), in fascicles. ....**123. *Strombosia scheffleri***
   Flowers 4-merous, in racemes. ....**124. *Strombosiopsis tetrandra***

Strombosia scheffleri **Engl. (123)** Olacaceae

Omuhika (ki); Chiusa, Lyuisa, Ruiunza (ms); Munyakasekuro, Munyakashekero, Munyankono (na); Mtora (to).
30 m. Trunk straight, sometimes with branches from near base. The lower branches tend to grow vertically. Crown short to long, not spreading greatly, casting a heavy shade. Small buttresses sometimes present. Trunk sometimes fluted. Bark thin and smooth, light brown, flaking in small pieces to large sheets (c. 5-30 x 1-20 cm), giving the trunk a mottled pattern of different colours. Slash fibrous, pink to red. Leaves simple, alternate, often large, c. 23 x 10 cm, with c. 5-7 main lateral veins on each side of the midrib, margin entire. Petiole c. 2 cm long, deeply channelled.
OCCURRENCE: U1-4. Abundant in Kibale, Kalinzu and Kasyoha-Kitomi forests and in Kigezi and on west Mt Elgon, ascending to 2500 m. Rare below 1200 m, except in some lake-shore forests.
CULTIVATION AND PROPAGATION: Moderately fast-growing. Probably best grown in mixed stands under shade. Collect fruits from the ground under mother trees and remove the pulp to release the seed. Soak seeds for a day in cold water and sow.

Strombosiopsis tetrandra **Engl. (124)** Olacaceae

30 m. Trunk straight for about half of its length and then branching, with a dense crown. Trunk slightly fluted at base, with occasional knobs. Bark thin and smooth, flaking in pieces c. 2-20 cm across to give a mottled appearance (the pattern is smaller than that of *Strombosia scheffleri*). Slash fibrous, dark red, layered, exuding a red exudate from near the wood (the exudate is produced in fairly small quantities, but is quite conspicuous). Leaves simple, alternate, similar to those of *Strombosia*.
OCCURRENCE: U2. Only recorded from Ishasha Gorge.
NOTE: Most reliably distinguished from *Strombosia scheffleri* by floral characters.

Heisteria parvifolia **Sm. (125)** Olacaceae

Small tree to 15 m. Leaves similar to those of *Strombosia*, but lateral veins tending to be more numerous (6-11 on each side of the midrib). Young branches slightly winged or ridged.
OCCURRENCE: U4. Only recorded from Entebbe Botanical Gardens, but believed to be indigenous.

Brazzeia longipedicellata **Verdc. (126)** Lecythidaceae

Omushabarara (ki).
Small tree to 7 m. Leaves simple, alternate, c. 14 x 6.6 cm, with c. 3-8 arcuate main lateral veins on each side of the midrib, apex acuminate, base unequal-sided. Petiole absent to very short. The fruit is large, orange and borne on trunks and larger branches.

OCCURRENCE: U2. Range-restricted species of narrow endemism, only recorded globally from Bwindi Forest (Ishasha Gorge) and eastern D.R. Congo.
CONSERVATION STATUS: Global EN (IUCN), LC (TOU); National EN (WCS).

Glyphaea brevis (Spreng.) Monach. (127) Malvaceae

Omukoma-nyadabito, Omukoma-nyakabita (no).
8 m. Shrub or weak-stemmed untidy small tree, branching low down, sometimes multi-stemmed. Bark fairly thin, brown, moderately rough on older stems. Slash fibrous, pink, sometimes with white lines, turning darker. Leaves simple, alternate, c. 15 x 7 cm, with 3 (to 5) main veins from the base, the laterals reaching half way or further up the lamina, margin with small to medium-sized teeth. Flowers conspicuous, yellow. Fruit a woody, ridged, capsule, c. 4-6 cm long.
OCCURRENCE: U2 and 4. Widespread. Mainly encountered in secondary forest, on forest edges and near water.

Key to *Desplatsia.*

1. Leaves with red-brown hairs. ....**129. *D. chrysochlamys***
   Leaves without red-brown hairs. 2
2. Leaves glabrous or only with scattered stellate hairs below. ....**128. *D. dewevrei***
   Leaves with fine stellate pubescence below; domatia absent. **129a. *D. mildbraedii***

Desplatsia dewevrei (De Wild. & T. Durand) Burret (128) Malvaceae

Njuli (am); Kitonto (ga); Omukoma-nyakabita (no).
15 m. Deciduous understorey tree, sometimes multi-stemmed. Bark light brown, quite thick and rough. Slash fibrous, white to pink, turning darker. Leaves simple, alternate, c. 22 x 10 cm, with 2 prominent lateral veins from base, these reaching over a quarter (but less than a half) way up the lamina, margin with large teeth. Fruit large, oblong, often c. 12 x 9 cm.
OCCURRENCE: U2. A secondary forest species, 1000-1220 m. Abundant in Budongo Forest.
NOTE: Chimpanzees eat the fruits.

Desplatsia chrysochlamys (Mildbr. & Burret) Mildbr. & Burret (129) Malvaceae

Shrub or weak-stemmed tree to 7 m. The leaves are similar to those of *Desplatsia dewevrei*, but larger (often c. 30 x 12 cm) and the young branches and petioles are covered with red-brown hairs. The fruit is also covered with red-brown hairs, ribbed and c. 4 cm long.
OCCURRENCE: U2 and 4. Occurs sparingly in Mengo and Bwamba, as well as in Budongo Forest, 760-1160 m.

Desplatsia mildbraedii Burret (129a) Malvaceae

SYNONYM: *Desplatsia lutea* Hutch. & Dalziel
18 m. Tree with stellate and simple hairs. Leaves simple, alternate, underside stellate pubescent. Fruit large, oblong, usually 6-9 cm long
OCCURRENCE: U2 (Tooro and Bunyoro). Moist forest, 1000-1130 m. A range-restricted species of narrow endemism. Only known from Uganda and eastern D.R. Congo.
CONSERVATION STATUS: Global NE (IUCN), VU (TOU); National NE (WCS)

Plate 12. Malvaceae, Olacaceae and others (123-131)

123. *Strombosia scheffleri* 126. *Brazzeia longipedicellata* 127. *Glyphaea brevis* 128. *Desplatsia dewevrei* 130. *Grewia pubescens* 131. *Grewia mildbraedii*

Actual sizes: leaves and fruits x 2; tree profile x 800.

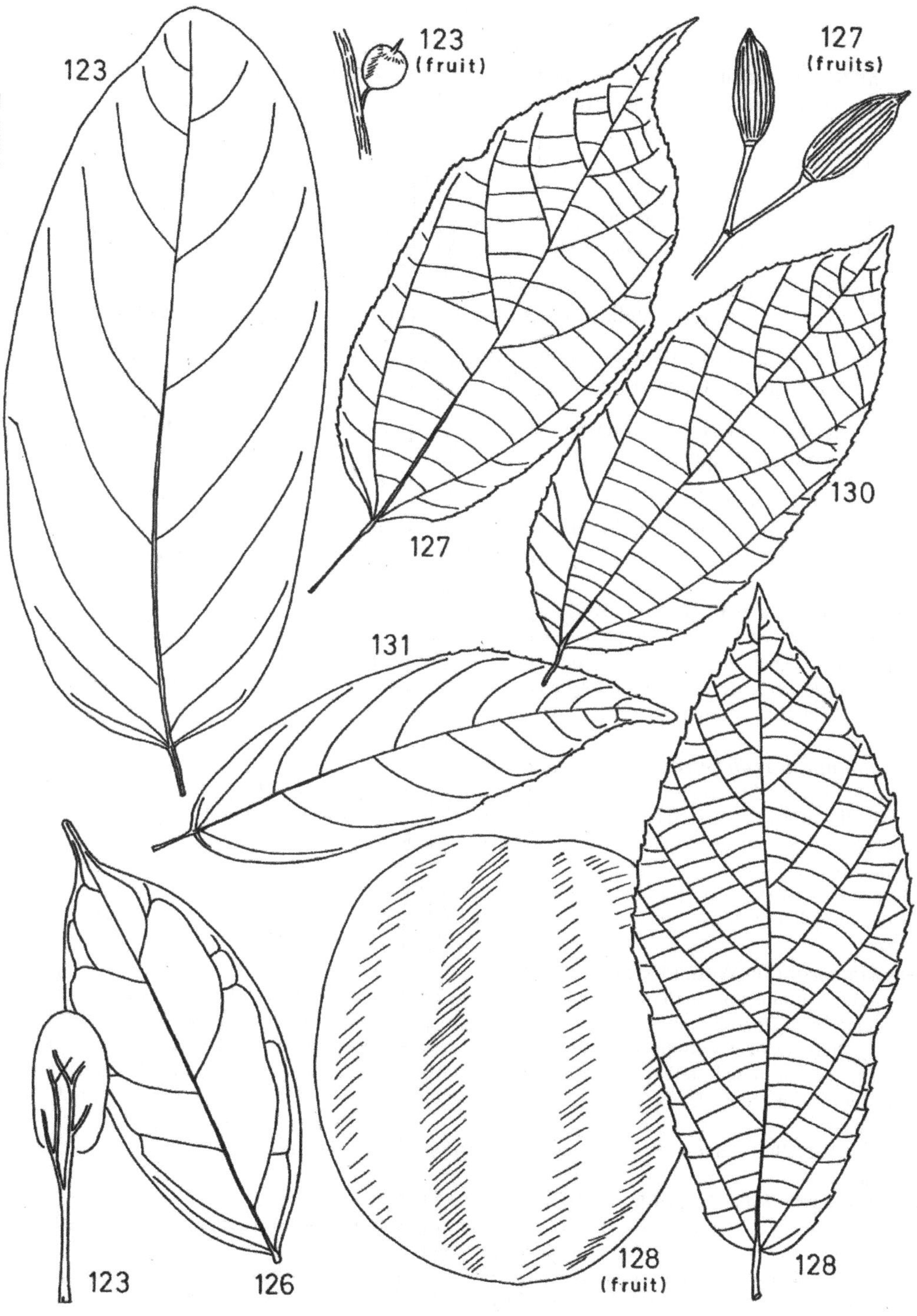

Plate 12. (123-131)

GREWIA

Grewia in Uganda includes a number of small trees, shrubs and stragglers. Most live in savanna. Species recorded from forest grow mainly on forest margins.

Key to *Grewia.*

1. Leaves rough (like sandpaper). ....**131b. *G. rugosifolia***
   Leaves not rough like sandpaper. 2
2. Leaves glabrous or nearly so above. 3
   Leaves with scattered stellate and simple hairs above. ....**131a. *G. seretii***
3. Petiole usually less than 4 mm long. ....**131c. *G.* sp. A of FTEA**
   Petiole usually at least 4 mm long. 4
4. Leaf margin slightly undulate. ....**131d. *G. ugandensis***
   Leaf margin not as above. 5
5. Leaf margin evenly finely serrate. ....**130. *G. pubescens***
   Leaf margin entire to shallowly serrate, often entire in lower two-thirds.
   ....**131. *G. mildbraedii***

Grewia pubescens P. Beauv. (130) Malvaceae
Tree to 8 m. Leaves c. 12 x 5 cm, with 2 main lateral veins from the base, these extending over half way up the lamina. The leaves are similar to those of *Glyphaea*, but are hairy below (those of *Glyphaea* and *Grewia mildbraedii* are glabrous or very nearly so) and tend to be more finely and evenly toothed.
OCCURRENCE: U2 and 4. Only common in Mabira Forest.

Grewia mildbraedii Burret (131) Malvaceae
12 m. Trunk straight and cylindrical. Bark smooth, very thin, grey to light brown, with orange lenticels and small vertical fissures. Slash fibrous, red. Leaves simple, alternate, elliptic-oblong, ovate to obovate-oblong, c. 16 x 6 cm, acuminate, base obtuse, margin entire to shallowly serrate towards apex, venation similar to *Grewia pubescens*, glabrous. Petiole 10-17 mm long, with scale-like hairs. Fruit unlobed, obovoid to pyriform, to 17 mm long, 8 mm wide, glabrous.
OCCURRENCE: U2. Only recorded from Ishasha Gorge.

Grewia seretii De Wild. (131a) Malvaceae
6 m shrub, scrambler or small tree. Trunk straight and cylindrical. Older branches dark red with yellowish lenticels. Bark smooth, very thin, light brown, with small vertical fissures. Slash fibrous, red. Leaves simple, alternate, ovate-elliptic, c. 16 x 6 cm, apex acuminate, base rounded, obtuse, truncate or shallowly cordate, with scattered stellate and simple hairs above, more or less toothed. Petiole 3-8 mm long, pubescent. Fruit unlobed, obovoid to subglobose or ellipsoid, 6-11 mm long, 8 mm wide, red, shiny, glabrous.
OCCURRENCE: U2 and 4. In forest, including swamp forest and on forest edges.

Grewia rugosifolia De Wild. (131b) Malvaceae
12 m. Shrub or tree, sometimes a scrambler. Branches stellate-pubescent when young. Leaves oblong to oblong-ovate, 6.5-13 cm long, 3-8.5 cm wide, apex obtuse to broadly acuminate, margin serrate, feeling sandpapery. Petiole 4-10 mm long, stellate pubescent. Androgynophore clearly cup-shaped, glabrous. Fruit indistinctly 1-4 lobed, subglobose, 9-14 mm wide.
OCCURRENCE: U2 and 4. On forest margins.

Grewia sp. A of FTEA (131c) Malvaceae
4 m. Young branches densely pubescent. Leaves ovate to ovate-oblong, 1.7-7.7 x 0.8-4.2 cm, apex variously obtuse, acute or slightly acuminate, base rounded, +/- asymmetric, margin distinctly serrate,

lamina glabrous to sparsely pubescent above, densely pubescent below. Petiole 1-4 mm long, densely pubescent. Fruit 2-lobed or unlobed (if abortive), sparsely pilose.
OCCURRENCE: U2 (Butiaba Flats, on forest edges). This is a range-restricted species only known from a single site in each of Uganda, eastern D.R. Congo and Kenya.

***Grewia ugandensis* Sprague (131d)** Malvaceae
5m. Shrub or small tree. Branches rusty brown-pubescent. Leaves ovate-elliptic, c. 10 x 4.5 cm, shortly acuminate, base obtuse, margin slightly undulate, virtually glabrous. Petiole 4-10 mm long, stellate pubescent. Fruit unlobed, c. 20 x 11 mm wide, sparsely pubescent.
OCCURRENCE: U2. This is a range-restricted species only known from Kibale and Itwara forests in Uganda and from Tanzania and D.R. Congo.

Key to Achariaceae, Salicaceae and similar-looking trees (132-162).

The species covered in this key belong to diverse taxa, namely the families Achariaceae, Ochnaceae and Salicaceae and the genera *Gymnosporia*, *Hugonia*, *Maytenus*, *Mystroxylon* and *Rinorea*. Several of these species were formerly placed in the botanical family Flacourtiaceae. All species included here tend to share a set of common features, being mostly small trees with simple alternate leaves which have more or less non-entire leaf margins. Another confusable species (not included in the key) is *Ficus asperifolia* (68). Additional keys are provided to species of *Dovyalis* (144-146), *Rinorea* (148-152a) and Ochnaceae (153-158a). Identification of some species of *Maytenus* and *Gymnosporia* in the sterile state can be particularly difficult, some species being variable. It has not proved possible to accommodate all forms of *Maytenus* in the key.

1. Leaves small (c. 7 x 3.5 cm), usually markedly obovate. ....**160. *Mystroxylon aethiopicum*; 162. *Gymnosporia heterophylla***
   Leaves not both small and markedly obovate. 2
2. Spines present on at least some stems. 3
   Spines absent (except sometimes on leaf margins). 9
3. Petiole long, often c. 7 cm or more. ....**132. *Caloncoba***
   Petiole shorter than above. 4
4. Dominant colour of slash a shade or red. ....**140-140a. *Scolopia***
   Not as above. 5
5. Leaves comparatively large (often c. 15 x 6 cm). ....**143. *Oncoba routledgei***
   Leaves smaller than above (except sometimes on epicormic shoots). 6
6. Lenticels more or less elongated on branchlets. 7
   Lenticels more or less rounded on branchlets. 8
7. Lateral leaf veins relatively few. ....**144-146. *Dovyalis***
   Lateral leaf veins relatively numerous (c. 4-7 on each side). ....**141. *Flacourtia***
8. Leaf margin entire or almost so. ....**144. *Dovyalis macrocalyx***
   Leaf margin toothed or crenate. ....**142. *Oncoba spinosa***

9. Axillary climbing hooks usually present on some stems (they are not present on young plants). ....**159. *Hugonia***
Axillary climbing hooks absent. 10

10. Teeth on leaf margin sharp and stiff; margin therefore spiny. 11
Teeth on leaf margin, if present, not spiny, though sometimes sharp. 12

11. Upper part of petiole normally deeply channelled. Petiole less than 3 cm long. ....**139. *Rawsonia*; 148. *Rinorea ilicifolia***
Upper part of petiole not deeply channelled. Petiole sometimes over 3 cm long. ....**133-134. *Lindackeria*** (particularly ***L. bukobensis***)

12. Leaves over 20 cm long (except near flowers), thick and leathery. ....**153. *Campylospermum densiflora***
Leaves less than 20 cm long or, if longer, then not thick and leathery. 13

13. Petiole long, often c. 7 cm or more. 14
Petiole shorter than above. 15

14. Leaf margin entire or with small teeth. ....**132. *Caloncoba*; 148-152a. *Rinorea***
Leaf margin conspicuously toothed. ....**133-134. *Lindackeria*; 148-152a. *Rinorea***

15. Leaves conspicuously 3-7 veined from base (see Plate 13). ....**147. *Trimeria***
Not as above. 16

16. Leaves small (c. 7.5 x 3 cm), ovate, apex acute to acuminate. A montane tree found above 1500 m. ....**163. *Maytenus acuminata***
Not as above. 17

17. Leaf margin with numerous, small and regularly arranged teeth (see Plate 14). Leaves rather narrow in proportion to length. Petiole often c. 0.3 cm long or less. ....**154. *Campylospermum vogelii*; 155-158a. *Ochna***
Leaf margin not as above. Petiole usually over 0.3 cm long. 18

18. Leaves small (usually to 9 cm long, occasionally to 11 cm), comparatively broad in proportion to length (see Plate 13) and with c. 4-8 main lateral veins on each side of the midrib, these borne at relatively steep angles to the midrib (both species included here are very variable). 19
Leaves either more than c. 9 cm long or, if smaller, then shape and venation not as above. 20

19. Slash colour a shade of red. ....**160. *Mystroxylon***
Slash colour not red. ....**141. *Flacourtia***

20. Petiole c. 0.4 cm long or less. Slash not red. ....**149. *Rinorea beniensis***
Petiole 0.5 cm long or more or, if shorter, then slash red. 21

21. Leaves large (often over 20 cm long), with a petiole often c. 3.5 cm long or more, leaf margin usually (but not always) rather obscurely toothed. ....**152. *Rinorea oblongifolia***
Leaves less than 20 cm long. 22

22. Slash with some shade of red as the dominant colour. 23
Slash not red (the species keying out here can be particularly difficult to separate from one another in the vegetative state). 24

23. Midrib raised and prominent on the upper surface of leaf, appearing about as wide as on the lower surface. ....**133-134. *Lindackeria*; 137-138. *Dasylepis***
Midrib narrower on upper surface than on under surface. ......**161. *Maytenus undata***

24. Petiole c. 0.5-1.3 cm long. ....**137-138. *Dasylepis***
Petiole on mature leaves usually over 1.3 cm long. 25

25. Acumen either absent or not very prominent, leaf apex rounded or obtuse. ....**150. *Rinorea brachypetala***
Acumen prominent, leaf apex more or less sharp-pointed. 26

26. Petiole conspicuously hairy. ....**133. *Lindackeria bukobensis***
Petiole not as above. 27

27. Upper part of petiole not deeply grooved. ....**134. *Lindackeria schweinfurthii***
Upper part of petiole deeply grooved. **150-151. *Rinorea brachypetala, R. dentata***

ACHARIACEAE AND SALICACEAE (132-147)

Species in these families were formerly placed in the family Flacourtiaceae (as in UFT), the taxonomy of which has been extensively revised in recent years. The re-assignment of some genera to families still remains uncertain. Achariaceae and Salicaceae in Uganda are small trees (except for *Casearia*, which is placed elsewhere in this book – 207-208; sometimes bearing spines; with simple, alternate leaves which usually have non-entire margins. Flowers various, often with numerous stamens. Fruits various.

Caloncoba crepiniana **(De Wild. & T. Durand) Gilg (132)** Achariaceae

SYNONYM: *Caloncoba schweinfurthii* Gilg
Okulukwer (ac); Tambe, Tambi (am); Oyanzu (md); Omuhekeheke (no).
10 m. Deciduous understorey tree, trunk branched near base, with a thin, spreading crown. Trunk sometimes bearing simple or branched spines. Bark thin, fairly smooth, brown. Slash light pink to light pink-brown, turning darker. Leaves simple, alternate, elliptic to ovate, c. 16 x 8 cm, apex obtusely acuminate, base cuneate to rounded, with c. 5-8 main lateral veins on each side of the midrib, margin entire (very rarely with small teeth). Petiole c. 7 cm long. Flowers large (c. 10 cm diameter), white and fragrant, borne after the tree has lost its leaves. Fruit smooth, green, mottled with yellow, c. 2.5-9 cm diameter. The fruits tend to persist on the tree.
OCCURRENCE: U1, 2 and 4. Abundant in secondary forest in Budongo Forest. Absent from lake-belt forests.

Lindackeria bukobensis **Gilg (133)** Achariaceae

SYNONYMS: *Lindackeria mildbraedii* Gilg, *Lindackeria bequaertii* De Wild.
9 m. Bark thin and rough. Readily recognized by its leaves. These are 5-20 x 2.5-12 cm, variable in shape, oblanceolate to oblong or elliptic, sometimes obovate, apex very shortly acuminate, acute or obtuse, base attenuate to obtuse, rarely rounded or sub-cordate, margin very coarsely toothed, almost spiny, chartaceous (paper-like in texture), at first densely hairy on both surfaces, turning nearly glabrous with age, lateral veins 6-8 on each side of the midrib. Petiole 1.5-7.2 cm long, pubescent. Flowers 3-7 in racemes. Peduncle 1.5-3(-5) cm long, pedicel 1.5-2.5(-3) cm long, articulate (segmented) at base. Capsule nearly globose, large, c. 1-3 cm diameter, including spines (that are pubescent), swollen at base.
OCCURRENCE: U2 and 4. From 700 m altitude (Semuliki National Park) to 1400 m (Bwindi-Impenetrable National Park). Uncommon.

Lindackeria schweinfurthii Gilg (134) Achariaceae
4.5 m. Leaves c. 17 x 6 cm, lanceolate-oblong to elliptic, apex long acuminate, base attenuate, margin sub-entire to crenate or toothed, papery, generally glabrous, main lateral veins 7-10 on each side of the midrib. Petiole c. 1.75 cm long, not or only slightly grooved. Flowers 3-6 in racemes. Peduncle c. 1.5 cm long, pedicel articulate (segmented) well above the base. Capsule subglobose, 1-1.5 cm diameter, including spines (that are pubescent), swollen at base.
OCCURRENCE: U2 and 4. From 750 m altitude (Semuliki National Park) to 1220 m (Mabira Forest).

Dasylepis eggelingii J.B. Gillett (137) Achariaceae
10 m. Understorey tree, with a crooked, often leaning, trunk and a long, narrow crown. Bark thin and smooth, brown. Slash red-brown on older trees. Leaves simple, alternate, oblanceolate to oblong, apex acuminate, base cuneate to rounded, c. 14 x 5 cm, rather thick, with c. 6-9 main lateral veins on each side of the midrib, margin regularly serrate with large sharp teeth. Petiole c. 1 cm long, grooved. Fruit c. 1.25 cm diameter, reddish, densely pubescent, without spines, hanging in clusters on stalks up to 12 cm long.
OCCURRENCE: U2 and 4. Perhaps most common in Kibale and Kayonza forests. This is a range-restricted species of narrow endemism, only known from Uganda and eastern D.R. Congo.
NOTE: The leaves are similar to those of *Dasylepis racemosa*, *Rinorea brachypetala*, *R. dentata* and some species of *Lindackeria*.

Dasylepis racemosa Oliv. (138) Achariaceae
Omusadya (ki).
Much-branched tree to 20 m. Bark thin and smooth. Slash pink. Leaves simple, alternate, c. 16 x 6 cm, elliptic to oblanceolate, apex shortly obtusely acuminate, base rounded, margin with a few widely separated teeth (mainly situated in the upper half of the lamina), c. 5-8 main lateral veins on each side of the midrib. Petiole c. 12 mm long (very rarely to 20 mm), grooved. Flowers on short, erect stalks, c. 5 cm long. Fruit nearly globose, c. 2.5 cm diameter, reddish, virtually glabrous.
OCCURRENCE: U2. Recorded from Bwindi Forest. Uncommon.
NOTE: Tends to occur at higher altitudes than *Dasylepis eggelingii*.

Rawsonia lucida Harv. & Sond. (139) Achariaceae
10 m. Understorey tree with a wavy trunk, often branched near base, with a dark-coloured crown. Bark thin and smooth, red-brown or grey, flaking in pieces c. 0.5-8 cm across to give a mosaic of different colours on the trunk. Phellogen sometimes green. Slash thin, yellow or brown, turning darker. Leaves simple, alternate, lanceolate to oblanceolate, occasionally to oblong-obovate, glabrous, c. 14 x 4.5 cm, with c. 7-9 main lateral veins on each side of the midrib, leathery, apex acuminate, base cuneate, margin spiny serrate. Petiole c. 0.6 cm long. Flowers white to pink, with c. 15 perianth parts. Fruit globular, orange, with persistent style remains.

Plate 13. Achariaceae and Salicaceae (132-147); see also Plate 14

132. *Caloncoba crepiniana* 133. *Lindackeria bukobensis* 136. *Lindackeria schweinfurthii*
137. *Dasylepis eggelingii* 138. *Dasylepis racemosa* 140. *Scolopia rhamniphylla*
141. *Flacourtia indica* 142. *Oncoba spinosa* 143. *Oncoba routledgei*
144. *Dovyalis macrocalyx* 145. *Dovyalis abyssinica* 147. *Trimeria grandifolia*

Actual sizes: leaves, flower and and fruits x 2.

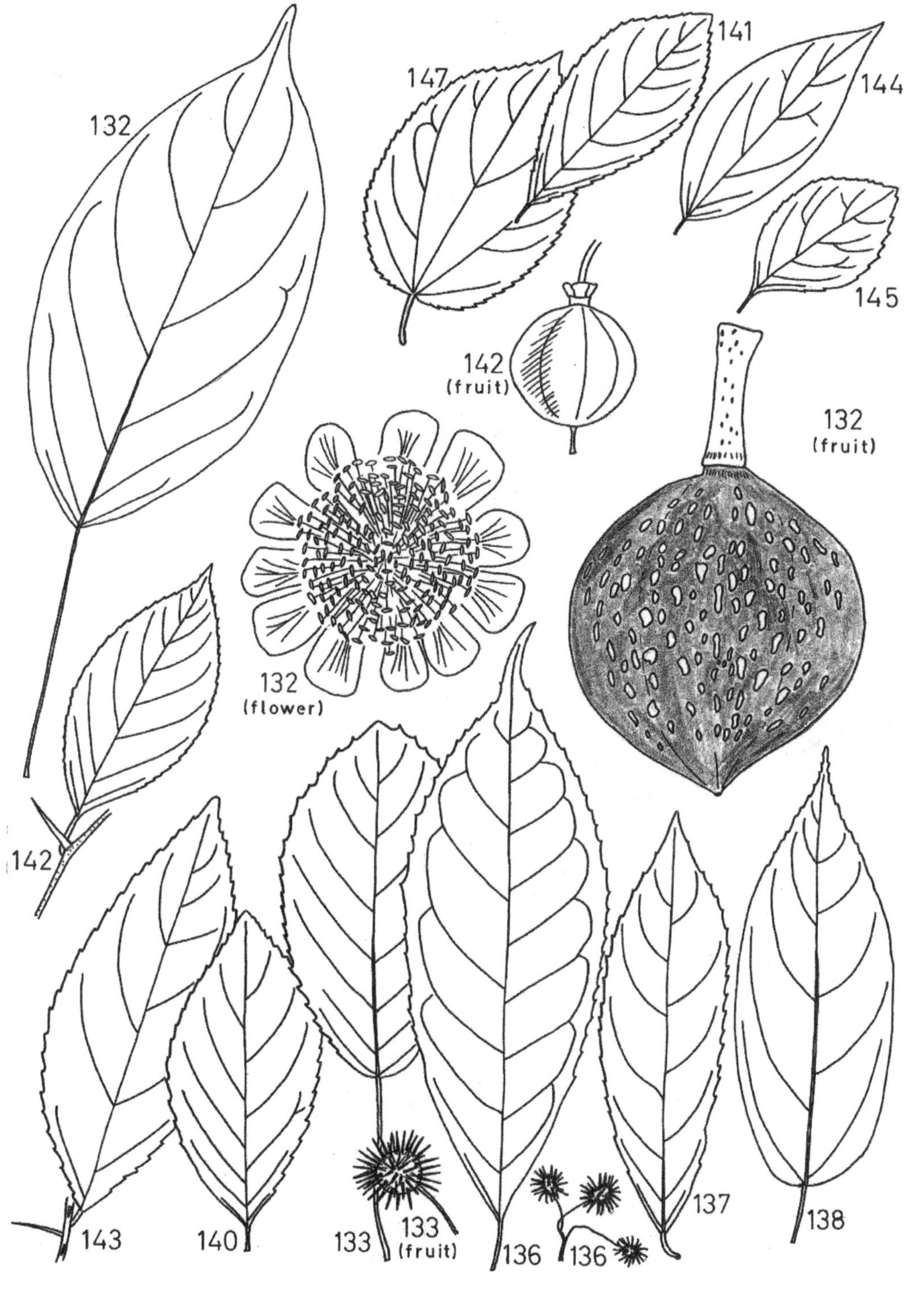

Plate 13. (132-147)

OCCURRENCE: U2 and 3. Mainly in drier types of forest. Abundant under *Cynometra* in South Maramagambo Forest and in South Kibale.
NOTE: *Rinorea ilicifolia* (148) is the only species with similarly shaped and spiny leaves.

Scolopia rhamniphylla Gilg (140) Salicaceae

Nkanaga (ga); Mubambaryobe, Musobya (na).
8 m. Weak-stemmed tree, often multi-stemmed and with epicormic shoots armed with long (to 15 cm) spines. When present, spines on young shoots axillary and to 8 cm long. Bark brown. Slash soft and thick, pink to purple-red. Lenticels prominent on young shoots. Leaves simple, alternate, c. 9 x 4.5 cm (longer on sucker shoots), with c. 5-7 main lateral veins on each side of the midrib, apex shortly acuminate, base cuneate, margin bluntly toothed. Petiole c. 0.5 cm long. Fruit ovoid-subglobular, 6-7 mm in diameter.
OCCURRENCE: U2 and 4. Uncommon, except possibly in damp places in forests in Mengo.

Scolopia zeyheri (Nees) Harv. (140a) Salicaceae

Low-branching, much branched shrub or tree, occasionally to 25 m. Trunk sometimes with branched spines. Bark thin, dark grey. Branchlets unarmed or with straight spines to 20 cm long. Spines sometimes bearing leaves or flowers. Leaves variable, 2-8 cm long, 1-3 cm wide, apex obtuse to rounded, base cuneate (rarely obtuse to cordate on coppice shoots), margin entire, wavy or bluntly crenate, with 4-6 lateral veins on each side of the midrib. Petiole to 1.5 cm long. Racemes axillary. Fruit sub-orbicular, 7-8 mm in diameter.
OCCURRENCE: U2 (Ankole).
NOTE: Differs from *S. rhamniphylla* in having a densely white-hairy receptacle and compact inflorescence. The receptacle in *S. rhamniphylla* is glabrous to only sparsely hairy and the inflorescence is open.

Flacourtia indica (Burm. f.) Merr. (141) Salicaceae

Ebelebelebwoit, Elebelebwoit (ka); Tungururu (ku); Kokowi (la); Singululurwe (ms); Muzhebazhebe (na).
10 m, usually less, rarely more. A tree, shrub or scandent plant. Spines usually (but not always) present on smaller branches and trunk. Bark light brown, rough on older stems. Slash off-white, often with orange or yellow markings. Leaves simple, alternate, very variable, ovate to elliptic, sometimes suborbicular, apex obtusely acuminate to rounded, base cuneate to rounded, margin crenate, toothed to almost entire, 2.5-12 x 1.3-7.5 cm, with c. 4-7 main lateral veins on each side of the midrib. Petiole c. 0.6 cm long (occasionally to 2 cm). Fruit globular, to 2.5 cm in diameter, with persistent stylar remains.
OCCURRENCE: U1-4. Widely distributed. On forest edges and in savanna.
NOTE: This highly variable species appears not to normally grow inside rainforest.

Oncoba spinosa Forssk. (142) Salicaceae

Omujebajebe (ki); Kimeswan (ku); Chibeye, Nabizima, Shibeye (ms).
10 m. Tree with sharp spines on shoots and trunk, up to 8 cm long. Bark thick and rough. Slash white with yellow lines, turning darker. Lenticels on branchlets not markedly elongate. Leaves simple, alternate, often red when young, elliptic to ovate-elliptic, c. 8.5 x 4.5 cm, apex acuminate, base cuneate, papyraceous to subleathery, glabrous, margin with small teeth or crenations, lateral veins 6-8 on each side of the midrib. Petiole c. 0.8 cm long. Flowers solitary, peduncle 1-2 cm long. Fruit globose, hard and woody, c. 4 cm diameter, smooth, with lines running from top to bottom, sepals persistent at base.
OCCURRENCE: U1-4. Widespread, often in riparian forest. Also, on forest edges.
NOTE: Leaves on epicormic shoots can be of the same size as the normal leaves of *Oncoba routledgei*.

Oncoba routledgei Sprague (143) Salicaceae
Omuzebajebe (ki), Munyege (to).
Small understorey tree to 8 m, with axillary spines 1-2.5 cm long. Leaves elliptic-ovate to oblong, 6-16 cm long, 3-6 cm wide, apex shortly acuminate, base obtuse to rounded, papyraceous to subleathery, glabrous and glossy, margin coarsely serrate-crenate, main lateral veins 4-6 on each side of the midrib. Petiole 0.4-1 cm long. Flowers solitary, peduncle 1-2 cm long. Fruit subglobose, c. 4.5 cm long, 5.5 cm across, smooth.
OCCURRENCE: U2 and 3. Recorded from Kigezi, Tooro and Mbale.
NOTE: The leaves are larger than those of *Scolopia* (except on epicormic shoots) and *Dovyalis*. This species differs from *Oncoba spinosa* in having larger leaves (to 15 x 6 cm) that are coarsely serrate-crenate and in having a stigma with several branches. The leaves of *O. spinosa* are only faintly serrate or crenate-serrate, and the stigma is peltate.

Key to *Dovyalis*.

1. Ovary with 2-3 (rarely 4) placentas, each placenta with one ovule; styles 2 or 3 (rarely 4). ....**144. *D. macrocalyx***
   Ovary with 4-8 placentas, each placenta with 2 ovules; styles (4-)8-20(-40). 2
2. Fruit apple-like, 10-15 cm across when fresh (6-10 cm when dry); seeds glabrous. ....**146. *D. spinosissima***
   Fruit subglobular, (2-)4-6 cm across when fresh (2-4 cm when dry); seeds densely covered with woolly hairs. ....**145. *D. abyssinica***

Dovyalis macrocalyx (Oliv.) Warb. (144) Salicaceae
Lucukucuko (ac); Mutunku (ga); Ntengenene (to).
6 m. Understorey tree with a weak trunk and spreading crown. Thin, straight, axillary spines present on branchlets and smaller stems. Bark light brown, vertically fissured. Slash yellow and layered (sometimes with some red colour) or white with orange markings. Lenticels on branchlets small and rounded. Leaves simple, alternate, c. 7 x 3.5 cm, elliptic to ovate, apex obtuse, base broadly cuneate to rounded, margin usually entire, glabrous. Petiole c. 0.3 cm long. The calyx bears conspicuous glands and enlarges during fruiting. Fruit ellipsoid, glabrous or sparsely hairy.
OCCURRENCE: U1-4. Widespread. On forest edges and in secondary and open forests.

Dovyalis abyssinica (A. Rich.) Warb. (145) Salicaceae
6 m. Understorey species. A similar-looking plant to *Dovyalis macrocalyx*, but lenticels on branchlets light brown and slightly elongated. Leaves simple, alternate, c. 9 x 4.5 cm, usually toothed or crenate (occasionally entire). Petiole c. 0.3 cm long.
OCCURRENCE: U1-4. Above c. 1500 m in Karamoja and on Mt Elgon, where it is common.

Dovyalis spinosissima Gilg (146) Salicaceae
SYNONYM: *Dovyalis macrocarpa* Bamps
Small understorey tree, similar in appearance to *Dovyalis macrocalyx*, but lenticels on branchlets light brown and slightly elongate. Leaves simple, alternate, ovate to elliptic, c. 9 x 4.5 cm, apex obtuse to shortly acuminate, base broadly cuneate to rounded, margin usually toothed or crenate, but occasionally entire, with pellucid points visible against the light. Lower surface of leaves hairy (rather than glabrous or only slightly hairy, as with *D. macrocalyx*). Petiole c. 1 cm long. Fruit large, c. 10-15 cm diameter.
OCCURRENCE: U1 and 2. On forest edges and in thickets in Kigezi. Common on the edge of Maramagambo Forest. Found above c. 1500 m on the mountains of Karamoja, where it is common.

Trimeria grandifolia (Hochst.) Warb. subsp. _tropica_ (Burkill) Sleumer (147) Salicaceae
Synonym: *Trimeria bakeri* Gilg
Jemberyambogo (ga): Omwatanshare (ki); Epelong (ku): Lusigi (so).

Shrub or small tree to 7 m. Leaves simple, alternate, ovate to elliptic or broadly reniform, c. 8 x 5 cm, apex shortly acuminate to obtuse, base rounded to cordate, densely softly hairy, 3-7 veined from base, with 1-3 other main veins on each side of the midrib, margin glandular, crenate or serrate, hairy below, at least on midrib. Petiole c. 1.3 cm long. Flowers small, sessile. Fruit a trigonous-ovoid capsule, c. 3 mm long.
OCCURRENCE: U1-4. On forest edges. May survive in secondary forest.

Key to *Rinorea.*

1. Leaves thick, margin spiny. .... **148. *R. ilicifolia***
   Leaves not spiny, sometimes with large teeth. 2

2. Petiole less than 0.5 cm long. .... **149. *R. beniensis***
   Petiole at least 0.5 cm long. 3

3. Leaf margin finely serrate-dentate. .... **151. *R. dentata***
   Leaf margin serrate-crenate to subentire. 4

4. Leaf apex acute or shortly and abruptly acuminate; capsule smooth. .... **150. *R. brachypetala***
   Leaf apex acuminate; capsule rough or with scales. 5

5. Petiole at least 1.5 cm long. .... **152. *R. oblongifolia***
   Petiole less than 1.5 cm long. .... **152a. *R. tshingandaensis***

***Rinorea ilicifolia* (Oliv.) Kuntze (148)** Violaceae

5 m. Understorey tree with a wavy trunk, often branched near base. Crown dark-coloured. Bark thin and smooth, dark-coloured. Slash hard, dark orange. Leaves simple, alternate, c. 15 x 5 cm, glabrous or rarely pubescent on midrib beneath, base cuneate to rounded or cordate, apex acute to acuminate, margin spiny. Petiole c. 1 cm long, glabrous or pubescent. Capsule 1.4-1.7 cm long, hairless but rugose.
OCCURRENCE: U1-4. A widely distributed species, sometimes abundant, e.g. in parts of Mabira Forest.
NOTES: Only var. *ilicifolia* occurs in Uganda. *Rawsonia lucida* is the only other species with similarly shaped and spiny leaves.

***Rinorea beniensis* Engl. (149)** Violaceae

SYNONYM: *Rinorea ardisiiflora* (Welw. ex Oliv.) Kuntze
Omuikaraheyere (no).
10 m. Understorey tree with a rather narrow crown. Bark thin, brown, flaking. Slash brittle, yellow. Leaves simple, alternate, c. 9 x 3 cm, apex long acuminate, base cuneate to subrounded, glabrous or with a few hairs on midrib and lateral veins beneath, with c. 5-7 main lateral veins on each side of the midrib, apex long acuminate, base cuneate to subrounded, margin toothed to undulate. Petiole c. 0.4 cm long, pubescent. Stipules thin and pointed or slightly curved, c. 0.4 cm long. Capsule c. 1 cm long, hairless.
OCCURRENCE: U1, 2 and 4. Very abundant beneath *Cynometra* in Budongo Forest.
NOTE: The leaves are smaller than those of other species of *Rinorea.*

***Rinorea brachypetala* (Turcz.) Kuntze (150)** Violaceae

Omuikaraheyere (no).
7 m. Understorey tree. Leaves simple, alternate, c. 14 x 5.5 cm, apex acute (sometimes acuminate), often rounded or obtuse right at the end, base subrounded to cuneate, fairly thick, usually glabrous when mature or with some hairs on midrib and lateral veins beneath, margin serrate-crenate or subentire, with 6-10 main lateral veins on each side of the midrib, margin

crenate, serrate or subentire. Petiole c. 1.5 cm long (sometimes rather longer), glabrous or rarely sparsely pubescent. Stipules rather broader than those of *Rinorea beniensis*, c. 0.6 cm long, straight or slightly curved.
OCCURRENCE: U1, 2 and 4. Abundant in Budongo Forest.

Rinorea dentata (P. Beauv.) Kuntze (151) Violaceae
5. Understorey tree. Leaves simple, alternate, c. 13 x 6 cm, apex acuminate, base cuneate to subrounded, glabrous above, usually with hairs on midrib and lateral veins below, margin finely serrate or dentate, with 8-12 main lateral veins on each side of the midrib. Petiole c. 1.2 cm long (variable, but not shorter than 0.5 cm), pubescent. Capsule 1-1.8 cm long, scaly.
OCCURRENCE: U2 and 4. Perhaps commonest in lake-shore *Piptadeniastrum* forests.
NOTE: A similar looking plant to *Rinorea brachypetala*. The leaves tend to be thinner and end in sharp points (the apex or acumen of *R. brachypetala* is usually rounded).

Rinorea oblongifolia (C.H. Wright) Chipp (152) Violaceae
13 m. Understorey tree. Bark thin, smooth to fairly rough. Slash fibrous, yellow to pale brown. Leaves simple, alternate, c. 22 x 8.5 cm, apex acuminate, base subrounded to cuneate, glabrous on both surfaces, margin sub-entire to slightly toothed or crenate (very rarely with large teeth or crenations), with 6-12 lateral veins on each side of the midrib. Petiole c. 4 cm long, glabrous. Stipules c. 0.8 cm long, rather broad. Capsule c. 1.8 cm long, rugose.
OCCURRENCE: U2 and 4. Mengo, Bunyoro, Masaka, Kigezi and Kitomi and Namatale forests. Uncommon.
NOTE: Distinguished from other species of *Rinorea* by the large leaves.

Rinorea tshingandaensis Taton (152a) Violaceae
13 m. Midstorey tree. Leaves simple, alternate, c. 10 x 3.5 cm, apex acuminate, base cuneate, glabrous, margin sub-entire to shallowly toothed or crenate, with 6-9 lateral veins on each side of the midrib. Petiole c. 1 cm long, glabrous. Capsule c. 1.6 cm long, rugose.
OCCURRENCE: U2 and 4. Kigezi, Ankole, Bunyoro, Masaka, Mengo. A range-restricted species, only known from Uganda and eastern D.R. Congo.

OCHNACEAE

Small trees. Leaves with short petioles and numerous small, evenly-spaced teeth or crenations. Fruit characteristic, consisting of a number of black drupes borne on an enlarged receptacle which is subtended by conspicuous red sepals. The genera *Campylospermum* and *Idertia* are characterized by having 10 stamens; petals yellow to orange. Species of *Ochna* have at least 13 stamens; flowers yellow to orange, white or pink.

Key to Ochnaceae.

1. Stamens at least 13; petals yellow to orange, white or pink; stipules not persistent. 2
   Stamens 10; petals yellow to orange; stipules persistent or deciduous. 5
2. Leaf tertiary venation not readily visible. ....**155. _Ochna membranacea_**
   Leaf tertiary venation readily visible, at least above. 3
3. Leaf margin spinulose-serrate or with minute bristles (setose); lateral veins 13-15. ....**158a. _Ochna insculpta_**
   Leaf margin densely curved serrulate; lateral veins 20 or more. 4

4. Leaf tertiary venation prominent only above; petiole often 1-3 mm long. ....**158. *Ochna holstii***
   Leaf tertiary venation prominent on both surfaces, though less so beneath; petiole 2.5-5 mm long. ....**157. *Ochna afzelii***

5. Leaf margin entire or nearly so. 6
   Leaf margin closely serrate with curved aculeate teeth. ....**154a. *Campylospermum likimiense***

6. Lamina often at least 6 cm wide. ....**153. *Campylospermum densiflorum***
   Lamina often less than 6 cm wide. 7

7. Stipules 3 mm long. ....**154. *Campylospermum vogelii***
   Stipules 4 mm long. ....**154b. *Idertia mildbraedii***

Campylospermum densiflorum (De Wild. & T. Durand) Farron (153)
Ochnaceae

SYNONYMS: *Gomphia densiflora* (De Wild. & T. Durand) Verdc.; *Ouratea densiflora* De Wild. & T. Dur.

Nsaggalanyi (ga).

5 m. Understorey tree. Bark flaking. Slash red. Leaves simple, alternate, thick and stiff, c. 25 x 9 cm, apex acute to rounded, base cuneate, rounded or truncate, venation difficult to see on upper surface, margin serrulate (with numerous small teeth), at least towards the base, or entire. Petiole thick, 5-10 mm long. Stipules narrowly triangular, mainly 6 mm long (range 5-10 mm), often deciduous. Flowers in solitary, raceme-like inflorescences or in terminal panicles. Flowers and fruits very attractive, the flowers yellow and the fruiting sepals red. Drupelet ellipsoid or subglobose, c. 8 x 5 mm.

OCCURRENCE: U2 and 4. Mengo, Masaka and Bunyoro; also Kibale Forest. Uncommon.

Campylospermum vogelii (Hook. f.) Farron (154) Ochnaceae

SYNONYMS: *Gomphia vogelii* Hook. f.; *Ouratea hiernii* (Tiegh.) Exell

Bitigandwa (ki); Mulyangabi (na).

10 m. Understorey tree. Bark smooth. Slash pink. Leaves simple, alternate, usually thin and papery, c. 15 x 3.5 cm, apex acute to acuminate, base narrowly cuneate, with c. 10 main lateral veins on each side of the midrib, margin serrate (with numerous small teeth) or (more often) entire. Petiole c. 0.3 cm long. Stipules 3 mm, striate, dropping early. Inflorescence subterminal. Drupelet ellipsoid, 7 x 5 cm.

OCCURRENCE: U2 and 4. Common in Kayonza and north Bwindi forests.

Campylospermum likimiense (De Wild.) I. Darbysh. & Kordofani (154a)
Ochnaceae

Plate 14. Achariaceae and Salicaceae (139-168); see also Plate 13

139. *Rawsonia lucida* 148. *Rinorea ilicifolia* 149. *Rinorea beniensis*
150. *Rinorea brachypetala* 151. *Rinorea dentata* 152. *Rinorea oblongifolia*
153. *Campylospermum densiflorum* 154. *Campylospermum vogelii*
155. *Ochna membranacea* 156. *Ochna bracteosa*
157. *Ochna afzelii* 61. *Maytenus undata* 162. *Gymnosporia heterophylla*
163. *Maytenus acuminata* 164. *Maesa lanceolata* 165. *Ilex mitis*
168. *Rhamnus prinoides* (a shrub - not described here)

Actual sizes: leaves and fruits x 2.

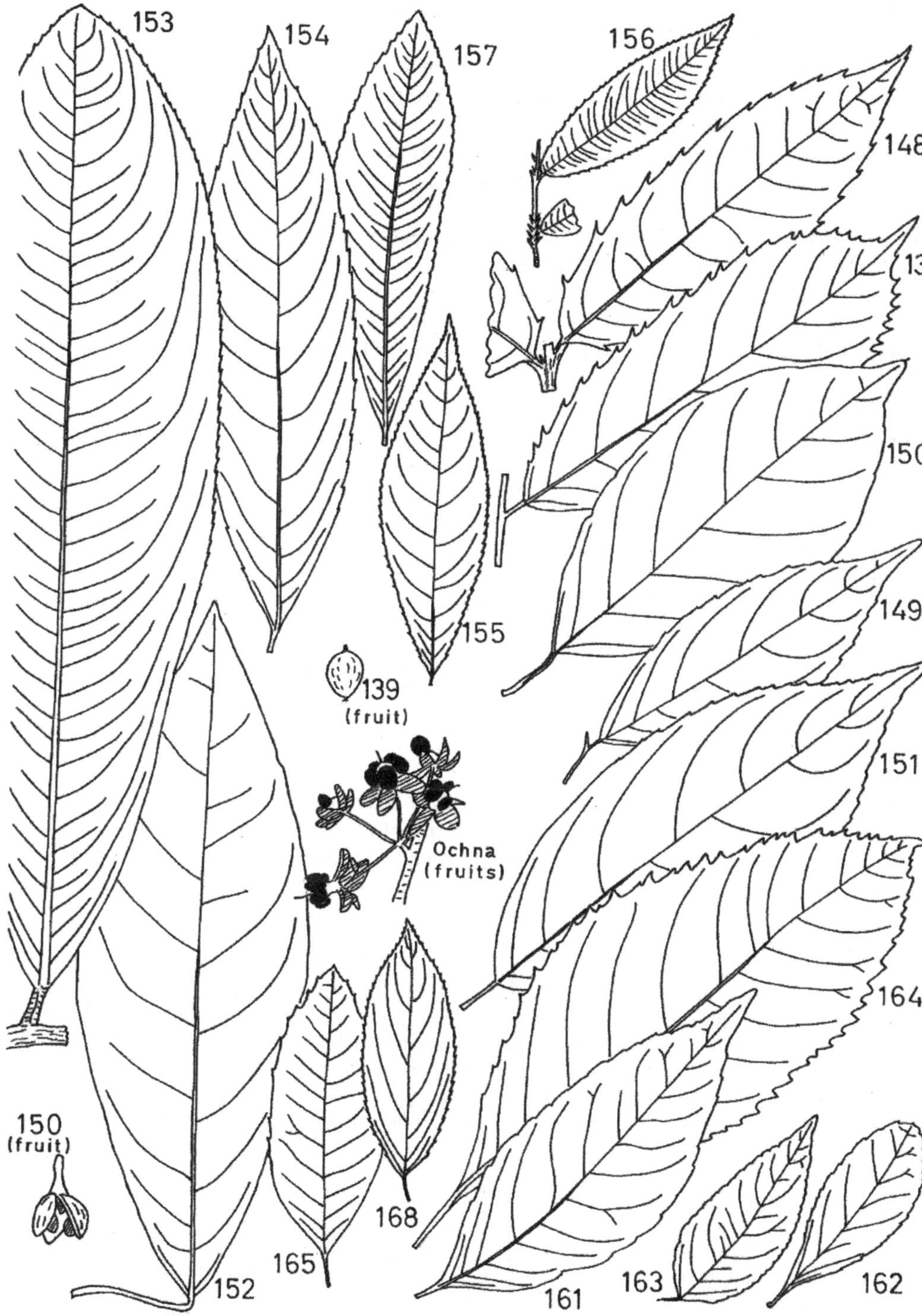

Plate 14. (139-168)

SYNONYMS: *Gomphia likimiense* (De Wild.) Verdc.; *Ouratea bukobensis* (Tiegh.) Exell. (ITU *pro parte*)
4 m. Bark smooth. Slash pink. Leaves simple, alternate, usually thin, width variable (to 4.8 cm), apex acute to acuminate, base cuneate, with 12-15 upwards-curving lateral veins, margin closely serrate with curved aculeate teeth. Petiole 1-4 mm long. Stipules linear, 5-10 mm long, aristate, dropping early. Inflorescence subterminal. Drupelet 5 x 4 mm.
OCCURRENCE: U2 and 4.

Idertia mildbraedii (Gilg) Farron (154b) Ochnaceae
SYNONYM: *Gomphia mildbraedii* (Gilg) Verdc.
6 m. Leaves simple, alternate, thin, 2.5-5.5 cm wide, apex abruptly acuminate, base cuneate, margin entire or (rarely) some leaves with 1-3 teeth. Petiole 4-7 mm long. Stipules very narrowly triangular, 4 mm, striate. Inflorescence axillary. Drupelet ellipsoid, c. 13 mm long, 6-7 mm wide.
OCCURRENCE: U2 and 4.
CONSERVATION STATUS: Global NE (IUCN), EN (TOU); National NE.

Ochna membranacea Oliv. (155) Ochnaceae
12 m. Understorey tree with a single trunk and much-branched crown (mini-tree habit). Bark thin. Slash whitish. Leaves simple, alternate, c. 11 x 3.5 cm, apex with blunt or pointed acumen, base cuneate to rounded, with c. 10-20 lateral veins on each side of the midrib, these veins prominent on both surfaces, tertiary veins not prominent, margin curved serrulate, with many small, regularly arranged teeth or crenations, widest in about the centre or sometimes in upper half. Petiole c. 0.3 cm long. Stipules linear, deciduous. Flowers in panicles 5-12 cm long (in other species of *Ochna*, the flowers are borne on short lateral branchlets). Drupelet reniform (kidney-shaped).
OCCURRENCE: U2 and 4. Mengo, Ankole, Tooro and South Maramagambo Forest (Kigezi). A common understorey tree in Zika Forest, Kisubi.
NOTES: *Ochna bracteosa* Robyns & Lawalrée (156) is a much-branched shrub to 3 m, with leaves similar in shape, toothing and venation to those of *Ochna membranacea*, but generally smaller (c. 9 x 2.5 cm). Petiole c. 0.2 cm long. It can be distinguished from other species of *Ochna* by the presence of persistent small brown scales (c. 0.2-0.4 cm long) on the young shoots. It is recorded from U2-4, being common under *Cynometra*.

Ochna afzelii Oliv. (157) Ochnaceae
12 m. Understorey tree with a single trunk and branches at right angles. Bark brown. Slash red, brown or orange. Leaves simple, alternate, variable in shape, often c. 12 x 3.5 cm, apex acuminate to rounded, base cuneate, lateral veins +/- 30 in total, tertiary venation prominent on both surfaces (though less so below), margin densely curved serrulate (with numerous regularly arranged teeth). Petiole c. 0.3 cm long. Flowers 2-6. Drupelet subglobose to subreniform, 6-7 mm in diameter.
OCCURRENCE: U1, 3 and 4. Recorded from Mengo, Masaka and Bugisu. Common in lake-shore forests.

Ochna holstii Engl. (158) Ochnaceae
Lokotono (ku); Siteti (ms)
25 m. Bark smooth, grey. Slash pink to red-brown. Leaves simple, alternate, c. 6.5 x 2.25 cm, apex acute to acuminate, base cuneate to rounded, lateral veins 20-25 in total, tertiary venation prominent above, margin with numerous teeth or crenations. Flowers in a racemiform inflorescence with rachis to 2 cm long. Drupelet ellipsoid, 9 x 7 mm. Petiole c. 0.1-0.2 cm long.
OCCURRENCE: U1 and 3. Recorded from near Suam valley (Mt Elgon), Kacagalau (Karamoja) and Imatong Mts at c. 2000 altitude.

***Ochna insculpta* Sleumer (158a)** Ochnaceae
SYNONYM: *Ochna* sp. near *O. macrocalyx* (*sensu* ITU)
Small tree to 9 m. Bark smooth, sometimes fissured. Leaves very glossy and bronze-tinged when young, 8 x 3.5 cm, apex acute, base cuneate, lateral veins 13-15, tertiary venation prominent above (less so beneath), margin spinulose-serrate or with minute bristles (setose). Petiole 1-3 mm long. Stipules 6 mm long. Flowers 1-6 in short raceme-like, subumbellate inflorescences, rachis to 7 mm long. Drupelets ellipsoid, 11 x 7 mm.
OCCURRENCE: U4. Mengo. Evergreen forest, forest margins and riverine forest.

***Hugonia platysepala* Oliv. (159)** Linaceae
Of variable habit – tree to 15 m, shrub to 3 m or climber with axillary, coiled, climbing shoots. Leaves simple, alternate, c. 11 x 3.6 cm, widest in upper half, with over 13 well-defined lateral veins on each side of the midrib, base cuneate to rounded, margin regularly to lightly toothed or crenate. Petiole c. 0.6 cm long. Flowers conspicuous, yellow.
OCCURRENCE: U2 and 4.
NOTE: Young leaves are similar in shape to those of *Argomuellera macrophylla* (174).

***Mystroxylon aethiopicum* (Thunb.) Loes. (160)** Celastraceae
SYNONYM: *Cassine aethiopica* Thunb.
Akado (al); Emus (at); Omushongati (ki); Musmobya (na); Lunyindi (nl).
15 m, usually less. Trunk short, branching near base, branches unarmed, branchlets pendulous, with a dense narrow crown. Latex absent. Bark thin and smooth, grey, flaking on older trees. Slash red to pink. Leaves simple, spirally-arranged, very variable in size and margin, often c. 9 x 4 cm, apex obtuse to rounded, base cuneate to subcordate, margin entire or with fairly regularly arranged teeth or crenations (undulate-crenate), with c. 5-8 main lateral veins on each side of the midrib. Petiole c. 0.5 cm long. Drupe globose to ovoid, 8-29 mm long, smooth or finely rugose.
OCCURRENCE: U1-4. On forest edges and in savanna. Up to 2400 m.
NOTES: A very variable species. The leaves are usually smaller than those of *Maytenus undata*.

GYMNOSPORIA AND MAYTENUS

Various species of *Gymnosporia* and *Maytenus* that are not included below are occasionally found in forest or on forest edges. *Maytenus* is a difficult genus taxonomically and some species are very variable. *Gymnosporia gracilipes* (Oliv.) Loes. (Syn.: *Maytenus gracilipes* (Welw. ex Oliv.) Exell) is a spiny shrub or small tree with very variably shaped leaves. *Gymnosporia senegalensis* (Lam.) Loes. (Syn.: *Maytenus senegalensis* (Lam.) Exell) is the common savanna species.

***Gymnosporia buchananii* Loes. (160a)** Celastraceae
8 m, sometimes sarmentose (with long thin runners or rhizomes). Spines to 2 cm long, terminating short axillary branches. No latex. Branches +/- flattened, striate, with numerous lenticels. Leaves not fasciculate (in bundles), lamina leathery, glabrous and pale green, elliptic to elliptic-oblong to ovate to oblanceolate, (3-)3.6-11.2(-17) x (1.4-)1.8-5(-8) cm, apex obtuse to rounded (rarely shortly acuminate on juvenile leaves), base attenuate to cuneate or (rarely) rounded, margin shallowly crenulate-serrulate, lateral veins and tertiary venation more prominent beneath than above. Petiole 2-9 mm long. Capsule obconical (cone-shaped, with narrower part towards base), glabrous, 4-10 mm long.
OCCURRENCE: U1 and 2. West Nile, Acholi and Ankole. In dry evergreen forest and riverine forest and on forest edges.

***Gymnosporia mossambicensis* (Klotzsch) Loes. (160b)** Celastraceae
SYNONYM: *Maytenus mossambicensis* (Klotzsch) Blakelock

8 m, sometimes sarmentose (with long thin runners or rhizomes). Spines to 8 mm long, axillary or terminating short axillary branches, without latex. Branches 4-lined, with numerous prominent lenticels. Leaves fasciculate (in bundles) or not, lamina bright green, paler below, ovate or lanceolate to elliptic or subcircular, 0.4-6(-9.7) x 0.4-3.5(-4.3) cm, apex acute (or rarely shortly and obtusely acuminate to acutely incurved-denticulate), base rounded (or rarely subcordate to cuneate or attenuate), lateral nerves and tertiary venation more prominent below than above. Petiole 1.5-6(-9) mm long. Capsule pyriform to obconical, glabrous, 7-14 mm long.
OCCURRENCE: U2. Ankole.
NOTE: Only var. *mossambicensis* occurs in Uganda.

Maytenus undata **(Thunb.) Blakelock (161)** Celastraceae

10 m. Trunk crooked, sometimes gnarled, bearing a spreading crown of dark-coloured leaves. Unarmed. Sucker shoots often present. Bark brown, quite thick and rough, flaking in small pieces c. 1-5 cm across. Slash hard, of even texture, pink to red, turning darker. Leaves simple, alternate, not in fascicles (bundles), variable in shape and margin, often c. 11 x 4.5 cm, apex acute to rounded, base usually cuneate (to truncate to subcordate), margin glandular-denticulate to spinulose-dentate (toothed), undersurface usually whitish (waxy) in colour. Petiole c. 0.5 cm long. Capsule reddish, smooth c. 0.6 cm diameter, with 3 valves.
OCCURRENCE: U1, 2 and 4. Widely distributed, found up to 2500 m. At lower altitudes, mainly in secondary forest and on forest edges.

Gymnosporia heterophylla **(Eckl. & Zeyh.) Loes (162)** Celastraceae

SYNONYM: *Maytenus heterophylla* (Eckl. & Zeyh.) N. Robson
Lusimamboli (so).
7 m. Small armed (occasionally unarmed), very variable tree. Spines to 24 cm long, axillary or terminating short axillary branches, without latex, often without visible lenticels. Leaves simple, alternate, often borne in clusters on short lateral shoots, 1-9 x 0.5-5 cm, glabrous, usually obovate, apex obtuse to emarginate, base cuneate to attenuate, margin shallowly and irregularly serrulate (particularly towards the apex) or crenate. Petiole 1-10 mm long. Capsule obovoid to globose, glabrous, 3-9 mm long.
OCCURRENCE: U1-4. On forest edges, especially where there has been disturbance. Also, an understorey tree in young secondary forest. Not in more mature forest types.

Maytenus acuminata **(L. f.) Loes. (163)** Celastraceae

Omurembwe (ki).
10 m. Small unarmed tree with latex threads in leaves, flowers and fruits. Branches 4-lined when young. Leaves simple, alternate, not in fascicles (bundles), c. 7.5 x 3 cm, dark green, glossy above, ovate, apex acuminate or acute, base attenuate, cuneate to rounded, margin glandular, denticulate to crenate (rarely entire), lateral and tertiary venation prominent beneath, but not or barely above. Petiole 2-7 mm long. Capsule 3-lobed or globose, smooth, 4-11 mm long.
OCCURRENCE: U2 and 4. Recorded from Masaka, Kigezi, Ankole and Rwenzori. In montane forest, 1500-3200 m.

Maesa lanceolata **Forssk. (164)** Primulaceae

Mutaka (am); Kiwondowondo (ga); Omuhanga (ki); Muhanga-honga (ko); Gororwo (ku); Kisangulia, Kisiangulu, Nabutwa, Naporo (ms); Muhanga, Muhanga-bagenyi (na); Muhanga-bagenzi (to).
15 m, usually less. Spreading small tree. Bark smooth. Slash usually brown and white, turning darker, usually with a dark red or orange resinous exudate. Leaves simple, alternate, variable in shape and size, c. 16 x 7.5 cm (particularly on higher altitude plants, but smaller at lower altitudes), apex obtuse to acuminate, base cuneate to rounded, with c. 8-16 main lateral veins on each side of the midrib, margin usually with large teeth. Petiole c. 2.5 cm long (particularly

on higher altitude plants, but shorter at lower altitudes). The broken petioles and young stems exude brown droplets.
OCCURRENCE: U1-4. Widely distributed, ascending to 2500 m. Sometimes abundant in secondary forest and on forest edges. Particularly common in Kigezi.
CULTIVATION AND PROPAGATION: Fast-growing. Can be grown on degraded soils, including those caused by the past presence of eucalyptus. Can be used as a pioneer to shelter slower growing species. Can be pruned and pollarded; suitable for the rapid production of fuelwood. Collect berries from mother trees and dry in the sun. Free the seeds by slightly crushing the dried fruits and plant as soon as possible.
NOTES: The exudate in the broken petiole is an excellent character. At least three varieties occur: (i) (at lower altitudes, e.g. in lake-belt forests) leaves small (c. 10 x 4 cm), inflorescence relatively small; (ii) (at higher altitudes) leaves and inflorescence large and (iii) (on the Bufumbira Volcanoes) a very small-leaved form (leaves c. 5.5 x 1.75 cm).

Maesa welwitschii **Gilg (164a)** Primulaceae

A related species to *Maesa lanceolata*, differing in having generally smaller leaves, with the lamina 4 times as long as the petiole (compared with 6 times as long in *M. lanceolata*).
OCCURRENCE: U3 and 4. Recorded from Mbale and Mengo (West Bugwe, Kajjansi and Mpanga forests). Highest altitude recorded 1200 m.

Ilex mitis **(L.) Radlk. (165)** Aquifoliaceae

Omunyambago, Omunyeiju (ki); Obwiso (ko); Segar, Seger, Sigara (ku); Mwandanda (ms).
15 m (occasionally 25 m). Trunk crooked. Crown dark-coloured, spreading to rounded (on taller trees). Bark smooth and thin, grey to nearly white. Slash white to yellow, turning darker. Leaves simple, alternate, very variable in shape, size and margin, often c. 7 x 3.5 cm, usually thick and leathery, with at least some leaves on each branch with sharp, distantly-spaced teeth, mainly in upper part of lamina.
OCCURRENCE: U1-4. On Mt Elgon, Mt Kadam, Rwenzori and in Kigezi, 1500-3100 m. Additional to higher altitude sites, it also occurs in Namalala Forest, Masaka. There is a record from Kibale Forest.
NOTE: The branchlets are used as tooth sticks. Only var. *mitis* is known to occur in Uganda.

Morella kandtiana **(Engl.) Verdc. & Polhill (166)** Myricaceae

SYNONYM: *Myrica kandtiana* Engl.
Nkikimbo (ga).
Small tree to 5 m with a dense crown. Leaves simple, alternate, small (c. 7.5 x 2.5 cm), toothed, widest in upper half, aromatic when crushed.
OCCURRENCE: U1, 2 and 4. In swamps, particularly with papyrus and *Syzygium cordatum*. Occasionally on river banks. Typically below 2000 m.
NOTE: Another species of *Morella*, *M. salicifolia* (258), grows in montane forest and woodland.

Aeglopsis eggelingii **M. Taylor (167)** Rutaceae

7 m. Spiny. Leaves uni- or sometimes bi- or trifoliolate, alternate, glabrous, gland-dotted, c. 25 x 10 cm, elliptic, apex acute or acuminate, base cuneate, margin crenate, lateral nerves 7-11 on each side of the midrib. Petiole 5-18 mm long. Fruit globose to pear-shaped, 9-11 cm long, aromatic and woody.
OCCURRENCE: U2 and 4. Mengo, Ankole, Tooro and Bunyoro. On forest edges. Uncommon. A range-restricted species of narrow endemism, found in Uganda and South Sudan.

ASTERACEAE

Formerly known as Compositae, this large family mainly consists of herbs and shrubs, with relatively few trees. Giant groundsels, *Dendrosenecio*, found at very high altitudes, are placed elsewhere in the descriptions (species numbers 19-21). Apart from the four small tree species described and numbered below, two shrubby species may be noted: (1) *Vernonia auriculifera* Hiern (*kikookooma* in Luganda), which is a shrub of forest edges. Leaves c. 15 x 4.5 cm, with rounded outgrowths at base of petiole. Flowers mauve; (2) *Solanecio mannii* (Hook. f.) C. Jeffrey (Syn.: *Crassocephalum mannii* (Hook. f.) Milne-Redh.), which is a common spindly shrub or tree to 10 m, found in forest clearings and on forest edges. Flowers yellow.

Brachylaena huillensis O. Hoffm., a tree to 25 m and typical of miombo savanna (a type of vegetation found today only south of the equator), has been recorded from Busoga and Mengo. It has been seen growing in Bubugo village on the east bank of the River Nile and may still be growing there. The locality is now an eco-tourism site, but the managers have been informed that they have a very rare and highly localized species for Uganda (Nkuutu David, pers. comm. to JK).

***Conyza vernonioides* (A. Rich.) Wild (169)** Asteraceae

SYNONYM: *Nidorella arborea* R.E. Fr.

10 m. Small tree. Leaves c. 12 x 2.5 cm, without a petiole, apex acute to acuminate, base cuneate, attenuate (but auriculate near stem), margin toothed, particularly in the upper half. Flowers conspicuous, bright yellow.

OCCURRENCE: U2 and 3. Common on Mt Elgon at 2750-3300 m, growing with bamboo, in the *Hagenia-Rapanea* Zone and on forest edges. There is a record from the Bufumbira Volcanoes.

***Vernonia conferta* Benth. (170)** Asteraceae

Tree to 13 m. Young stems with T-shaped and simple hairs. Bark smooth, grey. Slash off-white. Leaves elliptic to obovate, very large, c. 80 x 25 cm (but variable in size, total length range 10-90 cm, total width range 8-26 cm), apex acute, base cuneate to attenuate, margin sinuate (wavy), glabrous or nearly so above, tomentose beneath. Petiole a bit expanded, 2.5-7.5 cm. Flowers white.

OCCURRENCE: U2 and 4. In swamp forest in Masaka, Ankole, Kigezi and Tooro.

***Vernonia amygdalina* Delile (170a)** Asteraceae

Mululuuza (ga).

8 m. Shrub to small tree, much branched and spreading. Leaves elliptic to lanceolate, c. 10 x 4 cm, apex shortly acuminate or apiculate, base cuneate or rounded, margin minutely toothed to coarsely serrate, lamina nearly glabrous to sparsely pubescent.

OCCURRENCE: U1-4. Forest edges and clearings, savanna and farmland.

CULTIVATION AND PROPAGATION: Fast-growing. Can be planted on land that is much degraded. Resistant to termites and drought. Collect the fruiting heads and shake into a bag to collect the seeds

NOTES: Valued as a herbal medicine (anti-malarial) and widely grown or spared during clearing vegetation.

***Vernonia calvoana* (Hook. f.) Hook. f. (171)** Asteraceae

Tree to 10 m. Leaves c. 20 x 6 cm, with numerous sharp teeth. Flowers white, tinged with lilac.

OCCURRENCE: U2. On Rwenzori and the Bufumbira Volcanoes, at c. 2700 m.

CONSERVATION STATUS (of subsp. *ruwenzoriensisy*): Global NE (IUCN), DD (TOU); National NE.

NOTES: (1) It is the subsp. *ruwenzoriensis* C. Jeffrey that grows in forest. Another subsp., *V. calvoana* subsp. *adolfi-friderici* (Muschler) C. Jeffrey (Syn.: *Vernonia* sp. aff. *adolfi-friderici* Muschl. (*sensu* ITU and UFT), grows in woodland and bushland. (2) Attempts by some scholars to divide African *Vernonia* into split-genera (e.g. renaming this species *Baccharoides calvoana* (Hook. f.) Isawumi, El-Ghazaly & B. Nord.) have not been accepted by any of the African floras.

Plate 15. Euphorbiaceae, Phyllanthaceae and Putranjivaceae (172-184)

172. *Alchornea floribunda* 173. *Alchornea hirtella* 174. *Argomuellera macrophylla*
176. *Discoclaoxylon hexandrum* 179. *Shirakiopsis elliptica*
180. *Gymnanthes leonardii-crispii* 181. *Suregada procera* 182. *Drypetes gerrardii*
183. *Dryptes* sp. 184. *Drypetes ugandensis*

Actual sizes: leaves x 2; tree profile x 800.

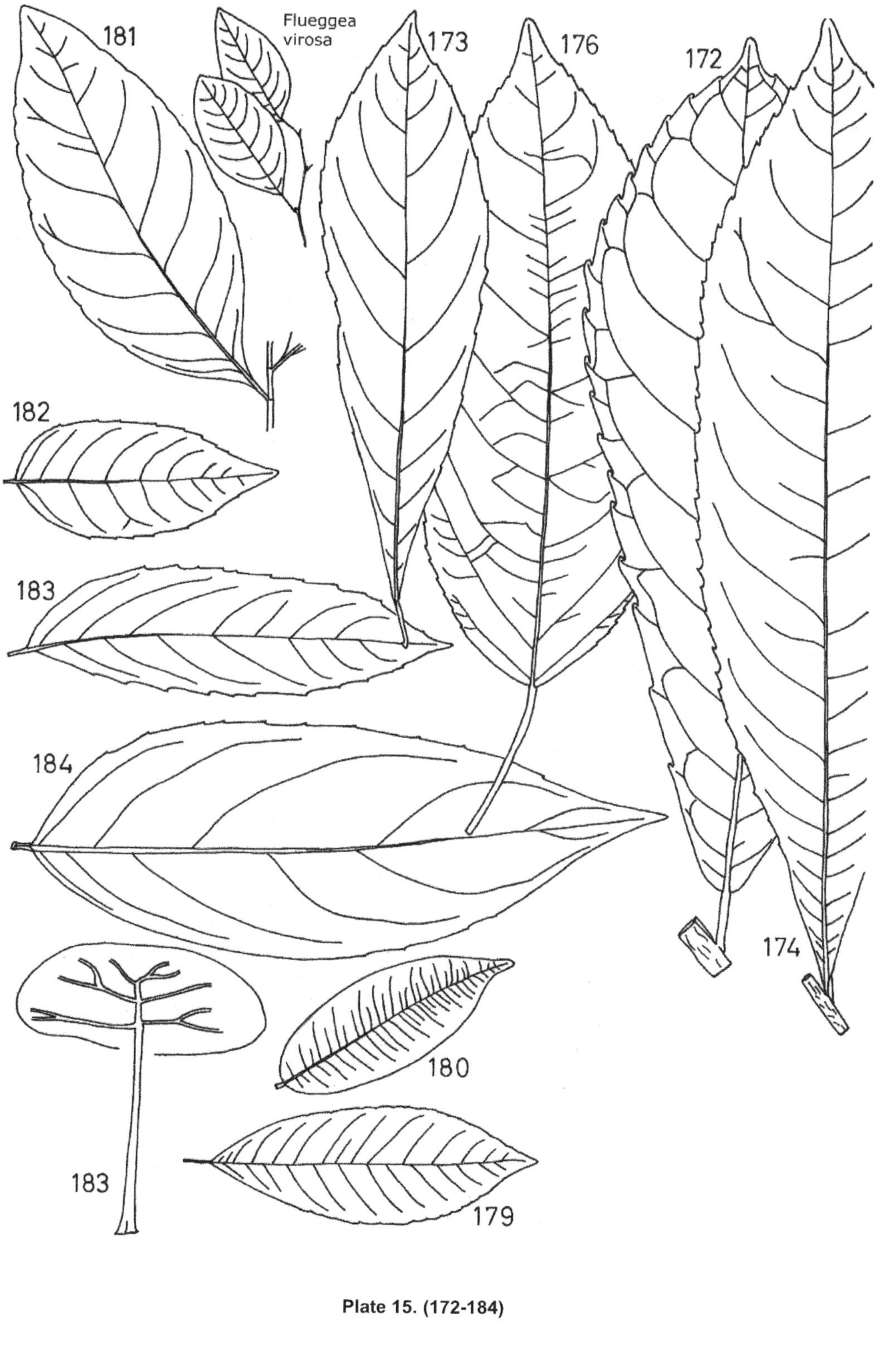

Plate 15. (172-184)

EUPHORBIACEAE AND RELATED FAMILIES (tree numbers 172-185)

Those Euphorbiaceae, Phyllanthaceae and Putranjivaceae with simple, alternate leaves, non-entire margins, lacking conspicuous veins from the base of the lamina and lacking white latex in the slash are included here. They are mostly understorey trees, but some species of *Drypetes* are medium-sized trees, as can be *Shirakiopsis elliptica.* See before tree 94 for an overview of Euphorbiaceae and related families.

Flueggea virosa (Willd.) Voigt (Syn.: *Securinega virosa* (Roxb. ex Willd.) Baill.), which is illustrated on Plate 15, is a much-branched shrub or small tree with white edible berries, c. 0.75 cm in diameter. It is a common light-requiring species, found on forest edges, in large forest clearings and in savanna (particularly on termite mounds).

Key to *Alchornea.*

1. Leaves distinctly cordate. ....**102. *A. cordifolia***
   Leaves not cordate or only slightly so. 2
2. Leaves with stipels. ....**103. *A. laxiflora***
   Leaves lacking stipels. 3
3. Leaves 14-37 cm long, lateral veins 11-21 on each side of the midrib. ....**172. *A. floribunda***
   Leaves 6-25 cm long, lateral veins 6-10 on each side of the midrib. ....**173. *A. hirtella***

Alchornea floribunda **Müll. Arg. (172)** Euphorbiaceae

5 m. Understorey tree. Leaves often clustered at ends of shoots, simple, alternate, oblanceolate, c. 30 x 10 cm, apex acuminate, with a long attenuate base which is rounded right at the end, margin toothed, lateral veins 11-21 on each side of the midrib. Petiole c. 2 cm long. Male inflorescence axillary; female inflorescence terminal. Fruit 3-lobed, 6 mm long, 1 cm across.
OCCURRENCE: U2 and 4. Mengo, Masaka and Bunyoro. Uncommon.
NOTES: Distinguished from *Argomuellera* by the leaf base, which is rounded right at the end.

Alchornea hirtella **Benth. (173)** Euphorbiaceae

Oruzogo (ki).
10 m. Understorey tree, branches numerous, at all heights, crown spreading and dense. Bark thin, light brown, fairly smooth. Slash red-brown. The leaves tend to be clustered at the ends of the branches. Leaves simple, alternate, elliptic to oblanceolate, c. 14 x 4 cm, apex obtusely or acutely acuminate, base attenuate, with c. 6-10 main lateral veins on each side of the midrib (not including the smallest veins near the base of the lamina), often with domatia in axils, margin toothed. Petiole c. 1 cm long. Male inflorescence terminal (sometimes axillary); female inflorescence terminal. Fruit 3(-4)-lobed, 5 mm long, 9 mm across.
OCCURRENCE: U1-4. Abundant in Bwindi, Kayonza and Kalinzu forests and probably also common in some lake-shore forests (Ssese, Buddu).
NOTES: (1) Only forma *glabrata* (Müll. Arg.) Pax & K. Hoffm. occurs in Uganda. The midrib and main veins of mature leaves in this forma are glabrous or nearly so beneath. Those of forma *hirtella* are evenly to sparsely bristly. (2) Distinguished from *Argomuellera* by the smaller leaf size.

Argomuellera macrophylla **Pax (174)** Euphorbiaceae

Laka (na).
Understorey shrub or small tree to 3 m, usually less. Stem weak. Leaves simple, alternate, elliptic to oblong, c. 30 x 10 cm, apex acuminate, base attenuate, margin with large teeth. Petiole c. 0.8 cm long. Fruit 3-lobed, 0.7 cm long, 1.3 cm across.

OCCURRENCE: U1-4. Very common.
NOTE: Young leaves of *Hugonia platysepala* (159) are similar in shape.

Pseudagrostistachys ugandensis **(Hutch.) Pax & K. Hoffm. (175)** Euphorbiaceae
Small tree to 5 m. Leaves large (c. 30 x 16 cm), apex acuminate, base rounded, margin serrulate to subentire, glabrous. Petiole c. 2.5 cm long. The leaves may be distinguished from the above species of *Alchornea* and *Argomuellera* (172-174) by the rounded to obtuse, not long attenuate, base. Fruit 3-lobed, 1.6 cm long, 2 cm across.
OCCURRENCE: U2 and 4. Only recorded from Bwindi Forest and the Sango Bay forests. A range-restricted species of narrow endemism, known from very few localities in Uganda, Tanzania and eastern D.R. Congo.

Discoclaoxylon hexandrum **(Müll. Arg.) Pax & K. Hoffm. (176)** Euphorbiaceae
SYNONYM: *Claoxylon hexandrum* Müll. Arg.
Understorey tree to 7 m. Leaves simple, alternate, c. 20 x 7.5 cm, elliptic oblong, apex acute to obtuse, base cuneate to rounded, margin toothed. Petiole c. 5 cm long. Fruit 0.5 cm long, 1 cm across.
OCCURRENCE: U2 and 4. Recorded from Mabira and Budongo forests. Uncommon.

Maesobotrya purseglovei **Verdc. (177)** Phyllanthaceae
Much-branched tree to 20 m. Bark brown, quite thick and rough, vertically fissured. Slash very thin, bright red. Leaves simple, alternate, c. 14 x 5.5 cm, elliptic to oblanceolate, with c. 6-8 main lateral veins on each side of the midrib, apex acute to shortly acuminate, base cuneate to rounded. A characteristic feature is the leaf margin, which is glandular-denticulate and bears small tufts of hairs. Petiole c. 2.5 cm long. Fruit ellipsoid, 7 mm long, 5 mm across.
OCCURRENCE: U2. Only recorded from Ishasha Gorge

Shirakiopsis elliptica **(Hochst.) Esser (179)** **Euphorbiaceae**
SYNONYM: *Sapium ellipticum* (Krauss) Pax
Bileri (al); Musasa (am, ga, na, to); Musanvuma (am, na, to); Elipilepo (at); Muzzanvuma, Muzzaŋŋanda (ga); Alokwe, Einiu (gb); Omushasha (ki); Mugustet (ku); Musaja (la); Musasia (ms); Mujasa (nl); Mujasajasa, Musadhasadha (so); Muchasa, Mukasa (sa).
25 m. Trunk crooked, usually branched low down, with a spreading crown and drooping branchlets. Bark thick and rough, vertically fissured, sometimes flaking, brown. Slash yellow to orange, sometimes with orange or whitish streaks, fibrous (brittle on some larger trees), very rarely exuding drops of white latex. White latex exuding from broken young parts. Old leaves turn red. Leaves simple, alternate, elliptic to oblanceolate, c. 10 x 3.5 cm, apex obtuse to oblanceolate, base cuneate to rounded, with c. 10-15 main lateral veins on each side of the midrib, margin with regularly arranged small teeth or crenations. Petiole c. 1 cm long. Fruit a 2-lobed capsule carrying the persistent styles, 6 mm long, 8 mm across.
OCCURRENCE: U1-4. Widely distributed and often abundant. It is a light-requiring species, found on forest edges, in larger gaps and in open forest.
NOTES: The leaves and fruits resemble those of *Prunus africana* (199), from which *Shirakiopsis* can easily be distinguished by the slash and the presence of white latex in young parts. Caterpillars are frequently found on this species.

Gymnanthes leonardii-crispi **(J. Léonard) Esser. (180)** Euphorbiaceae
SYNONYMS: *Sapium leonardii-crispi* J. Léonard; *Duvigneaudia leonardii-crispi* (J. Léonard) Kruijt & Roebers
Small understorey tree to 8 m. Leaves simple, alternate, elliptic to oblong-lanceolate, c. 8 x 3 cm, apex acuminate, base rounded, lateral veins 8-12 on each side of the midrib, rather indistinct above but prominent beneath, margin entire. Petiole 0.2 cm long. Fruit 3-lobed, 8 mm long, 1 cm across.
OCCURRENCE: U2. Only recorded from Ishasha Gorge and Kalinzu Forest. A range-restricted species of narrow endemism, only in Kigezi and eastern D.R. Congo.

Suregada procera (Prain) Croizat (181) Euphorbiaceae

SYNONYM: *Gelonium procerum* Prain

15 m (to 25 m according to herbarium specimens). Understorey tree, often with a weak stem and a spreading crown. Bark thin and smooth, dark green to light brown, with prominent lenticels. Phellogen green. Slash white to yellow, sometimes layered. Leaves simple, alternate, elliptic to elliptic-ovate, c. 14 x 6 cm, apex obtuse, base asymmetrically cuneate or rounded, decurrent onto petiole, with c. 7-12 main lateral veins on each side of the midrib, margin entire, undulate to toothed (particularly in upper half). Petiole very short, c. 3 mm. Fruit 3-locular, often 3-lobed, 0.8 cm long, 1 cm across.

OCCURRENCE: U1, 2 and 4. A widely distributed lower altitude understorey tree, abundant beneath *Cynometra* in South Kibale Forest and also common in Mabira Forest.

NOTES: This species resembles *Drypetes* in many characters, but can easily be distinguished by the prominent vein reticulum on the leaves, the smallest veins being of a different colour to the intervening tissue.

DRYPETES

A genus that can usually be recognized by the following combination of characters. Bark very thin and smooth, often dark green. Slash yellow or white, sometimes turning darker. Leaves simple, alternate, unequal-sided at base, with at least some irregularities along the margin. Only *Suregada* shares these features, but its leaves have a characteristic vein reticulum.

Key to *Drypetes*.

1. Flowers and fruits borne on main stem. ....**184. *D. ugandensis***
   Flowers and fruits borne on flowering branches. 2
2. Leaf base nearly symmetrically cuneate. ....**185. *D. bipindensis***
   Leaf base somewhat asymmetrically rounded-cuneate. ....**182. *D. gerrardii***

Drypetes gerrardii Hutch. (182) Putranjivaceae

SYNONYM: *Drypetes* sp. (183 of UFT)

Omushabarara (ki); Mushabarara (na).

30 m. Trunk straight, often cylindrical in upper part, with horizontal branches and a rounded to fairly spreading crown. Base of trunk usually fluted, sometimes with small buttresses. Bark dark green, smooth and thin, with prominent lenticels (in places, the bark may be dark brown and granular). Phellogen usually green. Slash brittle, yellow or white, sometimes turning slowly darker (at least in places), sometimes layered, more or less fibrous. Young stems with brown hairs. Leaves simple, alternate, dark green and rather thick, ovate, elliptic to lanceolate, small (c. 15 x 4.5 cm), apex acute to acuminate, base unequal-sided and rounded-cuneate, with c. 5-8 main lateral veins on each side of the midrib, margin almost entire to toothed (particularly in upper part). Petiole to 7 mm long. Fruit subglobose, 2-lobed, c. 1 cm long and wide.

OCCURRENCE: U2-4. An abundant species in many forests (e.g. in Mabira Forest and Kalinzu, where it reaches a large size). Rare in Budongo Forest. Ascending to 2400 m. Probably mainly on drier sites at lower altitudes.

NOTES: Distinguished from other species of *Drypetes* by the hairy young stems, smaller leaf size and sometimes large teeth or crenations on the leaves. All three varieties found in East Africa occur in Uganda: (1) var. *tomentosa* Radcl.-Sm., which has densely yellowish-brown tomentose young shoots; (2) var. *gerrardii*, which has its young shoots sparsely pubescent and leaves mostly 5-10 cm long; (3) var. *grandifolia* Radcl.-Sm., which has its young shoots sparsely pubescent or nearly glabrous and leaves mostly 10-17 cm long.

Drypetes ugandensis **(Rendle) Hutch. (184)** Putranjivaceae
Naliggwalimu (ga).
10 m. Understorey tree with a straight trunk and dark-coloured crown. Bark very thin and smooth, dark green, flaking. Slash light yellow to light brown, turning darker, of even texture. Leaves simple, alternate, c. 14 x 5 cm, with c. 4-7 main lateral veins on each side of the midrib, margin with medium-sized to small teeth (which may be few and widely scattered), apex often acuminate, base unequal-sided. Petiole c. 0.6 cm long.
OCCURRENCE: U2 and 4. Occasionally found in Bunyoro, Mengo and the Ssese forests.

Drypetes bipindensis **(Pax) Hutch. (185)** Putranjivaceae
Much-branched tree. The leaves are similar to those of *Drypetes ugandensis*, but the margin is slightly wavy to entire (rather than toothed). The ranges of the two species appear not to coincide.
OCCURRENCE: U2. Only recorded from Ishasha Gorge and Kalinzu Forest

Paropsia guineensis **Oliv. (186)** Passifloraceae
Seggwafu (ga).
20 m. Trunk straight or wavy, bearing a fairly open deciduous crown. Small buttresses sometimes present. Trunk sometimes slightly fluted. Bark light brown to almost white, quite thick, usually with conspicuous vertical channels (like young *Maesopsis*), sometimes splitting into rectangles. Slash brittle, with concentric rings of white to red and yellow. Branchlets simulating pinnate leaves, hairy. Leaves simple and alternate (not pinnate), c. 10 x 4 cm, elliptic-oblong, apex subacute to acuminate, base cuneate or nearly rounded and sometimes asymmetrical, margin toothed or crenate, lateral veins 7-8 on each side of the midrib, veins hairy on lower surface. Tufts of hairs along leaf margin. Petiole c. 0.3 cm long. Fruit c. 1.25 cm diameter, mostly borne on bare trees.
OCCURRENCE: U2 and 4. Mengo and Bunyoro. Uncommon.
NOTES: The slash and branchlets are characteristic.

Ficalhoa laurifolia **Hiern (187)** Sladeniaceae
Omuvumaga (ki).
25 m. Trunk straight and cylindrical. Main branches at right angles and curving up, smaller branches drooping. Bark brown, rough, with vertical fissures on older trees, smooth on young. Slash pink, with light yellow latex. Branchlets covered with long yellow hairs. Leaves simple, alternate, c. 8 x 2.5 cm, oblong to lanceolate, apex acuminate, base cuneate to rounded, margin with numerous small teeth, glabrous or sparsely pilose. Petiole c. 0.7 cm long. Fruit a capsule, c. 0.3 cm across.
OCCURRENCE: U2. Kigezi and Rwenzori, 2000-2500 m, often on or near ridges.

Balthasaria schliebenii **(Melch.) Verdc. (188) var. *intermedia* (Boutique & Troupin) Kobuski** Pentaphylacaceae
SYNONYM: *Melchiora schliebenii* (Melch.) Kobuski
Omukari (ki).
30 m. Trunk straight, either single or several together. Stilt roots often present, c. 1 m above ground. Bark rough and brown. Leaves simple, alternate, elliptic to oblanceolate, c. 10 x 3 cm, apex shortly acuminate, base cuneate, glabrous, with numerous lateral veins, margin with and numerous small teeth. Petiole c. 0.5 cm long. Fruit ovoid, c. 1 cm diameter.
OCCURRENCE: U2. Only recorded from Bwindi Forest, c. 2000 m. A range-restricted species of narrow endemism; worldwide only in Bwindi Forest, eastern D.R. Congo, Rwanda and Burundi.
CONSERVATION STATUS: Global NT (IUCN), LC (TOU); National NE.

Maesopsis eminii **Engl. (189)** Rhamnaceae

Musizi (ga, tn); Omuguruka (ki); Omuhongera (no).

40 m, but usually only to 25 m. Trunk straight and cylindrical, branches at right angles and curving up. Crown rounded on older trees, deciduous. Buttresses absent. Bark fairly thick, red-brown (to almost white on outside), with regularly arranged vertical channels, becoming fissured on larger trees. Slash rather brittle, red on outside, yellow within, sometimes with orange streaks, with a characteristic smell of cold cooked chicken. Leaves, simple, alternate to sub-opposite, ovate to oblong, c. 9 x 2.75 cm, apex acuminate, base rounded, margin with prominent, fairly widely spaced and blunt teeth, often glabrous. Petiole c. 1 cm long. Fruit a drupe, c. 2.5 cm long, obovoid, at first yellow, later black.

OCCURRENCE: U2, 3 and 4. Widely distributed, found up to c. 1400 m. Generally abundant in secondary forest and on forest edges. A light-requiring, fast-growing species.

CULTIVATION AND PROPAGATION: Fast-growing on better soils, slow-growing on poor or damp soils. Will not grow well under other trees. Can be planted in pure or mixed stands, growing best in widely-spaced pure stands. Collect fruits under mother trees, remove the pulp manually, crack the stones and dry the seeds in the sun. Soak the seeds in water for a day before sowing and plant as soon as possible. Protect the young plants from browsing animals.

NOTES: Only subsp. *eminii* occurs in East Africa. The wood is extensively used in Uganda for general purpose carpentry. The tree is quite widely planted, particularly in Mabira Forest. The fruits are eaten by chimpanzees and hornbills.

Prunus africana **(Hook. f.) Kalkman (199)** Rosaceae

Ngwabuzito, Ntaseesa (ga); Omumba (ki); Ngoti (ko, to); Oromoti (ku); Chiramat, Chirumandi, Gulumati, Gumwirumari, Namwini (ms); Mugote (na); Entasera (no); Mueri, Red stinkwood (tn).

35 m. Trunk usually straight in lower part, bearing a few large branches. Crown spreading. Small buttresses occasionally present. Bark dark brown, sometimes almost black, usually thick, rough, flaking raggedly in pieces (c. 1-15 x 1-10 cm in size). Slash fibrous, pink, turning darker, smelling of cyanide. Leaves simple, alternate, ovate to oblong, c. 15 x 5 cm, apex obtuse, acute or shortly acuminate, base cuneate to rounded, margin crenate, thick, glabrous (leaves on high altitude specimens are much smaller and more rounded than those at lower altitudes). Petiole c. 1 cm long, reddish. Fruit a red drupe, rounded-ellipsoid, 1.25 cm across, wider than long, bearing a persistent style.

OCCURRENCE: U1-4. Widely distributed, ascending to 3200 m. At lower altitudes (up to 1500 m), on forest edges. At higher altitudes, growing inside apparently mature forest and mainly in climatically wetter areas. Attains its largest dimensions in Bwindi and Kalinzu forests and on parts of Mt Elgon.

CONSERVATION STATUS: Global VU (IUCN), LC (TOU); National VU (WCS).

CITES: Appendix II.

Plate 16. Myristicaceae, Lauraceae, Irvingiaceae and others (186-210); see also Plate 17

186. *Paropsia guineensis* 187. *Ficalhoa laurifolia* 188. *Balthasaria schliebenii*
189. *Maesopsis eminii* 199. *Prunus africana* 200. *Parinari excelsa*
201. *Warburgia ugandensis* 202. *Pycnanthus angolensis* 203. *Staudtia kamerunensis*
204. *Beilschmiedia ugandensis* 205. *Ocotea usambarensis* 206. *Ocotea kenyensis*
207. *Casearia runssorica* 209. *Klainedoxa gabonensis* 210. *Irvingia gabonensis*

Actual sizes: leaves x 2; trunk base x 80; tree profiles x 800.

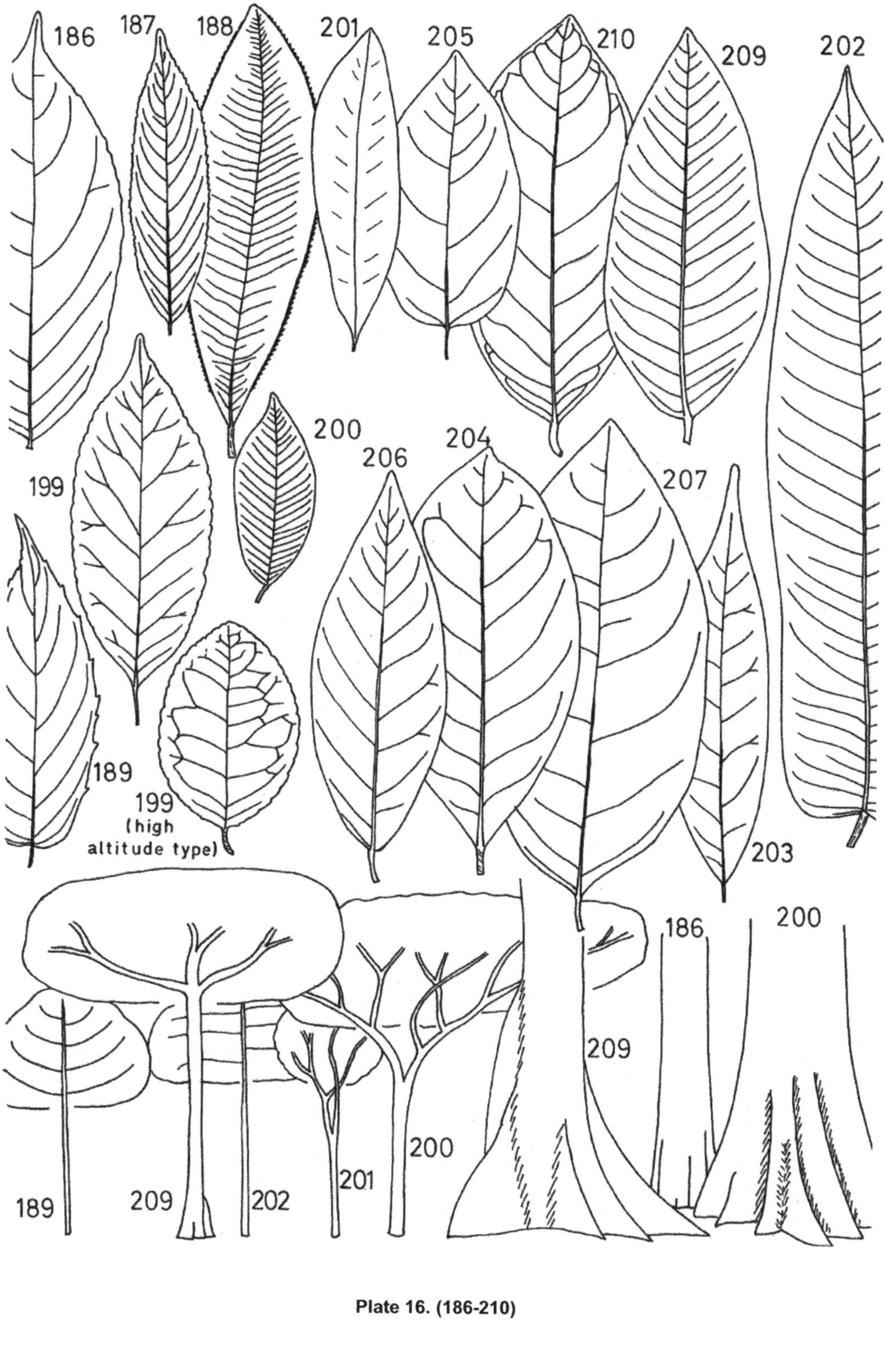

Plate 16. (186-210)

CULTIVATION AND PROPAGATION: Moderately fast-growing. Best grown on deep, well-drained soils under light shade. Can be grown in pure stands. Collect fruits from the ground under mother trees and soak in cold water for 24 hours to remove the pulp. Dry the seeds for a few hours and plant as soon as possible. Retain in the nursery for about 6 months. Protect transplanted plants until well-established.
NOTES: The wood is very strong and tough and can be used for flooring. The bark is medicinal.

Parinari excelsa Sabine (200) Chrysobalanaceae
Grey plum (en); Munazi, Namalambo (ga); Omushamba (ki); Mubura (na, tn, to); Omubura (no); Bula, Ebula, Ebura, Ibura (to).
45 m. Very large tree. Trunk thick, straight and cylindrical, with branches from relatively low down. Crown very large, spreading, thin. The leaves tend to be concentrated in non-overlapping clusters. Buttresses absent to medium-sized. Bark light brown, fairly thick, with vertical fissures c. 1-2 cm apart, usually also with horizontal fissures giving a granular texture. Phellogen more or less white. Slash brittle, red to dark red, sometimes with white lines, sometimes making a hissing noise when slashed hard. Leaves simple, alternate, c. 6.5 x 2 cm (much larger on small trees), with c. 20 main lateral veins on each side of the midrib, margin entire, dark green and glabrous above, covered with white or yellow-brown hairs below. Fruit c. 3.5 cm long, yellow.
OCCURRENCE: U1, 2 and 4. The dominant tree in Kalinzu and in North Kibale forests. Also abundant in Kayonza Forest and parts of Bwindi. Found up to 2200 m in Kigezi. It is common in Tero Forest (one of the Sango Bay forests), but rare in other lake-belt forests.
NOTE: Produces a very strong timber formerly much in demand for mine props.

Warburgia ugandensis Sprague subsp. _ugandensis_ (201) Canellaceae
East African greenheart (en); Mukuzannume, Muwiya (ga); Mwiha (na); Omusizambuzi (no); Balwegira (so); Abasi (tn); Muharami (to).
40 m (commonly to 30 m). Trunk cylindrical, straight to wavy, with a fairly small rounded crown. Buttresses and flutes absent. Bark thick and rough, brown, layered (with layers of lighter-coloured brown), fissured into rectangles c. 3 x 2 cm in size. Slash fibrous, red, tasting of pepper. Leaves simple, alternate, elliptic or oblong, c. 9 x 2.5 cm, apex acute and unequal-sided, base cuneate, thick and leathery, venation rather obscure, dark green and shiny above, paler beneath. Petiole 0.3 cm long. Fruit ovoid, green, turning purple, 3-5 cm long.
OCCURRENCE: U1-4. Widely distributed, often on dry sites. In Mabira Forest, Sango Bay forests, etc. Common in Kibale Forest.
CONSERVATION STATUS: Global NE (IUCN), LC (TOU); National VU (WCS).
CULTIVATION AND PROPAGATION: Tends to glow slowly when first planted, faster later. Collect the ripe (purple) fruits from mother trees or from the ground. Remove the seeds manually and sow as soon as possible. Retain seedlings in nursery for at least 6 months. Can be easily propagated from stem cuttings.
NOTES: The leaf shape is characteristic. The bark is used medicinally and in curries.

Pycnanthus angolensis (Welw.) Warb. (202) Myristicaceae
Muno (al); African nutmeg, False nutmeg (en); Lunaba (ga, tn); Munaba (ga); Ongono (sa); Ilomba (tn); Mukogoto (to).
35 m. Trunk straight and cylindrical, often with a shallow crown. Branches at right angles and curving upwards, branchlets drooping. Bark brown, moderately thick, with shallow fissures, but general effect smooth. Slash granular, pink, red to red-brown, often with numerous orange and/or white spots, turning darker, sometimes exuding red exudate (this may only become visible if the slash is deep). Young shoots, young leaves and fruits covered with orange-brown

hairs. Leaves simple, alternate, borne in two well-marked ranks. Leaves c. 20 x 7 cm, with c. 20-35 main lateral veins on each side of the midrib, apex acuminate. Fruit 2.5-4.5 cm long, brown, splitting in two to reveal a single seed with a lacy red aril.
OCCURRENCE: U2-4. Widely distributed in lower altitude forests, often on forest edges. Particularly common in lake-shore forests.
NOTES: The arils are eaten by monkeys, and hornbills. *Pycnanthus* is related to the nutmeg, (*Myristica fragrans* Houtt.), a native of the Spice Islands (Moluccas).

Staudtia kamerunensis Warb. (203) Myristicaceae
30 m. Trunk straight, with branches at right angles and a small crown. Bark smooth, flaking to leave concave scars, becoming rough on older trees. Slash white to pink, turning darker, exuding large quantities of red sap. Leaves simple, alternate, glabrous, c. 14 x 4 cm, acuminate.
OCCURRENCE: U2 and 4. Uncommon. Mengo and Bunyoro.
NOTE: Distinguished from *Pycnanthus* by the glabrous leaves.

Beilschmiedia ugandensis Rendle var. _ugandensis_ (204) Lauraceae
Mwasa (ga); Befe (gb); Omushoyo (ki); Mukalata, Mukarata (na).
25 m. Trunk usually crooked, with a dark-coloured, heavily foliaged, crown. Buttresses usually absent. Bark light brown, smooth, flaking in large sheets. Slash soft, dark red, with brown lines. Leaves simple, alternate, elliptic or oblong, c. 11 x 5.5 cm, apex acute to acuminate, base cuneate to rounded, rather thick, glabrous, with rather irregular venation. Petiole c. 1 cm long. Fruit ellipsoid, c. 2.5 cm long.
OCCURRENCE: U1, 2 and 4. Usually (but not always) in swamp forest or in other damp places.
CONSERVATION STATUS: Global VU (IUCN), LC (TOU); National VU (WCS).
CULTIVATION AND PROPAGATION: Collect fruits on the ground and leave to rot. Remove seeds manually from the decomposed flesh and dry in the sun for a few days. Germination is hastened by nicking the hard seed coat.

Ocotea usambarensis Engl. (205) Lauraceae
Omwiha (ki); East African camphor (tn).
25 m (to 45 m in Kenya). Trunk straight, with branches at right angles and a spreading crown. Buttresses absent. Bark brown, fairly smooth, scaling on older trees, with medium-sized vertical fissures on younger trees. Slash fibrous, pink, fragrant. Leaves simple, (sub-)opposite (but some alternate), elliptic to ovate, c. 10 x 4.5 cm, apex rounded to acuminate, base cuneate to rounded, margin recurved, with c. 4-9 main lateral veins on each side of the midrib, whitish below. Petiole 0.5-2 cm long. Fruit ellipsoid or globose, c. 1 cm long, 6 cm wide.
OCCURRENCE: U2. Very local in Uganda. Rwenzori Mts, Bwindi-Impenetrable Forest (Ishasha Gorge) and Kalinzu Forest (very rare in the latter).
CULTIVATION AND PROPAGATION: Moderately fast-growing. Best grown in montane areas with deep, well-drained, soils. Can be planted in mixed or pure stands. Collect fruits under mother trees, remove the pulp, keep the seeds moist and sow as soon as possible.
NOTE: The leaves are camphor-scented.

Ocotea kenyensis (Chiov.) Robyns & R. Wilczek (206) Lauraceae
20 m. Much-branched tree with a dark-coloured crown. Bark rough, dark-coloured. Leaves simple, alternate, red when young, aromatic when crushed, elliptic to ovate, c. 14 x 5.5 cm, apex acute or acuminate, base cuneate to rounded, margin often red and decurrent, more or less glabrous. Petiole c. 1 cm long. Fruit ellipsoid or ovoid, c. 2.5 cm long, 1 cm wide.
OCCURRENCE: U1 and 2. Recorded from Bwindi Forest (on ridge tops) and the Imatong Mountains, c. 2000-2500 m.
CONSERVATION STATUS: Global VU (IUCN), LC (TOU); National VU (WCS).

NOTE: The leaves are similar to those of *Beilschmiedia ugandensis* (204), which probably never occurs at such a high altitude.

Casearia runssorica Gilg (207) Salicaceae

SYNONYM: *Casearia engleri* Gilg (misapplied name in ITU and UFT)

35 m. Trunk straight and cylindrical. Bark light brown, fairly smooth, with prominent lenticels. Phellogen green. Slash variable in colour, red, pink, white to yellow/orange, sometimes turning slowly red. Leaves simple, alternate, c. 14 x 6 cm (sometimes considerably smaller), with c. 6-12 main lateral veins on each side of the midrib, apex acuminate, base cuneate or rounded, margin entire, with gland dots and short lines, particularly near the margin. Petiole c. 0.8 cm long. Fruit c. 1.25 cm long, yellow, splitting into three to reveal white seeds embedded in lacy red arils.

OCCURRENCE: U2 and 4. Widely distributed, but nowhere common. Not recorded from Mengo. To 2000 m.

CONSERVATION STATUS: Global NE (IUCN), LC (TOU); National VU (WCS).

NOTE: Distinguished by the gland dots and lines on the leaves.

Casearia battiscombei R.E. Fr. (208) Salicaceae

Seruwa (ku); Gusieruss, Sihuling (ms).

Tree very similar to *Casearia runssorica* in shape and leaves. This species differs from *C. runssorica* in having prominent reticulation on both surfaces of the leaves (it is prominent only on the undersurface in *C. runssorica*). Leaves simple, alternate, c. 15 x 5 cm, apex obtusely acuminate, base rounded to broadly cuneate, with c. 14-20 main lateral veins on each side of the midrib, margin undulate. Petiole c. 0.8 cm long. Fruit c. 1.1 cm long, yellow-orange.

OCCURRENCE: U2 and 3. Recorded from Mt Elgon (where it in common at 1800-2300 m) and Kalinzu and Kasyoha-Kitomi forests.

Klainedoxa gabonensis Engl. (209) Irvingiaceae

Mututtu (ga); Omukuzanyana (no).

40 m. Very large tree with a straight trunk and wide spreading crown with large branches. Buttresses very large. Surface roots prominent. Blunt thorns (to 2.5 cm long) sometimes present on trunk. Bark dark brown, flaking irregularly. Slash granular, yellow to orange. Leaves simple, alternate, ovate to elliptic or oblong-lanceolate c. 7 x 4 cm, thick and a bit leathery, glabrous, apex shortly acuminate, base cuneate to rounded, with up to 35 narrow lateral veins on each side of the midrib. Petiole 0.2-.8 cm long. Stipules remarkably long (5-10 cm), narrow, pointed. Fallen stipules carpet the ground beneath the tree. Flowers pink, beautiful. Fruit green, depressed-spherical, slightly 5-lobed, c. 6 cm diameter, falling off the tree without opening and rotting on the ground.

OCCURRENCE: U2 and 4. Uncommon, except in Namalala Forest (Sango Bay) and in Budongo Forest, where it is said to have the largest crown of any species of tree. Recorded from Bugoma Forest.

NOTE: The falling fruits can be dangerous for those standing beneath.

Irvingia gabonensis (Aubrey-LeComte ex O'Rorke) Baill. (210) Irvingiaceae

Wild mango (en).

20 m (occasionally 35 m). Tree with a spreading, dense crown (like that of a mango). Slash strongly scented of mangoes, densely reticulate, fibrous in lower part. Leaves simple, alternate, leathery, glabrous and glossy, elliptic to obovate- or ovate-elliptic c. 12 x 6 cm, apex shortly acuminate, base cuneate to rounded, with c. 7-10 main lateral veins on each side of the midrib. Stipules curved, to 1.5 cm long, narrow, covering the young leaves. Fallen stipules carpet the ground beneath the tree. Petiole c. 1 cm long, grooved. Fruits c. 7 cm long, 5 cm broad, yellow, with a single seed, resembling a mango.

OCCURRENCE: U2 and 4. Mengo, Tooro and Bunyoro. Common in Budongo Forest, but rare elsewhere.
CONSERVATION STATUS: Global LC (IUCN, TOU); National EN (WCS).
NOTE: The fruit is edible, but not very good.

Diospyros abyssinica (Hiern) F. White subsp. _abyssinica_ (211) Ebenaceae
Nkinga (am); Mpimbya (ga); Omuhoko, Omuwirute (ki); Cheptua (ku); Muhoko (na, to); Mubale (nl); Mayonjo, Miyonjo, Muyonja (sa); Lusui (tn).
30 m. Trunk cylindrical, thin, very straight, with a small rounded crown. Buttresses absent. Bark dark brown (to almost black on larger trees), thick and fibrous, vertically fissured, sometimes flaking. Slash fibrous (to brittle on older trees), yellow, usually with orange streaks, sometimes turning slightly darker, not or only slightly scented. Leaves simple, alternate, elliptic oblong to oblanceolate, c. 11 x 3 cm, apex obtuse to shortly subacuminate, base cuneate to rounded, margin often wavy, lateral veins 5-10 on each side of the midrib. Petiole c. 0.5 cm long. Fruit spherical, yellow to orange-red, c. 0.9 cm diameter.
OCCURRENCE: U1-4. Abundant on drier sites, particularly on upper slopes (e.g. in Mabira, Kibale and Maramagambo forests).

Diospyros katendei Verdc. (211a) Ebenaceae
Medium-sized tree. Bole irregularly fluted. Branchlets flattened and sharp-edged. Slash thin and orange. Leaves oblong, deep green, 6-12 cm wide, 2.2-4.2 cm wide, narrowly acuminate, base cuneate. Petiole 0.8-1.0 cm long. Fruit ellipsoid, c. 1 cm across.
OCCURRENCE: U2. Only known from a single collection in Kasyoha-Kitomi Forest. A range-restricted species, a Ugandan endemic.
CONSERVATION STATUS: Global CR (IUCN, TOU); National CR (WCS).
NOTE: The flattened branchlets are characteristic.

ANNONACEAE

Small to medium-sized trees. Bark on young trees reticulately fibrous, sometimes with prominent lenticels. The slash lacks any trace of red coloration and generally has an outer dark-coloured rim (phellogen). Leaves simple, alternate. This is an evolutionary primitive family, with perianth segments in threes and numerous stamens. The fruit consists of numerous free carpels (except in *Monodora* and *Isolona*).

Key to Annonaceae.

1. Plant often climber or straggler, sometimes tree; hairs often stellate, sometimes simple. ....**222-225. _Uvaria_**
   Plant often tree or shrub, rarely climber or strangler. 2
2. Leaves usually over 40 cm long. ....**221. _Uvariodendron_**
   Leaves usually under 40 cm long. 3
3. Young stems glaucous (with waxy bloom which rubs off easily). **219-220. _Monodora_**
   Young stems not glaucous. 4
4. Leaves silky hairy in young leaf bud; sepals 2 mm long. 5
   Leaves not silky, but, if silky, then sepals at least 3 mm long. 6
5. Young branchlets glabrous. ....**212 _Uvariopsis_**
   Young branchlets densely yellow hairy. ....**213. _Greenwayodendron_**

6. Young branchlets with hairs that are usually yellow or ferruginous. 7
   Young branchlets glabrous. ....**214. *Cleistopholis***

7. Petals 6, in 2 whorls; sepals 1.5-5 mm. long. ....**216-218. *Xylopia***
   Petals 6, in 1 whorl; sepals 3-3.5 mm long. ....**215. *Isolona***

Uvariopsis congensis **Robyns & Ghesq. (212)** Annonaceae

12 m. Understorey tree with a wavy trunk and spreading crown. Bark thin, smooth, brown, with vertical lines of lenticels. Slash white (very rarely brown or yellow), fibrous, turning darker. Phellogen black. If slashed thinly, a reticulate pattern of fibres can be seen. Young shoots glabrous. Leaves simple, alternate, elliptic or oblanceolate, c. 14 x 4 cm, apex obtuse or narrowly acuminate, base cuneate or rarely rounded, with c. 10-14 prominent main lateral veins on each side of the midrib, margin appearing wavy, glabrous at maturity. Petiole c. 0.4 cm long. Male flowers on leafy, but female flowers on leafless branches. Fruit ellipsoid or cylindrical, 1.7-4.5 cm long.
OCCURRENCE: U2-4. Abundant in Mabira, Kibale and probably other forests.
NOTE: The tree sometimes resembles *Diospyros abyssinica* (211), but is readily distinguished by the bark, tree shape and slash (if not the yellow variant).

Greenwayodendron suaveolens **(Engl. & Diels) Verdc. subsp. *suaveolens* (213)** Annonaceae

25 m. Trunk straight, cylindrical, with branches at right angles and curving upwards. Crown small, deciduous. Bark grey to brown, smooth. Phellogen black. Slash yellow, with reticulate fibres, turning brown after some time, scented. Young shoots with yellow hairs. Leaves simple, alternate, more or less glabrous when mature, elliptic or oblong-elliptic, c. 13 x 4.5 cm, apex acute or acuminate, base cuneate or rounded, main lateral veins 5-13 on each side of the midrib, these veins impressed above and prominent below. Petiole c. 0.3 cm long.
OCCURRENCE: U2 and 4. Recorded from Mabira and Budongo forests (rare in both) and the Ssese Islands (where it is locally common).
NOTES: It resembles *Diospyros abyssinica* (211), from which it can be distinguished by the scented, discolouring slash and smooth bark. The tree also resembles *Uvariopsis congensis* (212), but differs in having hairy young shoots. The leaf venation is more arcuate than in either of these two species or in *Cleistopholis patens* (214).

Cleistopholis patens **(Benth.) Engl. & Diels (214)** Annonaceae

20 m. Trunk straight and cylindrical, with horizontal branches. Buttresses absent. Bark greyish-white, fairly smooth, with vertical corrugations (like half-grown *Maesopsis*) or fissures. Phellogen black. Slash yellow-brown to white, reticulately fibrous, strongly scented. Leaves simple, alternate, elliptic or oblong-lanceolate, c. 13 x 3.75 cm, apex acuminate, base cuneate or rounded, main lateral veins c. 10-24 on each side of the midrib and prominent below, lamina very shiny above. Petiole c. 0.5 cm long.
OCCURRENCE: U1 and 2. A light-demanding species, found on the edges of swamps and margins of rivers. Uncommon, except in a patch of riverine forest in Queen Elizabeth National Park (north of Maramagambo Forest), and perhaps also in Siba Forest.

Plate 17. Annonaceae and others (199-222); see also Plate 16

199. *Prunus africana* 202. *Pycnanthus angolensis* 204. *Beilschmiedia ugandensis*
209. *Klainedoxa gabonensis* 211. *Diosypros abyssinica* 212. *Uvariopsis congensis*
213. *Greenwayodendron suaveolens* 214. *Cleistopholis patens* 216. *Xylopia aethiopica*
218. *Xylopia parviflora* 219. *Monodora myristica* 220. *Monodora angolensis*
222. *Uvaria angolensis*

Actual sizes: leaves, flower and fruits x 2; trunk base x 80.

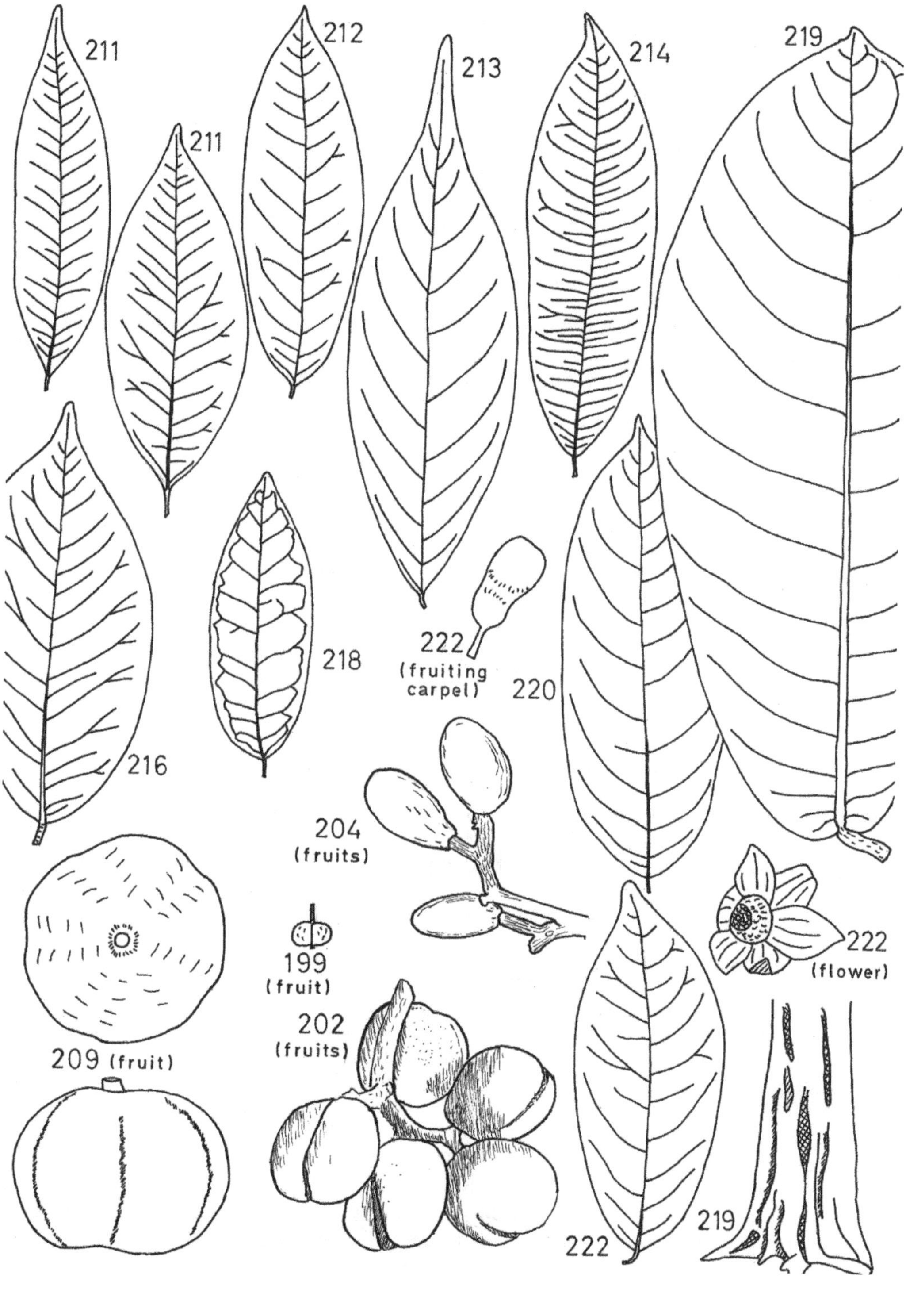

Plate 17. (199-222)

NOTE: The venation of *Greenwayodendron suaveolens* (213) is more arcuate than that of *Cleistopholis*.

***Isolona congolana* (De Wild. & T. Durand) Engl. & Diels (215)** Annonaceae
25 m. Bark grey, fissured. Young shoots pubescent. Leaves simple, alternate, c. 14 x 3.5 cm, apex acuminate, base rounded or cuneate, with c. 15-25 main lateral veins on each side of the midrib, glabrous at maturity (except for the midrib). Petiole c. 0.3 cm. Fruit ovoid or ellipsoid, 6.5-9 cm long, with longitudinal ribs.
OCCURRENCE: U2. Only known from Bwamba, where it grows with *Pterygota* by rivers.

Key to *Xylopia.*

1. Stilt roots usually present; leaves cuneate at base; only known from Bwindi Forest. ....**217. *X. staudtii***
   Stilt roots absent; leaves cuneate to truncate at base; distribution not as above. 2
2. Leaves leathery and tough, apex acuminate; indumentum on young shoots and leaves not visible to the naked eye. ....**216. *X. aethiopica***
   Leaves often thin, apex emarginate to acuminate; indumentum on young shoots and leaves spreading and visible to the naked eye. .... **218. *X. parviflora***

***Xylopia aethiopica* (Dunal) A. Rich. (216)** Annonaceae
Nsaggalanyi (ga).
30 m. Trunk thin, straight, with horizontal branches and a much-branched crown. Stilt roots probably sometimes present. Bark grey-brown with vertical channels. Phellogen black. Slash hard, fibrous, off-white. Leaves simple, alternate, elliptic to obovate, c. 12 x 5.5 cm, apex usually acuminate, base cuneate to rounded, thick and glabrous, shiny above, paler beneath, apex usually acuminate, base cuneate to rounded. Old leaves tend to turn red. Petiole c. 0.5 cm long, purple or black.
OCCURRENCE: U4. Lake-shore forests with *Piptadeniastrum*. Common in places.

***Xylopia staudtii* Engl. & Diels (217)** Annonaceae
Large tree to 45 m with a straight trunk. Short buttresses sometimes present. Stilt roots usually present. Bark rough, more or less fissured. Young branches with red-brown hairs. Leaves simple, alternate, elliptic to obovate, 4-12 x 2.5-6 cm, apex shortly acuminate, base cuneate, glabrous and shiny above, sparsely pubescent below, venation distinctly prominent on both surfaces. Petiole thick and grooved, 0.6-0.8 cm long.
OCCURRENCE: U2. Only known from Bwindi Forest.

***Xylopia parviflora* (A. Rich.) Benth. (218)** Annonaceae
Deciduous tree to 25 m. Trunk straight, with a small crown and whorls of short horizontal branches. Bark grey, usually smooth. Leaves simple, alternate, oblong to elliptic, c. 8 x 2.5 cm, thin, apex acute (rarely obtuse to acuminate), base cuneate or rounded, more or less hairy below.
OCCURRENCE: U1-3. Riverside forest.

***Monodora myristica* (Gaertn.) Dunal (219)** Annonaceae
Calabash nutmeg (en); Naggomola (ga); Muho (na); Mukoza (so); Mugema (to).
30 m. Trunk crooked, with a large spreading, deciduous crown of large leaves. The trunk is characteristically very uneven, rather gnarled, usually fluted, sometimes indistinctly buttressed. Bark brown, fairly thin, raised in places to form a more or less conspicuous reticulate pattern. Phellogen black. Slash with brown fibres forming an irregular reticulate pattern on a lighter coloured background. Leaves simple, alternate, pendulous, obovate or elliptic, large, often c. 25 x 10 cm (but much smaller on some branches, particularly near flowers), apex shortly acuminate, base rounded to cordate, with c. 10-23 main lateral veins on each side of the midrib

prominent on both surfaces, glabrous, margin entire. Petiole 0.5-1.5 cm long, channelled. Flowers hanging, large, greenish-yellowish with reddish spots. Flowers with three series of perianth segments. Fruit c. 16 cm diameter, spherical to slightly elongated, with numerous seeds.
OCCURRENCE: U2-4. Widely distributed and sometimes common. Particularly abundant at 1200-1600 m, e.g. in Central Kibale Forest and formerly on Mt Elgon (before extensive forest clearance since 1971).
NOTE: The seeds are edible and used medicinally.

***Monodora angolensis* Welw. (220)** Annonaceae
Mukufu (sa).
Small understorey tree to 15 m. Bark dark-coloured, shallowly vertically fissured. Shoots glabrous. Leaves simple, alternate, c. 11 x 4.5 cm, apex acuminate, base cuneate to rounded, with c. 8-11 main lateral veins on each side of the midrib, margin entire. Flower pendulous, green. Fruit 4-9 cm long, longitudinally ribbed.
OCCURRENCE: U2 and 4. Uncommon. Most records are from Budongo Forest.

***Uvariodendron magnificum* Verdc. (221)** Annonaceae
10 m. Tree branched near base, with a spreading crown. Bark thin, grey. Slash white, turning darker. Young shoots drooping, red. Leaves simple, alternate, oblanceolate, apex obtuse or shortly acuminate, base cuneate, very large (c. 50 x 14 cm), with c. 30 main lateral veins on each side of the midrib. Petiole c. 1-1.5 cm long, very thick, channelled. Fruit large, with numerous carpels.
OCCURRENCE: U2 and 4. Only known from Kasyoha-Kitomi Forest (in which it is gregarious and local) and Lutoboka Point in Ssese. A range-restricted species, a Ugandan endemic.
CONSERVATION STATUS: Global EN (IUCN, TOU); National VU (WCS).
NOTE: Easily distinguished by the large leaves.

UVARIA

A genus of small trees and scrambling shrubs, sometimes becoming large climbers. Leaves simple and alternate. Flowers yellowish. Fruit consisting of many free carpels. Two other genera of shrubby or climbing Annonaceae that occur in Uganda are *Artabotrys* and *Monanthotaxis*.

Key to *Uvaria*.

1. Young branches hairy, turning glabrous with age. 2
   Young branches glabrous or nearly so. ....**225. *U. scheffleri***
2. Nerves on leaves raised both above and below. ....**223. *U. welwitschii***
   Nerves impressed above, prominent below. 3
3. Lamina glabrous (except on nerves) or sparsely pubescent above and sparsely to densely hairy below. ....**222. *U. angolensis***
   Lamina very finely stellate puberulous (densely covered with short hairs) above, densely hairy below. ....**224. *U. schweinfurthii***

Uvaria angolensis* Oliv. (222) var. *angolensis Annonaceae
Shrub, large climber or small tree to 10 m. Bark with prominent lenticels. Phellogen black. Slash brown, reticulately fibrous. Branchlets at first hairy, becoming glabrous with age. Leaves oblong, elliptic or oblanceolate, c. 12 x 5.5 cm, apex obtuse or shortly acuminate, base rounded to slightly cordate. Petiole 3-8 mm long.
OCCURRENCE: U1, 2 and 4. Forest edges. The commonest species of *Uvaria*.

***Uvaria welwitschii* (Hiern) Engl. & Diels (223)** Annonaceae
Climber, shrub or tree to 6 m. Branches at first hairy, becoming glabrous with age. Leave mostly oblong, c. 13 x 4.5 cm, apex obtuse to shortly acuminate, base cuneate to rounded, margin and midrib ferruginous. Petiole 2-4 mm long.
OCCURRENCE: U2-4. Reported to occur on rocky outcrops in forest and in ravines.

***Uvaria schweinfurthii* Engl. & Diels (224)** Annonaceae
Shrub or straggling tree to 6 m. Leaves oblong or oblong-elliptic, c. 9 x 3.5 cm, apex acute to acuminate, base cuneate, rounded or slightly cordate, puberulous above with stellate hairs, markedly hairy below, nerves prominent below. Petiole 2-5 mm long.
OCCURRENCE: U2. In riverine forest in Murchison Falls National Park (probably in Rabongo Forest).

***Uvaria scheffleri* Diels (225)** Annonaceae
Climber, shrub or small tree to 3 m. Leaves oblong or oblong-elliptic, 1-14 x 0.7-5.8 cm, apex obtuse or acute, base cuneate to rounded, glabrous. Petiole 3 mm long.
OCCURRENCE: U1 (Karamoja).

TURRAEA

Turraea is an exceptional genus within the family Meliaceae in having simple (rather than pinnate) leaves. *Turraea* usually has tufts of hairs (domatia) in the axils of the veins on the undersurface of the leaves. Additional to the species below, *T. vogelii* Hook f. and *T. vogelioides* Bagsh. & Bak. f. (Plate 18) are shrubs found in U2 and 4. The former is particularly common in the undergrowth of lake-belt forests and the latter generally common. See before tree 390 for a description of the Meliaceae.

Key to *Turraea*.

1. Leaves usually less than 9 cm long; on Imatong Mts. ....**228. *T. holstii***
   Leaves at least 10 cm long. 2
2. Leaf base cuneate. 3
   Leaf base usually rounded (but may be subtruncate or cuneate). ..**226. *T. floribunda***
3. Leaf glabrous (except for scattered hairs on nerves below and in axils of main lateral veins); in Karamoja. ....**229. *T. abyssinica***
   Leaf surface pubescent. 4
4. Petiole to 0.7 cm long; capsule nearly as long as wide; in Budongo Forest. ....**229a. *T. pellegrianiana***
   Petiole to 1 cm long; capsule width double its length; more widely distributed. ....**227. *T. robusta***

***Turraea floribunda* Hochst. (226)** Meliaceae
Pogdliech (ac); Muhojole (nl); Murama (to).
10 m. Trunk straight or irregular, with a spreading crown. Bark brown, fairly thin and smooth, with vertical lines of lenticels. Slash fibrous, light yellow, turning darker in places near the bark. Leaves simple, alternate, ovate to lanceolate, c. 15 x 6 cm, apex acuminate, base subtruncate, rounded or broadly cuneate, with c. 9-17 main lateral veins on each side of the midrib, densely setose (bristly) when young, more sparsely so when older (except on the nerves), hairy below (at least on the midrib and main veins). Petiole c. 1 cm long. Petals greenish-white. Capsule usually obovoid-cylindric (rarely globose), to c. 2.5 cm x 1.5 cm, with red arillate seeds.
OCCURRENCE: U1-4. A light-demanding species found on forest edges.

***Turraea robusta* Gürke (227)** Meliaceae
Omukarakare (ki); Kivunambasa (so).

8 m. Bark light brown, smooth. Slash pink, sometimes with white lines, foul-smelling. Leaves elliptic or obovate, c. 10 x 5.5 cm, apex mostly acute, base cuneate, with c. 6-11 main lateral veins on each side of the midrib, often shortly pubescent below. Petiole c. 1 cm long. Inflorescence terminal or axillary. Petals creamy-white. Capsule 0.8 x 1.5 cm.
OCCURRENCE: U2-4. On termite mounds and forest edges and in young secondary forest. Very common on the edge of Maramagambo Forest.
NOTE: The leaf veins ascend at a steeper angle than those of *Turraea floribunda* (see Plate 18).

Turraea holstii Gürke (228) Meliaceae
Small, straight-boled, tree to 15 m. Leaves usually less than 9 x 4 cm, mostly elliptic, apex shortly and bluntly acuminate, base cuneate, lower surface nearly glabrous (except for scattered hairs on nerves and domatia). Petiole to 0.9 cm long. Inflorescence an axillary cyme. Petals white, turning yellow with age. Capsule depressed, globose, c. 0.7 x 1 cm.
OCCURRENCE: U1. Only known from the Imatong Mountains.

Turraea abyssinica Hochst. ex A. Rich. (229) Meliaceae
Small-sized tree to 8 m. Leaves c. 12 x 5 cm, lanceolate to lanceolate-elliptic, apex acutely acuminate, base cuneate, a little asymmetric, glabrous (except for domatia and scattered hairs on the nerves beneath). Petiole to 1 cm. Inflorescence a terminal or lateral cyme. Petals greenish-white or cream. Capsule depressed globose, c. 0.7 x 0.8 cm.
OCCURRENCE: U1. Only known from Karamoja.

Turraea pellegrianiana Keay (229a) Meliaceae
Small-sized tree to 8 m. Leaves c. 12 x 5.5 cm, elliptic or oblanceolate-elliptic, apex acuminate, base cuneate, lower surface sparsely covered with short hairs (puberulous). Petiole to 0.7 cm. Usually flowers when leafless. Inflorescence very rarely axillary. Petals greenish-white or cream. Capsule depressed globose, c. 0.9 x 1 cm.
OCCURRENCE: U2. Only known from Budongo Forest, on forest edges.

BAPHIA AND BAPHIOPSIS

Most Fabaceae (formerly Leguminosae) have obviously compound leaves, either trifoliolate (344-345) or pinnate or pinnatifid (423-447). Only *Baphia* and *Baphiopsis* (subfamily Faboideae) have apparently simple leaves (actually compound unifoliolate). See before tree 421 for an overview of the family.

Baphia wollastonii Baker f. (230) Fabaceae (Subfamily Faboideae)
Ndiabuturu, Njabituli (am); Omurungurungu (no).
20 m. Understorey tree, with an irregularly shaped trunk and branches at all heights. Trunk sometimes gnarled. Sometimes multi-stemmed. Crown fairly spreading. Bark thin and smooth, brown, sometimes with small vertical fissures and sometimes flaking in small pieces c. 1-3 x 0.2-1 cm in size. Slash of fairly even texture, yellow. Leaves apparently simple (actually compound unifoliolate), alternate, ovate or elliptic, c. 6.5 x 2.5 cm, with a long acumen, base a bit asymmetrical, often glabrous on both surfaces. Petiole c. 0.4 cm long. Flowers white with a yellow blotch near the base, the petals drying to brown. Pod 6-9 cm long, 1.5-2.2 cm wide.
OCCURRENCE: U1, 2 and 4. Abundant beneath *Cynometra* in South Maramagambo and Budongo forests, and in riverine environments in other forests.

Baphia capparidifolia Baker subsp. _multiflora_ (Harms) Brummitt (231)
Fabaceae (Subfamily Faboideae)
Munyamakanja (na).
Small tree to 5 m, often scandent, differing from *Baphia wollastonii* in the young stems and leaves being covered with yellow-brown hairs. Leaves apparently simple (actually compound unifoliolate), alternate, ovate to lanceolate, c. 9 x 3.5 cm, apex acute to obtuse or acuminate, base rounded to subcordate, upper surface glabrous, lower surface pubescent, with main veins very prominent. Petiole 1.25-3.75 cm long,

swollen at base and apex (like *Baphiopsis*). Petals white or yellowish, with an orange blotch near the base. Pod 4-9 cm long, 0.7-1.4 cm wide.
OCCURRENCE: U2. Only recorded from Kigezi and Tooro. Occasional in Bwamba.

Baphiopsis parviflora **Baker (232)** Fabaceae (subfamily Faboideae)

Mutoka (ga, na); Munyamakanja (na).
15 m. Spreading understorey tree, with a crooked (occasionally straight) trunk and irregular branching. Bark thin and smooth, dark green to dark brown, with vertically elongated and prominent lenticels, occasionally exfoliating in strips. Phellogen green to black. Slash fibrous, white to yellow, turning red in places. Leaves apparently simple (actually compound unifoliolate), alternate, c. 11 x 5 cm (but variable in size), with both the main lateral veins and the vein reticulum prominent on both surfaces. Petiole c. 1.5 cm long, markedly swollen at base and apex, characteristic. Fruit a swollen pod, c. 5 cm long, containing 1 to several seeds.
OCCURRENCE: U2-4. A widely distributed tree, abundant in many places, particularly in damp situations.

Maerua duchesnei **(De Wild.) F. White (233)** Capparaceae

Katombi (am); Muzikiza (ga); Munyirima, Mwirima (nyo).
8 m. Understorey tree with a widely spreading crown and an untidy appearance due to the presence of numerous thin branches. Bark thin and very smooth, almost black. Slash characteristic, very thin, bright red. Leaves simple, alternate, oblanceolate or elliptic, c. 10 x 4 cm, apex acutely acuminate, base cuneate, glabrous and leathery. Petiole c. 0.4 cm long. Fruit ovoid, c. 3 cm long.
OCCURRENCE: U1-4. Very abundant in many forests below 1300 m, particularly on drier sites.
NOTES: Easily recognized by the slash.

Tapura fischeri **Engl. (234)** Dichapetalaceae

Bererewa, Kaberero (am); Kazunganjuki (ga).
20 m (exceptionally 25 m). Understorey tree with a spreading, thin crown, often with layered foliage, sometimes with several trunks from base and often with epicormic shoots. Bark greenish to dark brown, thin, fairly smooth (to fairly rough with vertical fissures on old stems). Green phellogen usually prominent. Slash granular, light yellow, turning darker. Leaves simple, alternate, elliptic to obovate, apex shortly acuminate, base cuneate to rounded and often asymmetrical, c. 8.5 x 3.25 cm, with c. 4-7 main lateral veins on each side of the midrib, hairy below, with tufts of white hairs in the vein axils. Petiole c. 0.4 cm long. The small inflorescence is borne on the petiole. Fruit ovoid or ellipsoid.
OCCURRENCE: U1-3. Widely distributed, mainly in secondary forest. Quite common in Budongo Forest.
NOTES: The tufts of white hairs in the vein axils are a prominent feature. The Luganda name (*kazunganjuki*) implies that the plant is popular with bees.

Trichocladus ellipticus **Eckl. & Zeyh. subsp.** ***malosanus*** **(Baker) Verdc. (235)** Hamamelidaceae

Plate 18. ***Turraea, Baphia*** **and others (226-238)**

226. *Turraea floribunda* 227. *Turraea robusta* 228. *Turraea holstii*
230. *Baphia wollastonii* 232. *Baphiopsis parviflora* 233. *Maerua duchesnei*
234. *Tapura fischeri* 235. *Trichocladus ellipticus* 236. *Barteria nigritana*
237. *Apodytes dimidiata* 238. *Leptaulus daphnoides*

Actual sizes: leaves and fruits x 2.

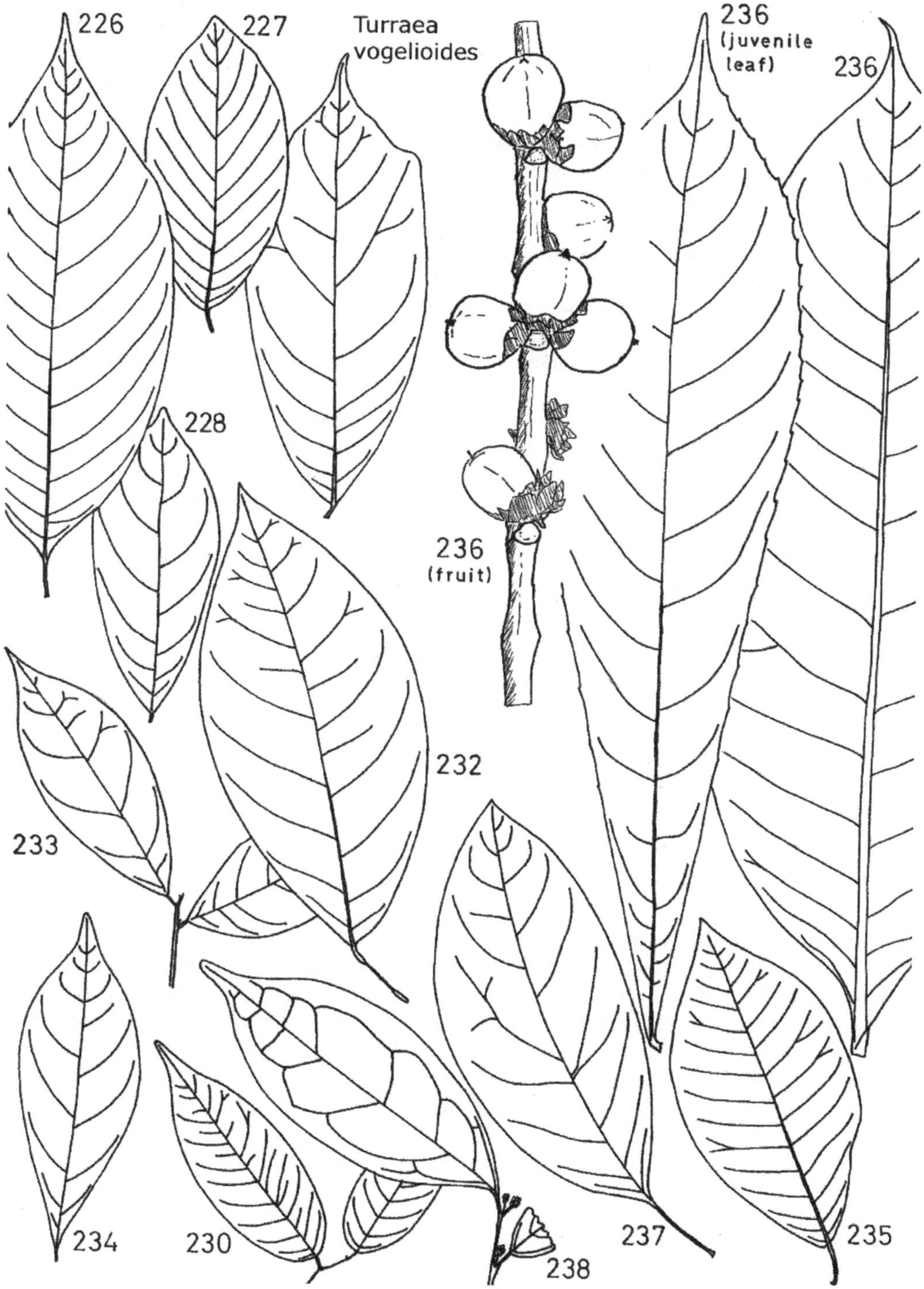

Plate 18. (226-238)

Berrakaya (ku); Bisoroko, Soroko (ms).
10 m. Weak stemmed or thicket-forming understorey tree. Bark light brown, thin and flaking. Slash fibrous, red. Leaves simple, alternate, often elliptic, c. 9 x 5 cm, apex acute or acuminate, base cuneate to rounded, margin entire, upper surface glabrous and glossy, lower surface covered with white to brown stellate hairs. Petiole 0.5-1.5 cm long.
OCCURRENCE: U1, 3 and 4. In damp places in lake-shore forests, being particularly common in *Podocarpus* forest in the Sango Bay area. Also common on Mts Moroto and Napak and at 2150 m on the north-eastern slopes of Mt Elgon.

Barteria nigritana **Hook. f. subsp. *fistulosa* (Maet.) Sleumer (236)**
Passifloraceae
SYNONYM: *Barteria acuminata* Baker f.
8 m. Trunk irregular, bearing wide spreading branches. Bark thin, red-brown, with prominent lenticels. Slash light yellow to red, sometimes with traces of orange, thin. Branchlets hollow. Leaves simple, alternate (in the sense of not being spirally arranged), variably elliptic to oblanceolate, large, c. 35 x 11 cm, apex obtuse or acuminate, base rounded or attenuate, margin entire, lateral veins prominent on both surfaces, 9-19 on each side of the midrib. Petiole c. 1 cm long. The young leaves (which persist on small trees growing in shade) are of a completely different shape, being long cuneate at the base and toothed (see illustration). Fruit spherical, c. 2.5 cm diameter, resembling a passion fruit.
OCCURRENCE: U4. Abundant in lake-belt forests in places. Usually near or on forest edges.

Apodytes dimidiata **Arn. (237)** Icacinaceae
White pear (en); Munyamazzi (ga).
25 m, but commonly less. Tree or shrub, very variable in habit (depending on situation). Bark thin and smooth, greenish to whitish. Slash pink to red, scented. Leaves simple, alternate, elliptic to oblong, c. 12 x 6.5 cm, (but variable in size), apex acuminate, acute or obtuse, base cuneate, with c. 5-9 main lateral veins on each side of the midrib, these veins being prominent and red in colour, margin entire or almost so. Petiole c. 2.5 cm long. The leaves dry to black. Flowers white. Fruit asymmetric, 5-11 mm long, 3-4 mm wide.
OCCURRENCE: U1-4. Found in forest and savanna. Rare, except possibly in lake-shore forests in Masaka.
NOTE: The red veins on the leaves are a good character.

Leptaulus daphnoides **Benth. (238)** Cardiopteridaceae
10 m. Understorey tree. Trunk irregular, with branches from near base. Bark thin and smooth, dark green to grey. Slash either yellow (sometimes with orange markings) to white with a yellow rim. The shoots elongate through the growth of an axillary bud at each node (this character may be difficult to see). Leaves simple, alternate, c. 13 x 4 cm, with c. 5-7 main lateral veins on each side of the midrib, these veins looping well within the margin and fusing with one another, glabrous, margin entire, apex acuminate. Petiole c. 0.8 cm long. Fruit red to orange at maturity.
OCCURRENCE: U2 and 4. Widely distributed. Recorded from Bunyoro, Mengo, Masaka and Kigezi.

Leptaulus holstii **(Engl.) Engl. (239)** Cardiopteridaceae
Small forest shrub or tree to 3 m. Leaves similar to those of *Leptaulus daphnoides*.
OCCURRENCE: U2 and 4. Recorded from Kasa Forest (Mengo) and Kalinzu Forest.

EUPHORBIACEAE AND RELATED FAMILIES (tree numbers 240-251)

Those Euphorbiaceae and members of related families included here have simple, alternate, leaves with entire margins, lack white latex in the slash and lack prominent veins from the leaf base. Small to medium-sized trees. See before tree 94 for an overview of Euphorbiaceae and related families.

Uapaca mole **Pax. (240)** Phyllanthaceae

SYNONYMS: *Uapaca paludosa* Aubrév. & Leandri; *Uapaca guineensis* Müll. Arg.
Freshwater mangrove (en); Mukusu (ga).
20 m. Trunk short, with a dense rounded crown of large leaves. Stilt roots present, very prominent. Spines sometimes present (these being young stilt roots). Bark light brown, with vertical lines of lenticels, usually flaking but general effect fairly smooth, becoming fissured with age. Slash white to light brown, with orange streaks, rather granular, rapidly turning red-brown. Leaves simple, alternate, clustered at ends of branches, obovate to oblanceolate, c. 25 x 15 cm, apex mostly rounded (sometimes obtuse), base cuneate or rounded, with c. 10-20 main lateral veins on each side of the midrib, prominent below, margin entire. Petiole c. 5 cm long (but sometimes much longer). Stipules large.
OCCURRENCE: U2 and 4. A swamp forest tree, e.g. in South Maramagambo Forest and on the Ssese Islands, where it is very common.
NOTES: It is named in Luganda after the grey parrot (*enkusu*), which disperses its seeds.

Uapaca sansibarica **Pax (240a)** Phyllanthaceae

Tree to 15 m. Differing from *U. mole* in lacking stipules. Also, the leaves are smaller (10-15 cm x 4-7 cm).
OCCURRENCE: U1. Northern Uganda. In riverine forest, woodland, wooded grassland and bushland.

Spondianthus preussii **Engl. subsp. *glaber* (Engl.) J. Léonard & Nkounkou (241)** Phyllanthaceae

SYNONYM: *Spondianthus preussii* var. *glaber* (Engl.) Engl.
Butwa, Mimbiri, Muttambuzi (ga).
15 m. Tree with irregular trunk, branched from near base and a spreading, dense, crown. Stilt roots occasionally present. Bark thick and rough, vertically fissured, brown. Slash coarsely fibrous, red with orange lines, sometimes exuding red sap. Young leaves red. Leaves simple, alternate, elliptic or elliptic-ovate, c. 25 x 13 cm, apex obtuse or subacute, base cuneate or rounded, with c. 7-11 main lateral veins on each side of the midrib, veins prominent below (but not above), margin entire. Petiole c. 8 cm long. Fruit ovoid-ellipsoid, 1.5-2 cm long, 1.2-1.5 cm diameter.
OCCURRENCE: U1, 2 and 4. Swamp forest and other damp places.
NOTES: Most parts of the tree are very poisonous, which is probably the basis for its Luganda name (which means 'the one that kills goats'). The bark is used medicinally, e.g. as an antidote to snake bite and as a cure for cancer.

Tetrorchidium didymostemon **(Baill.) Pax & K. Hoffm. (242)** Euphorbiaceae

Mukejekeje (ga); Emunywamaizi (ki); Myakahoko (na).
12 m. Trunk straight, with branches usually at right angles and curving up. Bark fairly thin and smooth, slightly vertically fissured, light brown. Phellogen green. Slash white to light brown, with orange spots or streaks, usually exuding large quantities of brown sap when slashed hard. Leaves simple, alternate, c. 12 x 5.5 cm, obovate to elliptic, apex acuminate, base cuneate, with c. 7 main lateral veins on each side of the midrib, widest in upper half, apex acuminate, base cuneate, margin entire. Petiole c. 0.5 cm long.

OCCURRENCE: U2 and 4. Widely distributed. Common in Kayonza, Kalinzu and lake-shore forests, often on forest edges.

Key to *Bridelia.*

1. Young shoots fulvous or ferruginous hairy; stipules soon caducous. 2
   Young shoots not fulvous or ferruginous hairy; stipules not caducous. 3

2. Shoots remain tomentose with age. ....**244a. *B. ndellensis***
   Older shoots glabrescent (glabrous or nearly so). ....**244. *B. brideliifolia***

3. Branches usually with woody thorns; lamina not shiny above. .....**243. *B. micrantha***
   Branches without thorns; lamina shiny above. ....**244b. *B. atroviridis***

***Bridelia micrantha* (Hochst.) Baill. (243)** Phyllanthaceae

Katazamiti (ga); Kumuholang, Shigakara (gb); Omujimbu (ki); Margalgalyet (ku); Odugu-kulo (la); Kataza, Mujiji (na); Muhangwe (nl); Lulongamombe, Mulondongombe (sa); Mwesende (so); Mubalagaza (to).

15 m. Small tree with trunk branched near base and a dense crown. Spines often present on trunk. Stilt roots sometimes present. Bark fairly thick, vertically fissured. Slash fibrous, pink to red, turning darker. Young shoots sparingly pubescent. Leaves simple, alternate, mostly elliptic, c. 12 x 4.5 cm, apex elliptic or acuminate, base cuneate or rounded, with c. 9-16 main lateral veins on each side of the midrib, these veins being parallel to one another and fusing with a marginal vein, somewhat prominent above, prominent below, margin entire, glossy above. Petiole c. 8 mm long. Stipules 5-10 mm long. Fruit obovoid-subglobose or ellipsoid, 8-10 mm long, 5.6 mm wide.

OCCURRENCE: U1-4. Common on forest edges and in large clearings. Recorded up to 2150 m.

NOTE: The leaf venation is characteristic of the genus.

***Bridelia brideliifolia* (Pax) Fedde subsp. *brideliifolia* (244)** Phyllanthaceae

Omujimbu (ki); Muanza (ko); Kataza, Mujiji (na).

30 m. Young shoots sparingly to densely ferruginous pubescent, later turning glabrescent. Lamina (4-)6-15(-21) cm long, (2-)3-7(-10) cm wide, apex obtuse, subacute to shortly acuminate, base cuneate-rounded, truncate or shallowly cordate, lateral veins (11-)12-18(-20) on each side of the midrib, these not prominent above, prominent below. Petiole 5-10 mm long. Stipules soon caducous. Fruit ellipsoid to ovoid-ellipsoid, 7- 12 mm long, 4-7 mm wide.

OCCURRENCE: U1 and 2. Ankole, Kigezi, Tooro and Acholi. On forest edges.

NOTE: A similar-looking tree to *Bridelia micrantha*, with similarly shaped leaves and leaf venation. The young shoots differ in being covered with red-brown hairs.

***Bridelia ndellensis* Beille (244a)** Phyllanthaceae

SYNONYM: *Bridelia ferruginea* Benth.

Plate 19. Euphorbiaceae, Phyllanthaceae and others (240-252)

240. *Uapaca mole* 241. *Spondianthus preussii* 242. *Tetrorchidium didymostemon* 243. *Bridelia micrantha* 245. *Antidesma laciniatum* 246. *Antidesma membranaceum* 247. *Margaritaria discoidea* 248. *Phyllanthus inflatus* 249. *Thecacoris lucida* 250. *Cleistanthus polystachyus* 251. *Microdesmis puberula* 252. *Chaetachme aristata*

Actual sizes: leaves x 2.

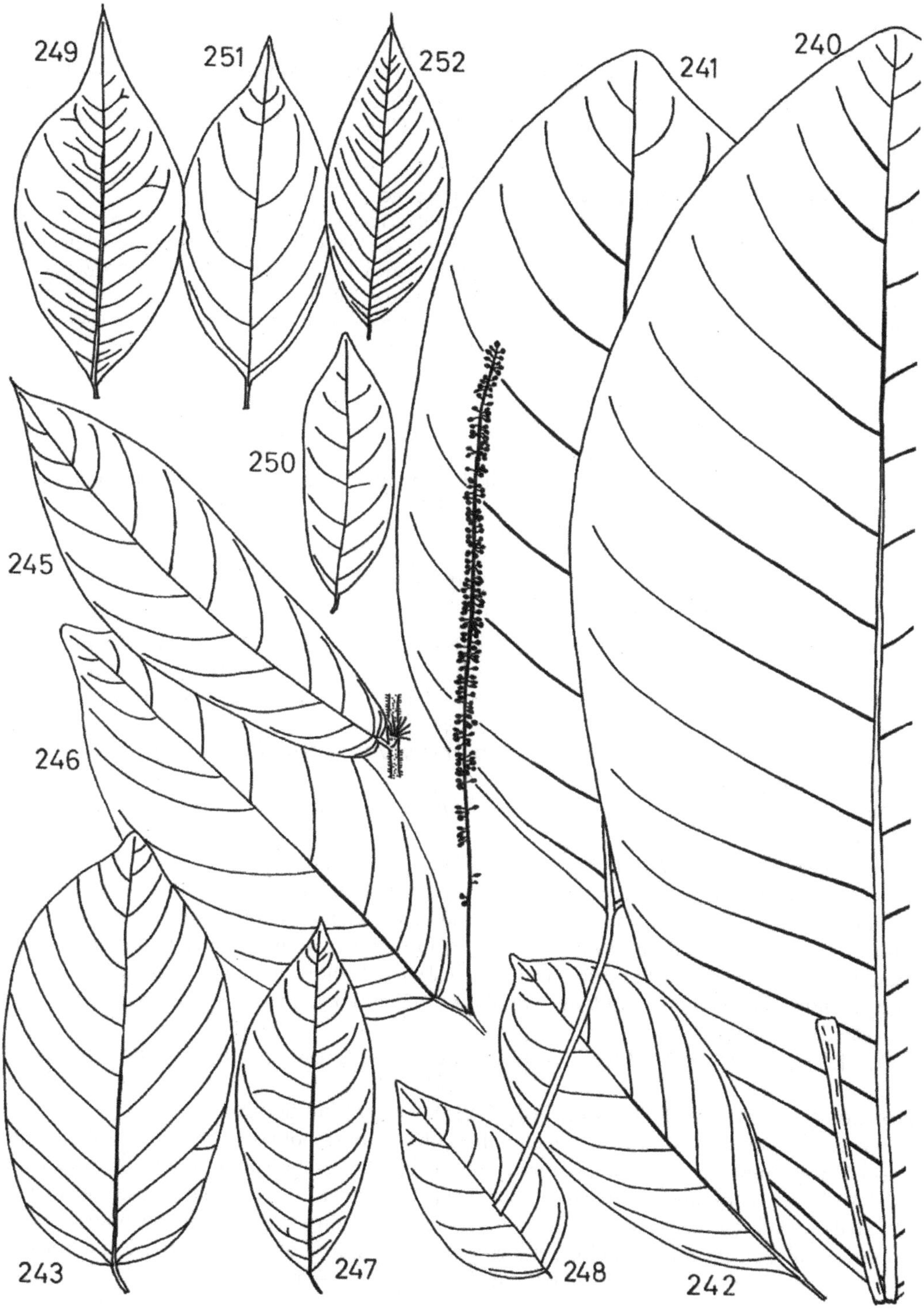

Plate 19. (240-252)

Much branched tree with a spiny trunk. Young shoots densely fulvous or ferruginous tomentose. Leaves simple, alternate, (5-)10-18 cm long, 4-8 cm wide, mostly elliptic, apex acuminate, base cuneate or rounded, lateral veins 9-13 on each side of the midrib, not prominent above, prominent below. Petiole 6-11 mm long. Stipules very quickly caducous. Fruit elliptic-oblong, 7-9 mm long, 5 mm wide.
OCCURRENCE: U1 and 2. In Budongo Forest and elsewhere. Evergreen forest, swamps, bushland, 1350-2000 m.

Bridelia atroviridis Müll. Arg. (244b) Phyllanthaceae
Much-branched shrub or tree to 12 m. Bark grey, reticulate. Young shoots sparingly pubescent, turning glabrescent with age. Leaves simple, alternate, (2-)6-11(-22) cm long, (1.5-)3-7(-10) cm wide, elliptic or oblanceolate, apex acuminate, base rounded-cuneate or rounded, lateral veins 10-22 on each wide of the midrib, these scarcely prominent above, fairly prominent below, lamina surfaces nearly glabrous, lamina shiny above, paler beneath. Petiole (2-)4-8 mm long. Stipules 5-8 mm long. Fruit obovoid-subglobose, 6-8 mm long, 5-6 mm wide.
OCCURRENCE: U2-4. On the edges of Budongo and Kyewaga forests.

Key to *Antidesma*.

1. Stipules divided into filiform or branched segments. ....**245. *A. laciniatum***
   Stipules simple and entire. 2
2. Leaf apex often rounded, obtuse, subacute or shortly acuminate; lamina sparingly pubescent to densely fulvous-ferruginous tomentose below. ...**245a. *A. venosum***
   Leaf apex usually distinctly acutely acuminate. 3
3. Leaves membranaceous, often pubescent below. ....**246. *A. membranaceum***
   Leaves coriaceous, often nearly glabrous both above and below (except along the midrib). ....**246a. *A. vogelianum***

Antidesma laciniatum Müll. Arg. subsp. _membranaceum_ (Müll. Arg.) J. Léonard (245) Phyllanthaceae
SYNONYM: *Antidesma laciniatum* var. *membranaceum* Müll. Arg.
Omusongi (no).
10 m. Understorey tree. Bark light brown, quite thin, vertically fissured. Slash fibrous, pink. Young stems covered with red-brown hairs. Leaves simple, alternate, elliptic-oblanceolate or elliptic-oblong, c. 16 x 6 cm, apex acuminate and mucronate, base rounded to slightly cordate, with c. 7-13 main lateral veins on each side of the midrib, veins slightly impressed above, prominent below, surface of midrib pilose both above and below, main veins also pilose below, margin entire. Petiole c. 0.6 cm long. Stipules c. 0.7 cm long, characteristically laciniate (cut into slender lobes).
OCCURRENCE: U2 and 4. Mengo and Bunyoro. Dense forests and edges of forest patches.

Antidesma venosum E. Mey ex Tul. (245a) Phyllanthaceae
Small tree, sometimes a straggler, branches drooping. Bark smooth or slightly fissured, flaking. Slash fibrous, pink-brown. Leaves simple, alternate, elliptic-obovate to oblong-oblanceolate, 4-14.3 x 2.9-7.6 cm, apex rounded, obtuse, subacute or shortly acuminate, base rounded or rounded-cuneate (sometimes cuneate), lateral veins 6-8 on each side of the midrib, these impressed above, prominent below, pubescent along midrib above (otherwise glabrous or sparingly pubescent), sparingly pubescent to densely fulvous- or ferruginous-tomentose below, shiny above, paler and dull below, margin entire. Petiole 3-7 mm long, pubescent to tomentose. Stipules simple, 4-8 mm long. Fruit 5-7(-8) mm long when dry.
OCCURRENCE: U1-4. Forest edges, riverine forest and associated savanna habitats.

Antidesma membranaceum Müll. Arg. (246) Phyllanthaceae
Kufora (gb).
10 m. Small tree with a spreading crown. Bark flaking. Slash fibrous, pink to yellow. Leaves simple, alternate, c. 16 x 8 cm (but sometimes rather smaller), apex acutely acuminate, base rounded to cuneate, with c. 5-12 main lateral veins on each side of the midrib, pubescent along midrib above (otherwise glabrous or sparingly pubescent), sparingly pubescent to densely fulvous- or ferruginous-tomentose below, margin entire, membranaceous (thin and semi-transparent).
OCCURRENCE: U1-4. In open forest types.
NOTE: *Antidesma venosum* and *A. vogelianum* Müll. Arg. are similar-looking.

Antidesma vogelianum Müll. Arg. (246a) Phyllanthaceae
9 m. Similar in nearly all respects to *Antidesma membranaceum*, except for the leaves that are coriaceous (leathery and tough) and usually nearly glabrous on both surfaces (except along the midrib).
OCCURRENCE: U2-4. Recorded from Budongo Forest and forests on the Ssese Islands. Forest edges, riverine forest and associated bushland.

Margaritaria discoidea (Baill.) G.L. Webster (247) Phyllanthaceae
SYNONYM: *Phyllanthus discoideus* (Baill.) Müll. Arg.
Otego (ac); Amakeke (am); Erionoi (at); Kamenyambazzi (ga); Lakamakambugo, Mutaigumbwa (gw); Omuhahara, Omukare (ki); Atego, Atigo (la); Odzeki (md); Muremamparigo (na); Kakazi, Katunganfulu, Luka (so).
Commonly to 5 m (exceptionally 25 m). Tree usually branched near base, with a spreading, deciduous crown with layered branches. Branches on young trees at right angles to the trunk. Branchlets pendulous. Bark thin and smooth, slightly vertically fissured, becoming thick and rough, fibrous, with large vertical fissures. Slash fibrous, pink, usually with orange streaks. Leaves simple, alternate, rather variable in size and shape (from elliptic-lanceolate to suborbicular-obovate), often c. 10 x 4 cm, apex acutely acuminate to rounded, often with a small mucronate tip, base cuneate to rounded, margin entire, translucent when held up to the light, with c. 10-16 main lateral veins on each side of the midrib, veins scarcely prominent above, a bit more prominent below. Petiole c. 0.6 cm long. Stipules linear-lanceolate, 2-5 mm long. Fruit a 3-valved subglobose capsule, c. 0.6 cm across.
OCCURRENCE: U1-4. Common on forest edges and in secondary forest. Also in thickets.
CULTIVATION AND PROPAGATION: Fast-growing. Can be grown in drier areas. Coppices easily and provides good quality poles. Collect the fruits near mother trees and crack open to obtain the seeds. Soak seeds for 2-3 hours in warm water before sowing.
NOTES: (1) The translucent leaf margin is a good character. (2) Three of the four East African varieties occur in Uganda, thus: (a) var. *nitida* (Pax) Radcl.-Sm. (with distal stipules usually 2-5 mm long); (b) var. *discoidea* (with distal stipules usually 5-10 mm long; newly formed shoots and petioles usually puberulous or pubescent); (c) var. *fagifolia* (Pax) Radcl.-Sm. (with distal stipules usually 5-10 mm long; newly formed shoots and petioles glabrous or only sparingly puberulous).

Phyllanthus inflatus Hutch. (248) Phyllanthaceae
SYNONYM: *Phyllanthus polyanthus* Pax (*sensu* ITU)
Mufuulanjuba (ga).
8 m. Small straggling understorey, sparingly spiny, tree. Branchlets resemble pinnate leaves. Bark grey. Slash red. Leaves simple, alternate, oblong-oblanceolate, apex acute to acuminate, rounded or truncate, base rounded or rounded-cuneate, c. 6.5 x 3.5 cm (but variable in size), often glabrous on both surfaces, lateral veins 7-10 on each side of the midrib, veins usually indistinct above, slightly more distinct below. Petiole c. 0.2 cm long. Fruit c. 2.5 cm diameter, inflated and bladder-like.
OCCURRENCE: U2 and 3. Recorded from Kalinzu, Budongo and Kasyoha-Kitomi forests.

NOTE: Easily distinguished by the young branches which simulate pinnate leaves. These branches (simulating pinnate leaves) have c. 10-15 leaves (simulating leaflets).

Thecacoris lucida **(Pax) Hutch. (249)** Phyllanthaceae

Bondabor (am).
10 m. Understorey tree with a spreading crown. Leaves simple, alternate, rather stiff, elliptic, c. 9.5 x 4 cm, apex acuminate with a small mucronate tip, base rounded or rounded-cuneate, margin entire, main lateral veins c. 10 on each side of the midrib, these veins not readily distinguishable from the secondary lateral veins on the lower surface of the leaf and often difficult to see on the upper surface. Petiole c. 0.6 long. Fruit trilobate, 0.5-0.7 cm diameter.
OCCURRENCE: U2 and 4. Common understorey tree beneath *Cynometra*, e.g. in Budongo, Bwamba and Maramagambo forests.

Cleistanthus polystachyus **Planch. (250)** Phyllanthaceae

Muhindi (to).
Understorey tree to 15 m. Trunk badly shaped, branching from near base, sometimes deeply fluted. Bark rough and fibrous, flaking. Slash fibrous, pink to red. Leaves simple, alternate, rather leathery and thick (coriaceous), glabrous on both surfaces (except for a few scattered hairs near the base of the midrib), mostly elliptic, c. 11 x 4 cm, apex acuminate, sometimes mucronate, base cuneate or rounded, main lateral veins c. 5-8 on each side of the midrib, veins indistinct above, slightly more distinct below. Petiole c. 0.4 cm long. Fruit trilobate-subglobose, 1.1-1.3 cm in diameter.
OCCURRENCE: U1, 2 and 4. Widely distributed tree, uncommon. In evergreen, riverine and semi-swamp forest.

Microdesmis puberula **Planch. (251)** Pandaceae

Understorey tree to 6 m. Leaves simple, alternate, c. 11 x 4 cm, with c. 5-8 main lateral veins on each side of the midrib, midrib running out into a mucronate tip, apex acute, base unequal-sided, margin entire or crenate. Petiole c. 0.5 cm long.
OCCURRENCE: U2. Only recorded from Budongo, Kibale and Rwoho forests.

Chaetachme aristata **Planch. (252)** Ulmaceae

Omubambanjobe (no); Mulere (sa).
10 m. Wide-spreading understorey tree, branches often zigzagging and arising from near base, often with several stems, usually with sucker shoots. Sucker shoots and young stems armed with spines which are sometimes branched. Bark often thick and rough, vertically fissured, light brown. Slash fibrous, yellow, turning darker and rather green. Leaves simple, alternate, leathery and thick, elliptic, c. 9 x 3.5 cm, apex acuminate and mucronate (at least on some leaves), base cuneate to rounded (or slightly cordate), unequal-sided, margin entire, upper surface glabrous, lower surface glabrous to densely pubescent, main lateral veins not easily distinguishable from secondary lateral veins. Petiole c. 0.5 cm long. Fruit a yellow drupe, c. 1.25 cm diameter, bearing two persistent styles.
OCCURRENCE: U1-4. Mainly on forest edges and in secondary forest, also in more open types of mature forest.

Pittosporum viridiflorum **Sims (255)** Pittosporaceae

SYNONYMS: *Pittosporum mannii* Hook. f.; *Pittosporum spathicalyx* De Wild. (tree numbers 253 and 254 in UFT)
Lapingyek (al); Omushekyera (ki); Chemwororia (ku); Mubaruka (na).
15 m. Shrub or small tree, with branches from near base. Bark thin, smooth, light brown, with prominent lenticels. Slash whitish, scented, slowly turning greenish. Leaves crowded at ends of branches, simple, alternate, c. 14 x 6 cm (c. 9.5 x 3 cm on Mt Elgon and the Karamoja mountains), widest in upper part, spatulate, obovate or oblanceolate, apex various, base cuneate, with c. 6-9 main lateral veins on each side of the midrib, venation inconspicuous on

upper surface, vein reticulum prominent on undersurface, glabrous. Petiole c. 2 cm long. Fruit a small capsule, dehiscing in two and exposing the red seeds.
OCCURRENCE: U1-4. Found up to 2400 m. On forest edges and termite mounds. Common in lake-shore forests and on Mt Elgon and the Karamoja mountains. Found in scrub and secondary forest in Echuya, Mafuga and Bwindi forests.
CULTIVATION AND PROPAGATION: Fast-growing, including on poor soils. Can provide shade for slower-growing species. Can be pruned and coppiced for firewood. Collect fruits from mother trees, remove seeds by hand and sow as soon as possible.

Pittosporum abyssinicum Delile (255a) Pittosporaceae
SYNONYM: *Pittosporum lanatum* Hutch. & E.A. Bruce
19 m. Leaves crowded at ends of branches, simple, alternate, obovate to oblanceolate, 6-10 x 2-3.5 cm, apex acuminate, upper surface glabrous, lower surface fulvous-tomentose, lateral veins indistinct on both surfaces, margin entire.
OCCURRENCE: U1. Karamjoja.
NOTE: This species differs from *Pittosporum viridiflorum* in having mature leaves that are densely fulvous tomentose below (those of *P. viridiflorum* are glabrous or only thinly pubescent).

Peddiea fischeri Engl. (256) Thymelaeaceae
Omushinya (ki).
10 m. Understorey tree, with branches from near base of trunk. Bark thin, smooth, brown. Slash pale pink to red. Leaves simple, alternate, lanceolate or elliptic, c. 11 x 4 cm, widest in about the centre, apex acute, base cuneate, venation rather indistinct on upper surface. Petiole c. 0.2 cm long. The claw-shaped scales that terminate the shoots are characteristic. Flowers bell-shaped, borne in axillary umbels. Perianth segments greenish.
OCCURRENCE: U2 and 4. Ascending to 2400 m.

Peddiea rapaneoides Engl. (256a) Thymelaeaceae
10 m. Much branched tree. Leaves simple, alternate, elliptic to lanceolate, 5-9 cm long, 2-4 cm wide, apex obtuse to acute, base cuneate, somewhat thick.
OCCURRENCE: U2-4. Upland forest. Recorded from Gahinga-Sabinio saddle.
NOTE: The fruit of this species is glabrous (that of *P. fischeri* is hairy at the top).

Dicranolepis incisa A. Robyns (256b) Thymelaeaceae
Small tree to 3 m. Leaves simple, alternate, oblong, 4-8 cm long, 2-3 cm wide, apex acuminate, base unequal-sided, glabrous above, sparsely pubescent below, midrib and main lateral veins prominent below. Petiole 1-3 mm long, glabrous or pubescent.
OCCURRENCE: U2-4. Recorded from Budongo and Sango Bay forests. A range-restricted species known from Uganda and eastern D.R. Congo.
CONSERVATION STATUS: Global NE (IUCN), VU (TOU); National NE.

Dicranolepis buchholzii Engl. & Gilg (256c) Thymelaeaceae
Small tree to 4 m, very similar to *Dicranolepis incisa*. Leaves simple, alternate, oblong or ovate-oblong, 4-8 cm long, 2-3 cm wide, apex acuminate, base unequal-sided, glabrous on both surfaces or slightly hairy below. Petiole 1-4 mm long, glabrous or pubescent.
OCCURRENCE: U2. Only recorded from Ishasha Gorge in Bwindi-Impenetrable Forest.
NOTE: The calyx tube of this species is less than 1.5 cm long (compared to 2-3 cm in *D. incisa*).

Erythroxylum fischeri Engl. (257) Erythroxylaceae
8 m. Understorey tree with a straight trunk and conical crown with numerous branches. Bark light brown, quite thick, rough, fissuring horizontally and vertically to give a pattern of small rectangles. Slash fibrous, pink, turning darker. Young shoots strongly flattened, oval in cross-section. Leaves simple, alternate, rather thick and leathery, elliptic to oblong, c. 14 x 5.5 cm,

apex acute to shortly acuminate, base cuneate. Petiole c. 0.8 cm long. Stipules quite large, placed between the petiole and the stem (intrapetiolar). Fruit an oblong drupe to 1.9 cm long.
OCCURRENCE: U1-4. Mainly in open types of forest. Abundant in W. Maramagambo.
NOTES: Easily recognized by the intrapetiolar stipules. The leaf shape is sometimes similar to that of *Peddiea* (256).

Morella salicifolia (A. Rich.) Verdc. & Polhill (258) Myricaceae
SYNONYM: *Myrica salicifolia* A. Rich.
Ekijeeje, Omugyegye, Omujeje (ki); Mukikembo, Mundrindi (ko); Segatetit (ku); Kiberassia, Maruss (ms).
10 m. Bark rough, dark-coloured. Slash red-brown with white lines. Leaves simple, alternate, very variable in shape and margin, varying from long and thin (c. 8 x 2 cm) to rather rounded (c. 3.5 x 1.75 cm), apex pointed to rounded, base mostly cuneate to truncate (or slightly cordate and unequal-sided), margin entire to toothed (in upper part), undersurface covered with small yellow glands. Petiole c. 1 cm long. Fruit elliptic to subglobose, 3-4 mm long, 2-4 mm wide.
OCCURRENCE: U1-3. A tree of montane forest and woodland, 2000-3000 m. Particularly common in secondary forest and derived woodland at 2000-2500 m.
NOTE: This species is represented by two subspecies in Uganda: (1) subsp. *salicifolia* (with leaf blades mostly more than 7 cm long) and (2) subsp. *mildbraedii* (Engl.) Verdc. & Polhill (with leaf blades less than 7 cm long).

Faurea wentzeliana Engl. (259) Proteaceae
SYNONYM: *Faurea saligna* Harv. (*sensu* UFT)
Omurengyere (ki); Mukaka (ko); Maiyokwo, Moyokwo (ku); Morororia (ms).
20 m. Trunk fairly straight. Bark thick and rough, dark brown to almost black. Slash fibrous, pink. Leaves simple, alternate, thick and leathery, lanceolate or elliptic, often c. 12 x 2.25 cm, with red veins, apex mucronate, base cuneate, margin undulate, midrib on older leaves entirely glabrous except for a few hairs near the base when young. Petiole 1-1.5 cm long. Flowers in spikes. Seeds hairy.
OCCURRENCE: U1-3. Montane forest, 1200-3000 m. Common on ridges in Bwindi Forest.

Protea caffra Meisn. subsp. _kilimandscharica_ (Engl.) Chisumpa & Brummitt (260)
Proteaceae
SYNONYM: *Protea kilimandscharica* Engl.
Small tree to 5 m. Leaves simple, alternate, thick and leathery, linear-lanceolate, c. 12 x 2 cm, base cuneate, margin distinctly undulate in lower half. Flowers borne in conspicuous large heads, c. 5 cm diameter, with stiff bracts.
OCCURRENCE: U1 and 3. Found on forest edges on rocky sites in the Ericaceous Belt on Mt Elgon and near the summit of Mt Kadam.

Agarista salicifolia (Lam.) G. Don (261) Ericaceae
SYNONYM: *Agauria salicifolia* (Lam.) Oliv.
Musegewa, Musengulu (ki).

Plate 20. Various families (253-266)

253-255. *Pittosporum viridiflorum* 256. *Peddiea fischeri* 257. *Erythroxylum fischeri*
258. *Morella salicifolia* 259. *Faurea wentzeliana* 260. *Protea caffra*
261. *Agarista salicifolia* 262. *Rapanea melanophloeos* 263. *Euclea schimperi*
264. *Nuxia congesta* 266. *Premna angolensis*

Actual sizes: leaves x 2.

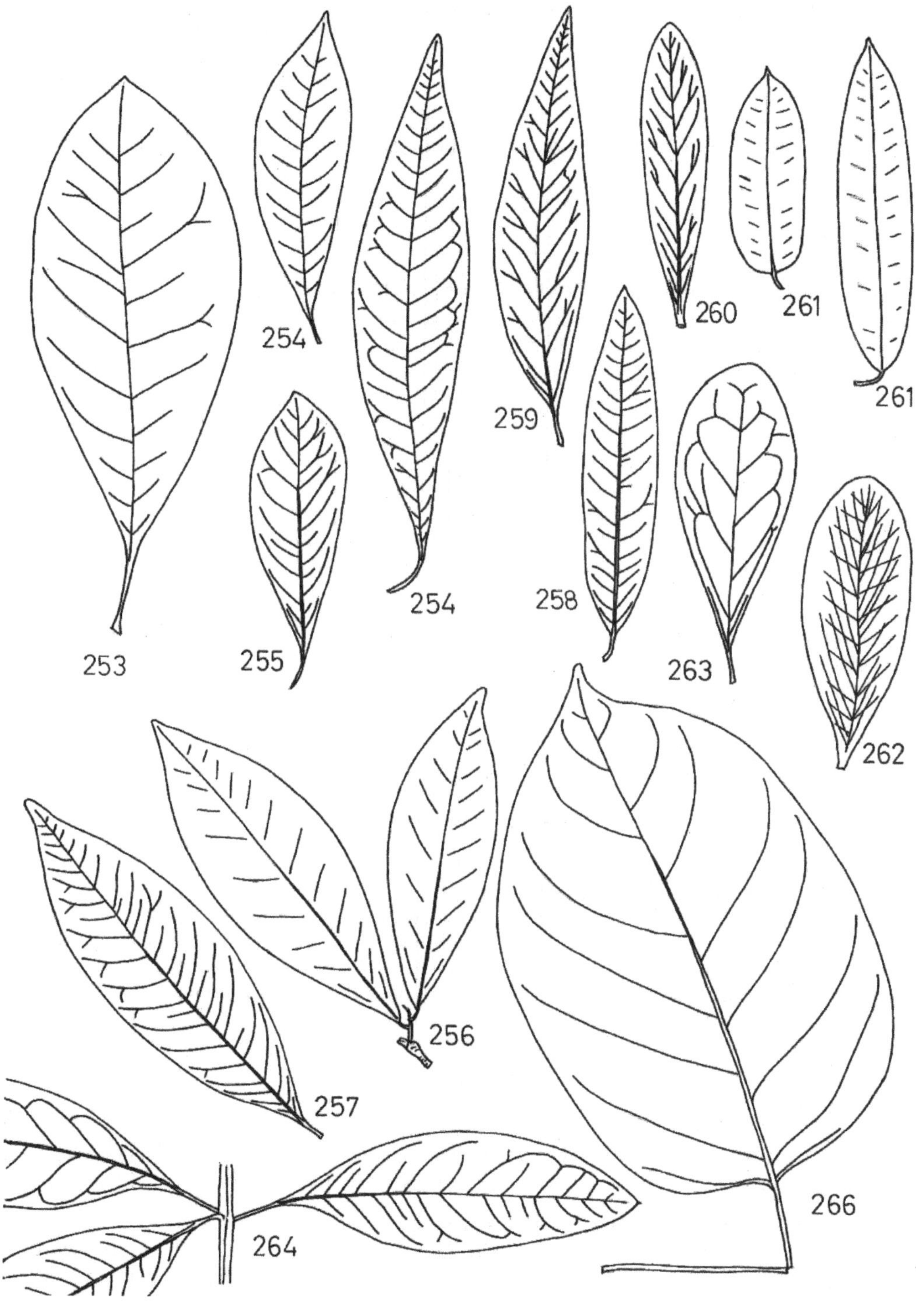

Plate 20. (253-266)

10 m. Small tree. Bark very thick and rough, fissured. Slash red to red-brown, sometimes with white lines. Leaves simple, alternate, mostly elliptic, c. 9.5 x 2 cm (but very variable in size), apex rounded or acuminate, base cuneate to subcordate, thick and leathery, margin entire, venation very indistinct on upper surface, glabrous except for the midrib below. Petiole 0.5-1 cm long, glabrous or pubescent. Fruit a capsule, 0.4-0.7 cm long, with persistent style.
OCCURRENCE: U1-3. In montane forest (both moist and dry), secondary forest and on forest edges, 1800-3200 m.
NOTE: A number of forms are recognized varying in leaf size, flower colour and hairiness of branchlets.

***Rapanea melanophloeos* (L.) Mez (262)** Primulaceae

SYNONYM: *Rapanea rhododendroides* (Gilg) Mez
Omukoni (ki); Musongonyonye (ko); Sitoto (ku); Mugaita, Mulimangombe (tn).
5 m. Trunk straight and cylindrical. Bark whitish, with small fissures. Slash pink with brown lines, granular. Leaves simple, alternate, clustered at ends of branches in pseudo-whorls, elliptic, oblong or obovate, c. 8.5 x 3 cm (sometimes considerably bigger), apex obtuse to shortly acuminate, base tapering into petiole, margin entire, glabrous, glossy above, midrib red when young, with numerous translucent streaks. Petiole 2-11 mm long.
OCCURRENCE: U1-4. Mainly in montane forest, 2300-3500 m. A dominant species of the *Hagenia-Rapanea* Zone (c. 3000-3200 m). In swamp forest at c. 2000 m in Kigezi and in Jubiya Forest in Masaka (1140 m).

***Euclea schimperi* (A. DC.) Dandy (263)** Ebenaceae

SYNONYMS: *Euclea racemosa* Murray subsp. *schimperi* (A. DC.) F. White; *Euclea latidens* Stapf (*sensu* ITU and UFT)
Emus (at); Sitsantsassi (ms).
Small tree to 15 m. Bark fairly thin, dark brown, with small vertical fissures. Slash fibrous, red to pink. Leaves simple, alternate, rather thick and leathery, variable in shape and size, often obovate to oblanceolate, c. 8 x 3 cm, widest in upper half, apex rounded, base cuneate, margin entire (sometimes undulate), glabrous below. Petiole c. 0.4 cm long. Fruit 6-8 mm long in diameter, glabrous.
OCCURRENCE: U1-4. On forest edges.

***Euclea divinorum* Hiern (263a)** Ebenaceae

Small tree to 9 m. Bark rough, grey-brown, with longitudinal fissures, flaking. Slash dark red/pink to crimson, pale yellow to orange towards wood. Leaves simple, subopposite, often rhombic, variable in size, widest near the middle, 1.6-12 cm long, 0.6-4.5 cm wide, apex rounded or emarginate, base attenuate to rounded, margin usually wrinkled, glabrous except for rusty peltate scales below. Petiole. c. 0.4 cm long. Fruit globose, 0.5-0.7 cm in diameter, with short white hairs.
OCCURRENCE: U1-4. On forest edges and in secondary forest.

Key to *Nuxia*.

1. Leaves entire; fruit barely longer than the calyx, densely hairy. ....**264. *N. congesta***
   Leaves usually distantly denticulate or bluntly serrate (rarely entire). 2
2. Leaf apex acute or acuminate; fruit nearly twice the length of the calyx.
   ....**265. *N. floribunda***
   Leaf apex rounded and usually mucronulate; fruit barely longer than the calyx, hirsute.
   ....**265a. *N. oppositifolia***

Nuxia congesta **Fresen. (264)** Stilbaceae

Umwesa (fu); Omubuzije (ki); Chorowa (ku).

15 m. Trunk irregular. Bark rather thin and smooth, sometimes flaking. Slash whitish to green, turning darker. Leaves usually in whorls of 3, elliptic to obovate, c. 10 x 4 cm (but variable in size), apex acute, rounded, often with a mucro, base cuneate, margin of mature leaves entire, glabrous. Fruit barely longer than the calyx, densely hairy.

OCCURRENCE: U1-3. In open places in montane forest (including in Kigezi and Tooro and on Mt Elgon and the Imatong Mts), particularly in secondary forest, 1500-2800 m.

Nuxia floribunda **Benth. (265)** Stilbaceae

Omubuzije (ki).

20 m. Bark brownish-grey. Leaves elliptic, to 14 cm long, 2-4 cm wide, apex acute or acuminate, base cuneate, margin distantly denticulate to entire, glabrous. Fruit nearly twice the length of the calyx, glabrous.

OCCURRENCE: U2. Only recorded from Kigezi, where it is common in secondary forest and derived woodland between 1600 and 2400 m.

NOTE: Similar looking to *Nuxia congesta*, but leaves differing in sometimes being toothed (rather than entire). It is not always easy to distinguish between the two species in the vegetative state. However, the inflorescences of the two species are different, that of *N. congesta* being a dense, terminal cyme or corymb up to 12 cm across and that of *N. floribunda* being a lax, repeatedly dichotomous cyme, up to 25 cm across.

Nuxia oppositifolia **(Hochst.) Benth. (265a)** Stilbaceae

12 m. Bark reddish brown. Leaves elliptic, to 12 x 2.5 cm, apex rounded and usually mucronulate, base cuneate, margin bluntly serrate (rarely entire), glabrous. Fruit barely longer than the calyx, hirsute.

OCCURRENCE: U1. Riverine forest.

Premna angolensis **Gürke (266)** Lamiaceae

Mutala (ga); Baniamunkiro (ki); Muhororo (na, to); Nkubwe (na); Mukomati (sa).

25 m. Trunk crooked, bearing a spreading crown with fairly large leaves. Bark light brown, of medium thickness, rough with vertical fissures. Slash soft, whitish, with small golden-yellow to yellow-brown lines. Leaves in whorls of 4 (occasionally 3). Lamina ovate, oblong or elliptic, c. 15 x 10 cm, apex acuminate, base rounded, cuneate or subcordate, main lateral veins c. 5-8 on each side of the midrib, glabrous above, pubescent below on main veins. Petiole c. 8 cm long. Fruits borne in clusters, globose, fleshy, blue.

OCCURRENCE: U1-4. On forest edges and in open forest. Widely distributed.

NOTES: Easily recognized by the whorled leaves with long petioles. Leaves usually severely damaged by insects.

APOCYNACEAE

A family easily recognized by the opposite or whorled, simple leaves and the presence of white latex in young stems. With the exception of older trees of *Pleiocarpa*, *Rauvolfia* and *Voacanga*, white latex is also present in the slash. Bark usually smooth. Slash usually white, orange or yellow, or a combination of these colours. The flowers are usually white or yellow, large and fragrant. The fruits are typically paired, often indehiscent (e.g. *Tabernaemontana*) or dehiscent and producing plumed seeds (e.g. *Funtumia*). *Rauvolfia* has small berries. Many large forest climbers belong to this family. A well-known example is *Landolphia owariensis* P. Beauv., which produces attractive, fragrant, white flowers and which is abundant on sandy soils near Lake Victoria. This climber, as well as the tree *Funtumia elastica*, are sources of high-quality rubber latex. The first botanical investigation in Uganda, made in 1905 by Mr. M.T. Dawe, was a survey of the indigenous rubber resources of the country.

Key to Apocynaceae.

1. Tree repeatedly dichotomously branched. 2
   Plant not repeatedly dichotomously branched. 3

2. Latex present in most parts. ....273-276. ***Tabernaemontana***
   Latex in bark and branchlets; small tree of damp places. ....278-278a. ***Voacanga***

3. Leaf blade with domatia beneath; trunk typically very straight. **271-272. *Funtumia***
   Leaf blade without domatia. 4

4. Colleters present in leaf axils; many parts not glabrous. 5
   Colleters absent from leaf axils; all parts glabrous except inside corolla tube. 6

5. Leaf blade coriaceous; secondary veins 25-50 pairs; tall tree with straight trunk.
   .... **267. *Alstonia***
   Leaf blade not coriaceous; secondary veins 6-35 pairs; small trees.
   ....**268-269a. *Rauvolfia***

6. Leaves always opposite; base rounded to obtuse. .... **277. *Picralima***
   Leaves opposite or whorled; base cuneate or decurrent into petiole. **270. *Pleiocarpa***

***Alstonia boonei* De Wild. (267)** Apocynaceae

Kigima (am); Mubajjangalabi, Musoga (ga); Omujwa (no); Nsiwa (so); Cheese wood, Mujwa, Pattern wood, Stool wood (tn).

40 m. Straight trunked, deciduous tree, with a spreading, dark-coloured crown. Branches whorled on young trees. Trunk very deeply fluted, the folds resembling those of a curtain. Bark thin to moderately thick, light brown, fairly smooth to granular, with prominent lenticels, sometimes with small vertical fissures. Slash granular, white to yellow, sometimes with orange stone cells, producing abundant white latex. Leaves whorled, often 6 or 8 in a whorl, c. 20 x 5.5 cm, widest in upper half, apex acuminate (rarely obtuse or retuse), decurrent into petiole, with many prominent lateral veins on each side of the midrib, thick and leathery. Fruit c. 20-40 cm long, with wind-borne seeds with tufts of hairs at both ends.

OCCURRENCE: U2-4. Widespread. Abundant in Budongo and Bugoma forests and frequent in valley forests in Mabira Forest. Rare in Bukedi and in Kalinzu Forest. A light-demanding species.

CULTIVATION AND PROPAGATION: Fast-growing. Prefers moist places, but can grow on well-drained soils and on drier hillsides. Can be planted alone or in mixed stands. Preferably collect seeds from fruits before they split. Soak seeds for 24 hours before planting. Scarification is reported to increase the germination rate. Germination in 2-4 weeks.

NOTE: The timber is soft and light.

Key to *Rauvolfia*.

1. Leaf apex apiculate; at least some inflorescence branches puberulous.
   .... **269. *R. vomitoria***
   Leaf apex acute, acuminate or cuspidate; inflorescence branches glabrous. 2

Plate 21. Apocynaceae (267-273); see also Plate 22

267. *Alstonia boonei* 269. *Rauvolfia vomitoria* 272. *Funtumia elastica*
273. *Tabentaemontana pachysiphon*

Actual sizes: leaves and fruits x 2; trunk base x 80; tree profiles x 800.

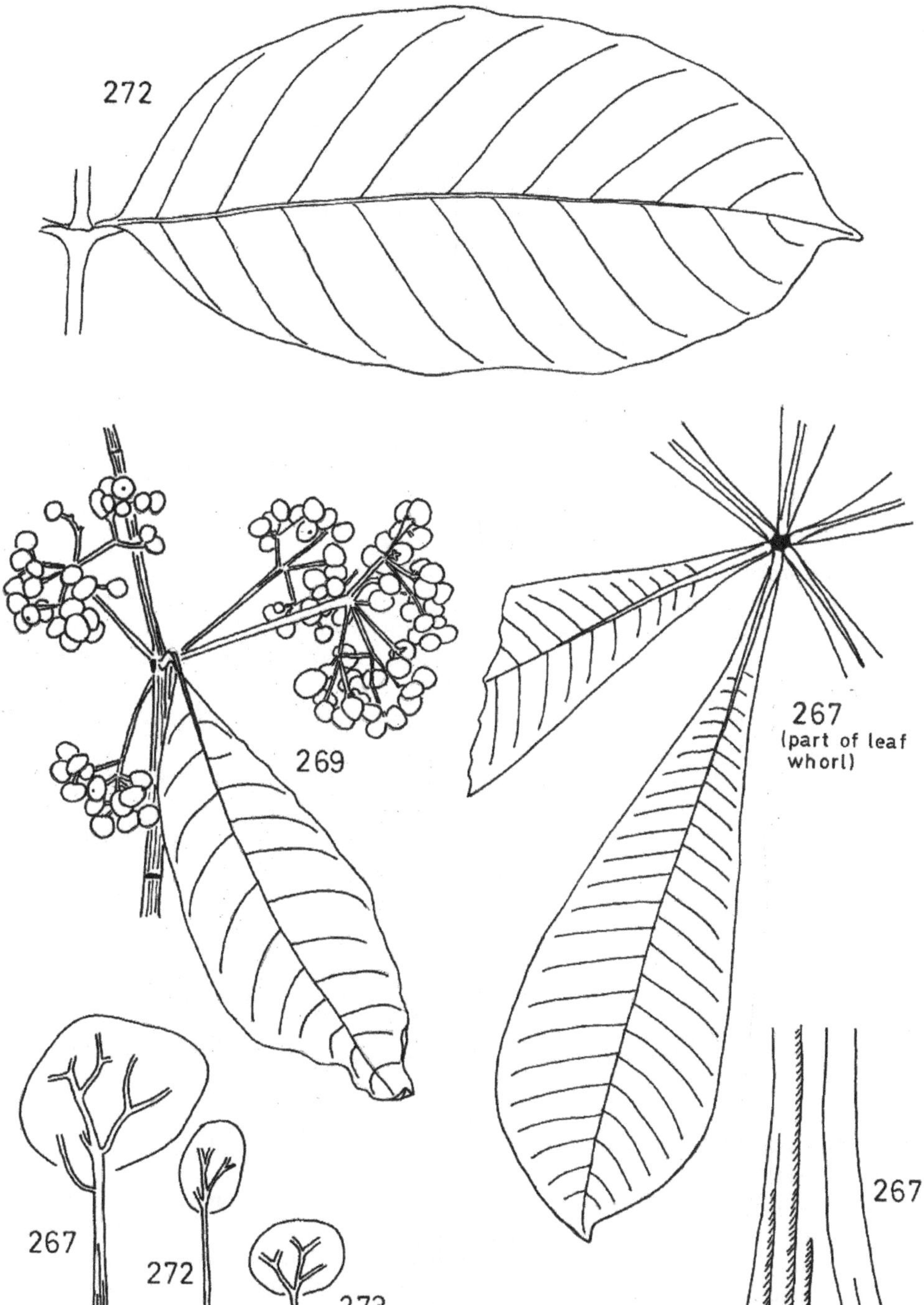

Plate 21. (267-273)

2. Bark very thick; leaf apex acute. ....**268. *R. caffra***
Bark scaly and peeling, not thick; leaf apex cuspidate. ....**269a. *R. mannii***

Rauvolfia caffra Sond. (268) Apocynaceae
SYNONYM: *Rauvolfia oxyphylla* Stapf
Kisalako (am); Munyamazzi (ga); Mutongo (ko).
20 m. Tree with a spreading, umbrella-shaped, crown. Bark quite thick, light brown to greenish, with prominent lenticels, rather granular, flaking. Slash white to yellow, with orange to brown markings, without latex on older trees (latex is present in young parts). Leaves in whorls of 3-6, leaf-whorls crowded together on stout branches. Leaves elliptic, c. 25 x 6.5 cm, widest in upper half, apex acute, base cuneate or attenuate, sometimes decurrent into petiole, with 12-35 main lateral veins on each side of the midrib, midrib yellow, lamina glossy and glabrous
OCCURRENCE: U1-4. Widespread. In riverine forest and open swamp forest.
CULTIVATION AND PROPAGATION: Moderately fast-growing. Grows best near water. Requires shade when young and full light once established. Can be grown in pure or mixed stands. Suitable for land reclamation. Collect fruits from mother trees or from the ground beneath. Extract the seeds manually and sow the seeds as soon as possible.
NOTE: The leaves are larger and the leaf-whorls more crowded than those of *Rauvolfia vomitoria*.

Rauvolfia vomitoria Afzel. (269) Apocynaceae
Kamwanyimwanyi (ga); Kawule (so).
10 m. Small untidy tree, branching from near base, with an open crown. Bark thin, dark brown. Slash white and yellow, without white latex on older stems. Leaves usually in whorls of 3 or 4, the leaf-whorls being borne on slender branches and not crowded together. Leaves c. 13 x 4.5 cm, apex apiculate, base cuneate to attenuate, lateral veins 8-17 on each side of the midrib, glabrous. Petiole c. 2 cm long. Fruit a berry, ripening to yellow and finally red.
OCCURRENCE: U2-4. On forest edges and in young secondary forest. Common in Mengo and in Budongo Forest. Recorded from Kayoha-Kitomi Forest.
NOTES: The Luganda name refers to its resemblance to the coffee tree. The root bark contains a valuable drug.

Rauvolfia mannii Stapf (269a) Apocynaceae
Shrub or tree to 8 m. Bark scaly, peeling, lenticellate. Leaves in whorls of 3-6 (but opposite at some nodes), elliptic or slightly obovate, to 25 cm long and 10 cm wide, apex acuminate or cuspidate, base cuneate, with 6-19 main lateral veins on each side of the midrib, glabrous. Petiole to 2.5 cm long, glabrous.
OCCURRENCE: U2. Recorded from Ishasha Gorge in Bwindi Impenetrable National Park. Moist forest, 2150-2250 m.

Pleiocarpa pycnantha (K. Schum.) Stapf (270) Apocynaceae
Mutoma, Nyakatoma (na).
10 m. Understorey tree with a short trunk. Bark smooth, dark-coloured. Slash white, with scattered yellow-brown lines, no latex in older stems. Leaves simple, opposite or in threes, elliptic to oblong, c. 11 x 3.5 cm, apex acute to acuminate, base cuneate or decurrent into petiole, dark green above, paler below. Petiole 0.5-2 cm long. Flowers white, fragrant or not. Berries yellow, c. 2 cm long.
OCCURRENCE: U2 and 4. Recorded from Kalinzu Forest and common in Kibale Forest. Once probably common on the Ssese Islands, but now less so due to deforestation for oil palm growing.

Funtumia africana **(Benth.) Stapf (271)** Apocynaceae

Bastard wild rubber (en); Nkago (ga, so); Nnamukago (ga); Ekinyamaate (ki); Munyamatunga, Nyamukago (na); Omusanda (no); Mujwamata (to).
30 m, but usually, less. Trunk straight and cylindrical, bearing a small, narrow, dark-coloured crown. Bark brown to dark-coloured, thin, with small vertical fissures, becoming granular on old trees, general effect smooth. Slash granular, orange/yellow to white, exuding copious white latex which does not coagulate into balls when rubbed between the fingers, but remains sticky. Leaves simple, opposite, ovate or elliptic, c. 20 x 9 cm, apex shortly acuminate, base cuneate, main lateral veins c. 8-14 on each side of the midrib, margin wavy. Usually, there are no pits (domatia) in the axils of the main lateral veins beneath. Flowers yellow-white and fragrant. Fruit up to 30 cm long, containing wind-borne seeds with hairs mainly at one end.
OCCURRENCE: U1-4. Common second storey species. In moist, riverine and swamp forest.
CULTIVATION AND PROPAGATION: Fast- and straight-growing. Can be planted as single specimens or in pure or mixed stands. Preferably collect fruits on mother trees before opening (as the seeds are very widely dispersed, once released). Sow seeds as soon as possible.
NOTES: The latex is useless as a source of rubber. Can be difficult to distinguish from *Funtumia elastica* in the field.

Funtumia elastica **(P. Preuss) Stapf (272)** Apocynaceae

African wild rubber, Lagos rubber tree (en); Nkago, Nnamukago (ga); Omusanda (no).
30 m. Tree similar to *Funtumia africana*, differing in the following respects: when rubbed between the fingers, the latex coagulates into balls and comes away cleanly from the skin; there are usually pits (domatia) in the vein axils on the underside of the leaves.
OCCURRENCE: U2 and 4. Generally a rarer tree than *Funtumia africana*, but relatively abundant in Budongo, Bugoma and Mabira forests. Recorded from forest in Semuliki National Park.
NOTES: The latex makes a high-quality rubber. The tree was tapped during the Second World War. The Runyoro name refers to the latex.

Key to *Tabernaemontana.*

1. Leaf petiole at least 3 cm long. 2
   Leaf petiole not more than 2.5 cm long. 3
2. Lamina with scattered dots below. .... **274. *T. stapfiana***
   Lamina without scattered dots below. .... **275. *T. ventricosa***
3. Leaf length nearly double its width. .... **273. *T. pachysiphon***
   Leaf length nearly thrice its width. .... **276. *T. odoratissima***

Tabernaemontana pachysiphon **Stapf (273)** Apocynaceae

SYNONYMS: *Conopharyngia holstii* (K. Schum.) Stapf; *Tabernaemontana holstii* K. Schum.
Bbeerelyankima, Kitwekyankima (ga); Ekinyamagosi (ki); Ikidehe (na); Mungogwenkende, Mwogogwenkende (to).
10 m. Spreading understorey tree with a wavy trunk and dense crown of large, dark-coloured, leaves. Bark thin, fairly smooth, light brown to greenish, with large, light brown, lenticels. Slash yellow to white, exuding copious white latex. Leaves simple, opposite, elliptic to obovate, c. 25 x 12 cm, apex acuminate to acute, base cuneate, with c. 9-21 main lateral veins on each side of the midrib, margin entire, glabrous. Petiole c. 2 cm long. Flowers white, fragrant. Corolla tube up to 3 cm long. Fruits paired, spherical, c. 10 cm diameter.

OCCURRENCE: U2-4. An abundant tree. Found on Bugala Island and up to 2200 m on Rwenzori and in Kigezi.
NOTE: The Rukiga name refers to the resemblance of the paired fruits to human testicles.

Tabernaemontana stapfiana **Britten (274)** Apocynaceae

SYNONYM: *Tabernaemontana johnstonii* (Stapf) Pichon
Namaondu, Namatumagali (ms).
18 m. Very similar to *Tabernaemontana pachysiphon*, but tree rather larger, leaves larger (up to 40 cm long) and corolla tube 2-2.5 cm long. The leaf apex in this species is acuminate, apiculate or rounded; the leaf base is cuneate or decurrent. The lamina has scattered dots below. Petiole 3 cm long.
OCCURRENCE: U2 and 3. Common on lower slopes of Mt Elgon and reported to occur in Kibale Forest and Kigezi.

Tabernaemontana ventricosa **A. DC. (275)** Apocynaceae

SYNONYMS: *Conopharyngia usambarensis* (Engl.) Stapf; *Tabernaemontana usambarensis* Engl.
Mwongogwenkende (ms).
15 m. Understorey tree with a wavy trunk and fairly spreading crown of dark-coloured leaves. Bark light brown, flaking on outside, of medium thickness, rough. Slash light yellow, exuding white latex. Leaves simple, opposite, elliptic, c. 20 x 5.5 cm, more than 3 times as long as wide, apex obtusely acuminate, acute or obtuse, base cuneate, margin undulate. Petiole c. 1 cm long, grooved. Fruits green, paired, ellipsoid, c. 5 cm long or less.
OCCURRENCE: U2-4. Widespread, usually on drier sites than *Tabernaemontana pachysiphon*. Below 1500 m. Forest margins; riverine and groundwater forest.

Tabernaemontana odoratissima **(Stapf) Leeuwenb. (276)** Apocynaceae

Ekinyamagosi (ki).
10 m. Understorey tree. Leaves simple, opposite, elliptic, c. 22 x 8 cm, apex apiculate to obtuse, base cuneate, main lateral veins c. 9-15 on each side of the midrib, usually with scattered black dots below. Petiole 0.5-1.5 cm long. Corolla tube 7.5-10 cm long, flowers open at night. Fruits paired, up to 15 cm long, elongated.
OCCURRENCE: U2 and 4. Common in Kalinzu Forest. Recorded from Kibale Forest (Tooro) and Mpanga Forest (Mengo).
NOTES: Sterile specimens may be difficult to identify. The leaves of *Tabernaemontana pachysiphon* tend to be broader.

Picralima nitida **(Stapf) T. Durand & H. Durand (277)** Apocynaceae

Uncommon understorey tree to 30 m, sometimes with a straight trunk. Crown spreading, dark-coloured. Leaves elliptic to oblong, c. 22 x 8 cm, apex abruptly acuminate, base rounded to obtuse, venation rather obscure, lateral veins running into a submarginal vein. Corolla tube 1.5-2 cm long. Fruits yellow or orange, smooth, obovoid to ellipsoid, c. 12 cm long.
OCCURRENCE: U2 and 4. Recorded from Mabira, Budongo and Bwamba forests.

Voacanga thouarsii **Roem. & Schult. (278)** Apocynaceae

Musanvuma (ga); Entoma (na).

Plate 22. Apocynaceae (271-278); see also Plate 21

271. *Funtumia elastica* 277. *Picralima nitida* 278. *Voacanga thouarsii*
Actual sizes: leaves, flower, fruits and seed x 2.

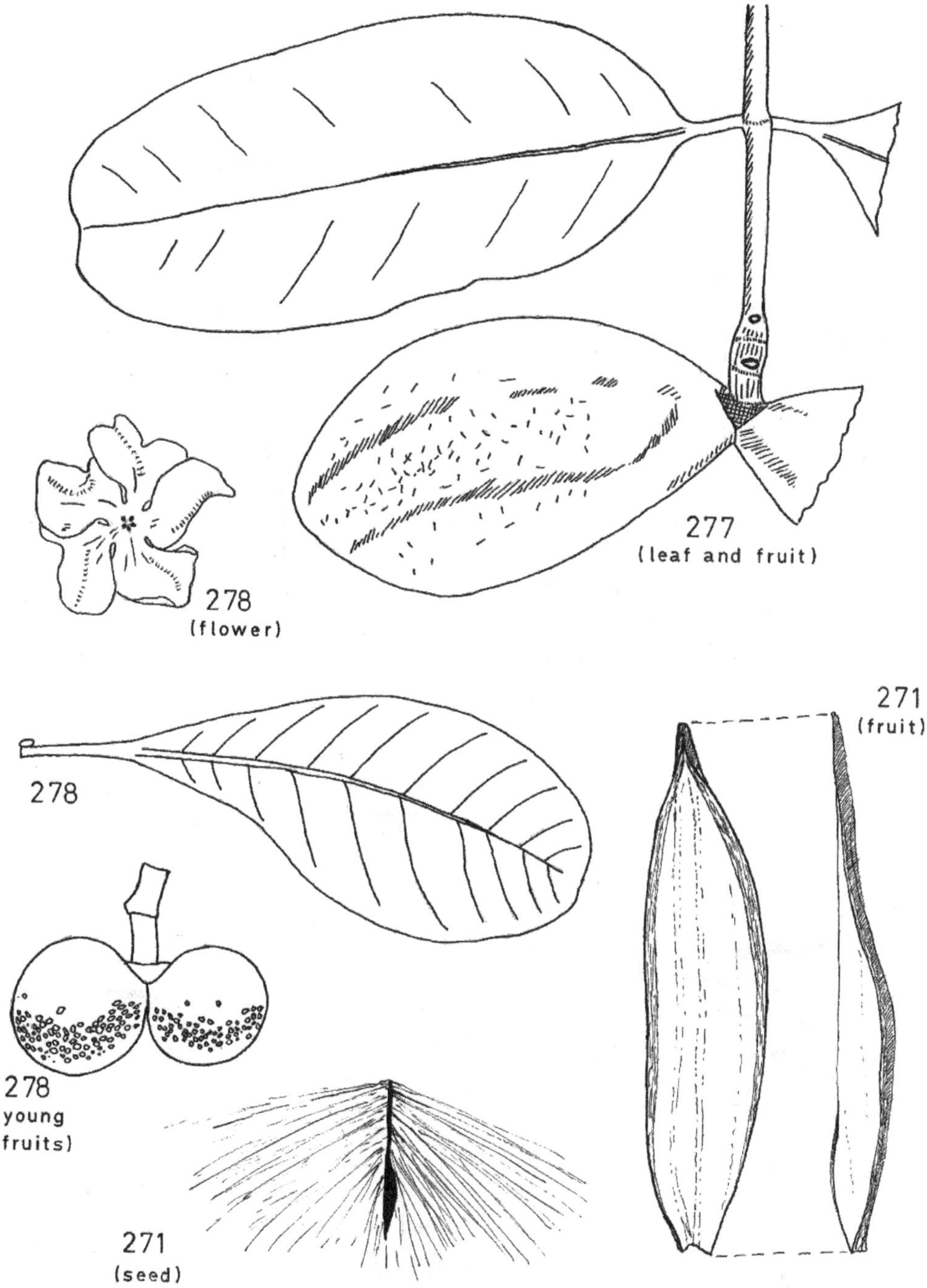

Plate 22. (271-278)

12 m. Spreading tree with a crooked trunk and fairly open crown. Bark light brown, thin, granular, but otherwise smooth. Slash thick and soft, white, with conspicuous orange stone cells and lines, slowly turning darker, without white latex on older trees. Leaves simple, opposite, markedly obovate, c. 13 x 5 cm, apex obtuse or rounded, base attenuate, main lateral veins c. 6-14 on each side of the midrib, glabrous, both surfaces with numerous pits. Petiole c. 1.5 cm long, glabrous or minutely puberulous at base. Flowers yellow, fragrant. Drupes paired, 4-7 cm diameter.
OCCURRENCE: U1-4. Common and widely distributed tree in swamps and valley bottoms.

Voacanga africana Stapf (278a) Apocynaceae
10 m. Shrub-like tree. Bark pale grey-brown, smooth or shallowly fissured. Slash yellowish-white. Leaves simple, opposite, elliptic, to 40 cm long and 15 cm wide, apex bluntly acuminate (rarely acute or obtuse), base cuneate or decurrent into petiole, glabrous (but sometimes pubescent below and on midrib above). Petiole 0-2 cm long, glabrous to pubescent.
OCCURRENCE: U1. Laropi Forest (West Nile).
NOTE: Differs from *Voacanga thouarsii* in having leaves usually sessile and elliptic, with acuminate apices.

CLUSIACEAE, CALOPHYLLACEAE AND HYPERICACEAE

Trees or shrubs (as with *Hypericum* and sometimes *Harungana madagascariensis*) with simple opposite leaves, yellow or orange coloured latex and brightly coloured flowers. The taller trees in this family have thick, leathery leaves. *Allanblackia*, *Garcinia*, *Mammea* and *Symphonia* were placed in the family Guttiferae in ITU and UFT.

Symphonia globulifera L. f. (279) Clusiaceae
Muyanja (ga); Omusisi (ki); Munimba (ko); Munyansungu, Musandasanda (na); Mukarangeye, Munyankwansi, Munyenye, Nkwasi (to).
30 m. Large tree with a straight cylindrical trunk and small crown. Branches at right angles, the larger curving up, the smaller drooping. Buttresses and flutes absent. Stilt roots probably sometimes present. Bark very thin and smooth, light brown, sometimes with small vertical fissures. Slash light pink, orange to yellow, exuding drops of yellow latex in rings. Leaves simple, opposite, shape variably lanceolate, elliptic, obovate, c. 11 x 3.5 cm, apex acuminate, base cuneate, lateral veins numerous, leathery and glossy. Petiole c. 0.8 cm long. Flowers bright red, waxy, very conspicuous. Fruit c. 2 cm diameter.
OCCURRENCE: U2 and 4. Mainly in swamp forest and valley bottoms in Mengo and Masaka. Common on hillslopes in Kalinzu and Bwindi forests and on Ruwenzori. Found up to 2500 m.

Garcinia buchananii Baker (280) Clusiaceae
SYNONYM: *Garcinia huillensis* Oliv. (*sensu* UFT)
Atenum, Ekwalakwala (at); Musaali, Nsaali (ga, so).
15 m. Understorey tree with a thick, dark-coloured, crown. Bark dark grey to brown, flaking. The underside of the bark is bright red-brown. Slash of even texture, white to yellow, exuding drops of bright yellow latex in rings. Leaves simple, opposite, elliptic to ovate, c. 13 x 5.5 cm, apex long acuminate, base cuneate, obtuse (sometimes rounded), lamina with numerous lateral veins which are prominent on both surfaces, smooth and glabrous. Petiole c. 0.5 cm long. Flowers yellow. Fruit c. 2.5 cm diameter.
OCCURRENCE: U1, 3 and 4. Uncommon, except in lake-shore forests. Usually on forest edges.
NOTE: The fruit (*ensaali* in Luganda) is edible.

***Harungana madagascariensis* Poir. (281)** Hypericaceae

Njuli (am); Mukaabiransiko, Mulirira (ga); Omunyananga, Omwongorero (ki); Asonbere, Serubele (md); Mutaha (na); Musoga (to).

12 m (exceptionally 25 m). Usually a much-branched tree or shrub, but sometimes with a straight, cylindrical trunk. Bark red-brown, scaling. Slash pink to brown, exuding orange latex. Leaves simple, opposite, c. 16 x 7 cm, upper surface dark green, undersurface light brown. The young leaves are clasped together and covered with red-brown hairs. Petiole c. 2 cm long. Flowers white, in dense clusters, fragrant. Fruits small, orange to yellow-brown.

OCCURRENCE: U1-4. Widely distributed. At lower altitudes, on forest edges and termite mounds and in forest clearings. In forest interiors in Kibale, Kalinzu, Kayonza and Bwindi forests. It attains its largest dimensions in Kigezi.

CULTIVATION AND PROPAGATION: Fast-growing. Can be used as a pioneer species on degraded land, including where soil degradation has been caused by the past presence of eucalyptus. Collect fruits from mother trees, remove the pulp and dry the seeds in the sun. Sow seeds as soon as possible. Can also be propagated by transplanting root suckers from mother trees.

***Allanblackia kimbiliensis* Spirlet (282)** Clusiaceae

Orutaka (ki).

Tall tree to 35 m. The shape and exudate are similar to *Symphonia*, from which it differs in having rough brown bark, which is fissured vertically and horizontally, and a pink-brown slash, which turns darker. Leaves simple, opposite, oblong, c. 15 x 4.5 cm, apex acuminate, base acute or obtuse, main lateral veins numerous, prominent on both surfaces, lamina leathery and glabrous. The fruit is large (c. 14 x 11 cm), has 5 rounded valves, falls off the tree before dehiscing and has many light brown seeds.

OCCURRENCE: U2. Only recorded from the vicinity of Ishasha Gorge. A range-restricted species of narrow endemism, known only from Bwindi Forest and eastern D.R. Congo.

CONSERVATION STATUS: Global NE (IUCN), LC (TOU); National CR (WCS).

***Mammea africana* Sabine (283)** Calophyllaceae

Large tree to 30 m with yellow latex. It differs from *Symphonia* in having rough, scaling, dark brown bark and a red to red-brown slash. Leaves simple, opposite, oblong to elliptic, c. 20 x 7 cm, apex acuminate, base cuneate, lateral veins numerous, slightly prominent on both surfaces, lamina leathery and glabrous. Petiole 1 cm long. Fruit orange, subglobose, c. 14 cm diameter.

OCCURRENCE: U2. Occasional in Budongo Forest.

HYPERICUM

A genus of small trees, shrubs and herbs, found mainly above 2000 m. Up to 4000 m in favourable sites. The larger species tend to grow at higher altitudes and in more open habitats. Trees and shrubs in *Hypericum* are readily distinguished by their small, narrow (less than 2 cm broad), opposite leaves and by their large yellow (rarely red to orange) flowers. Species of tree and shrub *Hypericum* in Uganda probably have yellow or orange latex, but no records of whether this is so have been seen. Apart from the two species below, two other species found in Uganda may be noted. They are illustrated as numbered species on Plate 23. One is *H. roeperianum* A. Rich. (287), a shrub or small tree to 6 m with orange flowers, recorded from Rwenzori, Mt Elgon and Mt Kadam. It is reported to occur in Kenya in riverine thicket, rocky sites near water and less commonly on forest margins away from water (Beentje 1994). The other is *Hypericum quartinianum* A. Rich. (288), a shrub with yellow flowers growing to 2.5 m and found on some mountains in Karamoja (perhaps also on Mt Elgon). It is reported to occur on rocky streambanks in Kenya (Beentje 1994).

Hypericum bequaertii De Wild. (285) Hypericaceae
Much-branched shrub or small tree to 12 m. Leaves lanceolate or oblong-lanceolate, to 4 cm long and 1 cm wide, apex acute, base clasping, one or two pairs of main lateral veins from base of lamina on each side of the midrib, these extending to near the apex, lamina with many longitudinal glands parallel to the midrib. Flowers solitary, at end of branches, red, cup-shaped at anthesis. Fruit a 5-valved capsule.
OCCURRENCE: U2 and 3. A range-restricted species of narrow endemism, only on Ruwenzori and Mt Elgon, 3150-4300 m.

Hypericum revolutum Vahl (284, 286) Hypericaceae
SYNONYM: *Hypericum leucoptychodes* Steud. ex A. Rich.
Cheborokorok (ku).
Much-branched shrub or small tree to 12 m. Leaves lanceolate or oblong-lanceolate, 1.25-3 cm long (but see note below), pinnately and reticulately veined, lamina with many longitudinal linear glands which are not translucent. Flowers yellow.
OCCURRENCE: U2 and 3. The commonest species in Kigezi and in the montane forest and Ericaceous belts of Elgon.
NOTES: This species is similar to *Hypericum bequaertii*, but differs in having flowers expanded (not cup-shaped) at anthesis. *Hypericum revolutum* Vahl subsp. *keniense* (Schweinf.) Robson (Syn.: *H. keniense* Schweinf.) is a recognized subspecies. Its leaf is illustrated on Plate 23 (numbered 284). The leaves tend to be long (to 5 cm) and have 3 conspicuous veins (including the midrib) running from near the base to near the apex. Recorded from Rwenzori, Mt Elgon and the Karamoja mountains.

Key to *Anthocleista.*

1. Branches armed with short paired spines. ....**290. *A. vogelii***
   Branches unarmed. 2
2. Leaves sessile or subsessile, often thin, margin wrinkled. ....**289. *A. grandiflora***
   Leaves (at least the upper ones) petiolate, often subleathery, margin somewhat undulate-crenulate. ....**291. *A. schweinfurthii***

Anthocleista grandiflora Gilg (289) Gentianaceae
SYNONYMS: *Anthocleista pulcherrima* Gilg; *Anthocleista zambesiaca* Baker
Cabbage tree (en); Omunyangabo, Omuzibiziba (ki); Gumsiwaniwiwa, Kisigewa, Kumisigewa, Namatumagali (ms).
25 m. Trunk straight and cylindrical, branches few, crown thin and spreading. Bark thin, greenish to brown, fairly smooth, with small vertical fissures c. 2 cm apart. Phellogen green. Slash hard and very granular, white, often with orange granules, turning darker. Branchlets unarmed. Leaves simple, opposite, usually sessile, oblanceolate or obovate, large (c. 45 x 25 cm, but much larger on young plants), apex rounded or acute, base cuneate, +/- inconspicuously auriculate, margin wrinkled, lateral veins 9-14 on each side of the petiole, prominent below, lamina thin (but sometimes a bit leathery).
OCCURRENCE: U1-3. Mainly in highland areas, ascending to 2300 m. In swamps, valley forest and persistent in damp secondary forest.

Plate 23. Clusiaceae and others (279-297)

279. *Symphonia globulifera* 280. *Garcinia buchananii*
281. *Harungana madagascariensis* 282. *Allanblackia kimbiliensis*
284. *Hypericum revolutum* subsp. *keniense* 285. *Hypericum bequaertii*
286. *Hypericum revolutum* 287. *Hypericum roeperianum* 288. *Hypericum quartinianum*
291. *Anthocleista schweinfurthii* 292. *Strychnos mitis* 293. *Afrocrania volkensii*
294. *Lijndenia jasminoides* 296. *Dichaetanthera corymbosa* 297. *Mallotus oppositifolius*
Actual sizes: leaves and fruit x 2; trunk base x 80; tree profiles x 800.

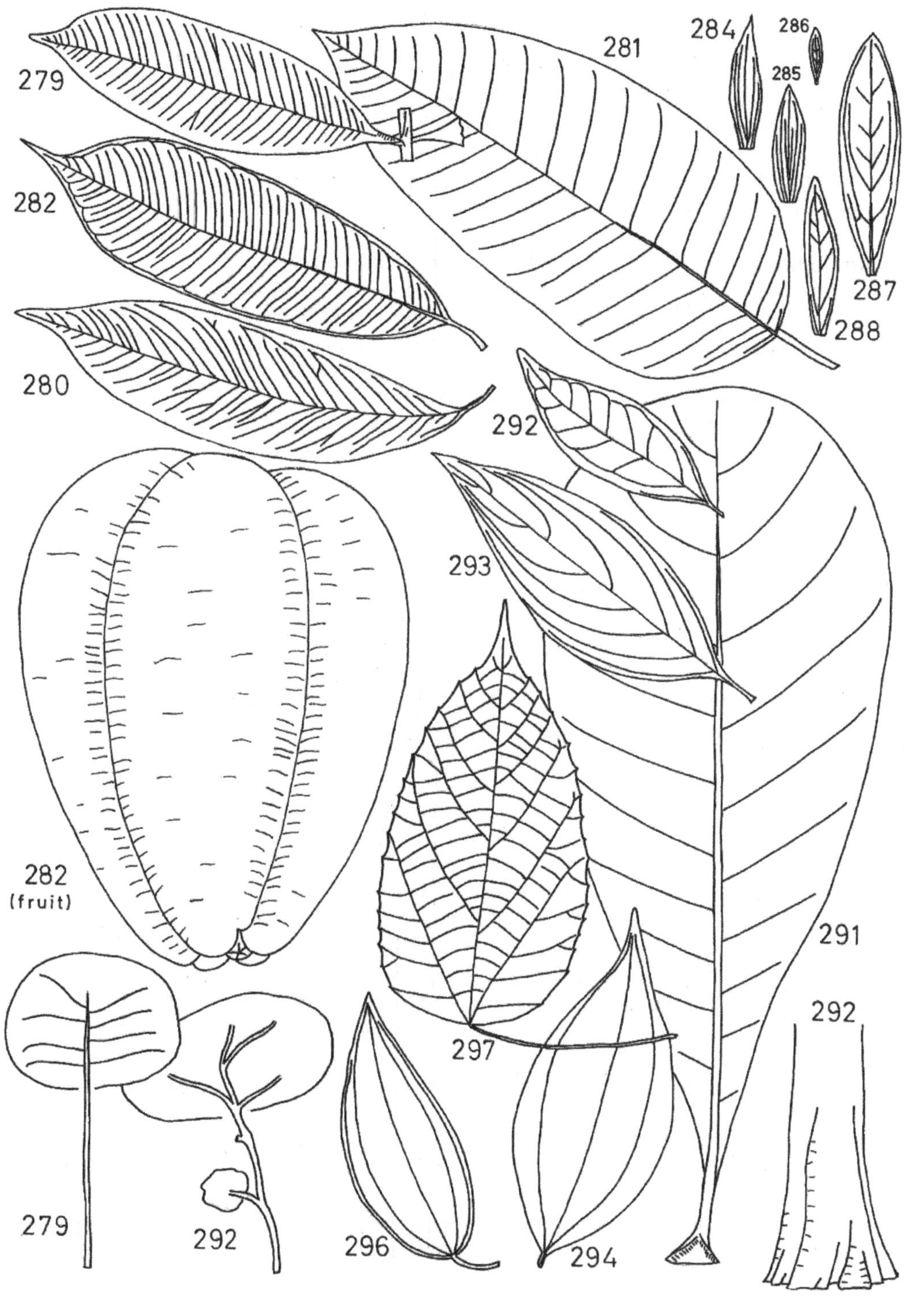

Plate 23. (279-297)

Anthocleista vogelii Planch. (290) Gentianaceae

Cabbage tree (en); Mubakampungu (na).

15 m. Shape and slash as for *Anthocleista grandiflora*. Branchlets armed, the spines paired, persisting on the trunk on older trees and becoming woody. Leaves normally sessile (rarely with a petiole to 2.5 cm long), usually obovate, large (c. 45 x 23 cm, but much larger on young plants.

OCCURRENCE: U1, 2 and 4. In primary and secondary swamp forests, 1200-1400 m.

NOTE: Distinguished from other species of *Anthocleista* by the presence of spines.

Anthocleista schweinfurthii Gilg (291) Gentianaceae

Cabbage tree (en); Mugabogabo (ga); Omunyangabo (ki).

20 m. Shape and slash as for *Anthocleista grandiflora*. Branchlets unarmed. Leaves simple, opposite, large, lateral veins 6-13 on each side of the midrib, prominent below, margin somewhat undulate-crenulate, lamina often subleathery. Upper leaves with a petiole (to 2 cm long), lower leaves subsessile.

OCCURRENCE: U2 and 4. Mainly secondary forest, 1200-1400 m. It used to be very common on the Ssese Islands and in their neighbourhood.

NOTE: This species may be confused with *Anthocleista grandiflora*, from which it may usually be distinguished by the petiolate upper leaves. *A. grandiflora* generally occurs at higher altitudes.

Strychnos mitis S. Moore (292) Loganiaceae

Awukebu (am); Mukusakusa (ga); Akomya (ms); Mungangara (to).

35 m. Trunk usually crooked, with branches from low down and a wide-spreading crown. The trunks of older trees can appear twisted from a distance (like eucalyptus). The trunk may flare out at the base, but lacks true buttresses. Bark very thin, very smooth, slightly flaking, light brown. Phellogen often green. Slash brittle, yellow, quite thin, characteristically coming away easily from the wood. The external face of the exposed wood is smooth and shiny and has conspicuous white lines. Leaves simple, opposite, elliptic to ovate-lanceolate, c. 7 x 2.5 cm, apex usually acuminate, base cuneate, with two relatively faint lateral veins from the base and a more conspicuous pair of lateral veins arising asymmetrically c. 1 cm up the midrib, lamina subleathery, glabrous above, glabrous or pubescent below. Petiole c. 0.2 cm long

OCCURRENCE: U1-4. Abundant on upper slopes in Kibale, Mabira and other forests. Common in riverine forest in drier parts of Uganda, e.g. at the base of Mts Kadam and Moroto.

NOTES: (1) Other species of *Strychnos* occur in Uganda, mainly as trees of dry country or as climbers in forest. All have the characteristic leaf venation described above. (2) *Strychnos congolana* Gilg is a lowland rainforest species that has been recorded along the Kampala-Entebbe Road. It may no longer exist in Uganda. (3) Species of *Strychnos* contain alkaloids and the bark, seeds and other parts of some are used as poisons and medicines. Strychnine is obtained from the seeds of *Strychnos nux-vomica* L., a native of India and Malaysia.

Afrocrania volkensii (Harms) Hutch. (293) Cornaceae

Musonganyonyi (ko); Tasakia (ku).

20 m. Canopy or understorey tree. Trunk straight to crooked. Bark granular, grey to black, with vertically elongated lenticels. Slash soft, yellow-brown, with darker streaks. Leaves simple, opposite, elliptic, c. 11 x 4 cm, apex acuminate, base cuneate, with c. 4-5 main lateral arcuate veins (curving like a bow) on each side of the midrib. Petiole c. 0.8 cm long, grooved. If a leaf is gently pulled in two, the two parts remain attached by the extended veins.

OCCURRENCE: U1-3. Montane forest. Kigezi, Rwenzori and on Mts Elgon and Moroto, 2500-3000 m.

Lijndenia jasminoides **(Gilg) Borhidi (294)** Melastomataceae
SYNONYMS: *Memecylon jasminoides* Gilg; *Warneckea jasminoides* (Gilg) Jacq.-Fél.
Nabbumba (ga).
7 m. Understorey tree with a small crown. Bark smooth, grey, flaking. Phellogen green. Slash off-white to brown. Leaves simple, opposite, c. 11 x 4.5 cm, ovate to elliptic, apex acuminate, glabrous, margin entire, with 2 main lateral veins from the base, these curving around and joining the midrib at the apex but not forming conspicuous loops. Petiole c. 0.5 cm long. Berries blue, globose, c. 0.7 cm diameter.
OCCURRENCE: U2 and 4. Widely distributed. Most abundant in lake-shore forests.
NOTES: The venation in the leaves of this species and the two below is typical of the family Melastomataceae. Some members of the family have striking blue or purple flowers.

Lijndenia bequaertii **(De Wild.) Borhidi (295)** Melastomataceae
SYNONYMS: *Memocylon bequaertii* De Wild.; *Memocylon* sp. (of UFT)
15 m. Leaves simple, opposite, elliptic, lateral veins prominent below, two from the base and two from about midway up the lamina forming loops and reaching the apex. Petiole c. 0.2 cm long. Berries globose, 0.9-1 cm diameter.
OCCURRENCE: U2. Only recorded from Bwindi Forest, c. 1500 m. Abundant in the understorey. A range-restricted species of narrow endemism, only in Bwindi Forest and eastern D.R. Congo.
CONSERVATION STATUS: Global NE (IUCN), LC (TOU); CR (WCS).

Dichaetanthera corymbosa **(Cogn.) Jacq.-Fél. (296)** Melastomataceae
15 m. Tree branching near base, with a spreading crown. The base of the trunk may become gnarled. Bark light brown, fairly smooth, flaking (papery). Slash of even texture, pinky-orange to light yellow, turning darker. Leaves simple, opposite, ovate-elliptic, c. 11 x 4.5 cm, apex acute, base rounded, with 2 main lateral veins from the base on each side of the midrib, these veins curving around and extending to the apex, midrib and 2 basal pairs impressed above, prominent below, lamina strigose on both surfaces, more densely so below. Petiole c. 1.2 cm long. Flowers large, purple, attractive.
OCCURRENCE: U2. Kigezi, Ankole (including Kalinzu Forest), Tooro. In open forests, 1400-2000 m.

Mallotus oppositifolius **(Geiseler) Müll. Arg. (297)** Euphorbiaceae
5 m. Understorey tree or bush, with an open crown. Bark smooth, light-coloured. Phellogen green. Slash white. Leaves simple, opposite, ovate to ovate-lanceolate, c. 12 x 6 cm, more or less heart-shaped, apex acuminate, base shallowly cordate, truncate or rounded-cuneate, with a pair of lateral veins extending steeply from the base of the lamina (see illustration), margin with small teeth. Petiole c. 6 cm long.
OCCURRENCE: U2 and 4. Raised and swamp forests. In Budongo Forest.
NOTES: *Mallotus* is unusual among Euphorbiaceae in having opposite leaves. See before tree 94 for an overview of the family.

CASSIPOUREA AND LASIODISCUS

Most species that have simple, opposite leaves together with interpetiolar stipules belong to the family Rubiaceae (303-327e). The only other species with these features are in the genera *Cassipourea* and *Lasiodiscus*. A sure character for separation of *Cassipourea* and *Lasiodiscus* from Rubiaceae is the superior ovary. Additionally, *Cassipourea* and *Lasiodiscus* can usually be readily separated from Rubiaceae because they have non-entire (toothed, crenate or wavy) leaf margins, rather than entire, as is normal with Rubiaceae. Mangroves belong to the family Rhizophoraceae.

Key to *Cassipourea.*

1. Leaves entirely glabrous. ....**301. *Cassipourea gummiflua***
   Leaves hairy below, at least on midrib. 2
2. Leaves less than 10 cm long. ....**298. *Cassipourea malosana***
   Mature leaves over 10 cm long. 3
3. Young stems densely covered with yellow hairs. ...**299. *Cassipourea ruwensorensis***
   Young stems with only a few scattered hairs. ....**300. *Cassipourea congoensis***

Cassipourea malosana (Baker) Alston (298) Rhizophoraceae
Aganiya (ku); Chizanzasi, Kusiu (ms); Pillarwood (tn).
20 m. Understorey tree with a very straight, cylindrical trunk, branches at right angles and a small rounded crown. Bark thin, smooth, with ring marks. Slash white to orange, reddish on outside. Leaves simple, opposite, elliptic to obovate, c. 5.5 x 3 cm, apex acuminate or obtuse, base cuneate or rounded, margin widely bluntly serrate or with at least some leaves on a branch toothed. Petiole c. 0.5 cm long. Fruit an ovoid capsule, c. 0.7 cm long, hairy, longer than style remains.
OCCURRENCE: U1 and 3. Mt Elgon, mountains in Karamoja and the Imatong Mountains, 1700-2500 m. Abundant on Mts Elgon and Kadam.

Cassipourea ruwensorensis (Engl) Alston (299) Rhizophoraceae
Kobwo (to).
15 m. Understorey tree with a very straight trunk and branches at right angles. Bark brown, very thin and smooth, with ring marks. Slash of even texture, yellow, sometimes layered. The young stems are densely covered with yellow hairs, which are also present on the leaf veins below and on the leaf margin. Leaves simple, opposite, elliptic to obovate, c. 14 x 5.5 cm, apex acuminate, base acute or cuneate, margin bluntly serrate, glabrous above. Petiole 1 cm long. Fruit an ovoid capsule, 0.5 cm long, hairy.
OCCURRENCE: U1-4. Widespread, common in Kibale and Bwindi forests. Found up to 2400 m.

Cassipourea congoensis DC. (300) Rhizophoraceae
10 m. Trunk straight. Bark smooth. Slash yellow and red, layered. Leaves simple, opposite, c. 11 x 5 cm, elliptic, apex subacuminate to obtuse, base cuneate or rounded, midrib hairy on lower surface, margin entire or bluntly toothed. Petiole 0.5 cm long. Fruit a capsule, 0.5 cm long, often glabrous.
OCCURRENCE: U2. Kayonza, South Bwindi and Kalinzu forests. Uncommon.

Cassipourea gummiflua Tul. (301) Rhizophoraceae
Engongwe (ki).
25 m. Understorey tree, with a straight trunk and small crown. Branches at right angles, drooping. Bark thin, smooth, brown, grey to black. Slash off-white to light red-brown, granular. Leaves simple, in twos or threes, rather stiff and thick, elliptic, c. 11 x 5.5 cm, margin with small teeth or entire, completely glabrous.
OCCURRENCE: U2 and 4. Widely distributed, but rare except in Bwindi Forest.
NOTE: Only var. *ugandensis* (Stapf) J. Lewis occurs in Uganda

Lasiodiscus pervillei Baill. (302) Rhamnaceae
SYNONYM: *Lasiodiscus mildbraedii* Engl. (of UFT, a misapplied name)
Bulindi (am); Omunyamaija (no).

15 m. Understorey tree, with a crooked trunk (rarely more or less straight) and a deep, untidy, crown. Trunk fluted on older trees. Bark light brown, of medium thickness, rather fibrous, with small vertical fissures, flaking in places. Slash brittle to fibrous, red to red-brown, often with white lines or rings. Leaves simple, opposite, c. 16 x 6 cm, with prominent lateral veins (as seen on lower surface), margin toothed, crenate or wavy. Petiole c. 0.5 cm long.
OCCURRENCE: U1-4. Very abundant under *Cynometra* in Budongo, South Kibale and Maramagambo forests. Also abundant in swamp forest in Masaka. Common in parts of Mabira Forest.

RUBIACEAE

Usually small to medium-sized, understorey or forest-edge trees, rarely big trees (*Fleroya*, *Nauclea*). Leaves opposite or (rarely) in whorls of three, with interpetiolar stipules. These characters are shared among Ugandan forest trees only by *Cassipourea* and *Lasiodiscus*, both of which usually have non-entire leaf margins, contrasting with the normally entire leaf margins of Rubiaceae. The flowers of Rubiaceae have fused petals and inferior ovaries. The flowers are often sweet-scented and vary from small to very large.

There are many species of Rubiaceae in Ugandan forests and sometimes they can be difficult to identify in the vegetative state. In common with UFT, we have excluded *Gardenia imperialis* K. Schum. (a handsome tree with large flowers, sometimes found in swamps), *Lasianthus kilimandscharicus* K. Schum. and *Psychotria riparia* (K. Schum. & K. Krause) E.M.A. Petit. The forests of the Kalinzu-Kayonza-Bwindi area appear to be particularly rich in members of this family.

On the other hand, we have included six species that were not in UFT, namely *Chassalia subochreata*, *Ixora seretii*, *Rytigynia acuminatissima*, *R. kigezieneis*, *R. ruwenzoriensis* and *Tricalysia bagshawei.* Apart from *Ixora seretii* (318a), which is morphologically close to *Pavetta molundensis* (318), we have not had access to sufficient information to place these species confidently in the key below. They are numbered 327a-327e.

Key to Rubiaceae.

1. Stipules relatively broad and long (normally larger than 2 x 1.5 cm), widest in about the centre (rather than near the base). 2
   Stipules less than 1.5 cm broad or, if broader, then widest near base. 4

2. Small tree (to 6 m), with a markedly asymmetric leaf base and short petiole (to c. 1 cm long). ....**306. *Oxyanthus unilocularis***
   Leaf base more or less symmetrical. 3

3. Wood bright yellow. A small tree to 12 m. ....**312. *Morinda***
   Wood not bright yellow. Tree taller than above. ....**303-305. *Fleroya, Nauclea***

4. Leaves relatively large (c. 30 cm or more long and/or c. 12 cm or more wide). Note: *Belonophora* and *Rothmannia whitfieldii* may occasionally key out here. 5
   Leaves less than c. 28 cm long and c. 12 cm broad. 8

5. Leaf base markedly asymmetric. Small tree to 6 m. .....**306. *Oxyanthus unilocularis***
   Leaf base more or less symmetrical. 6

6. Medium-sized tree (to 20 m tall) with a straight trunk and horizontal branches. In wet highland forests in south-west Uganda, 1300-1650 m. ....**307. *Pauridiantha callicarpoides***
   Combination of characters not as above. 7

7. Leaves _ 12 cm broad. ....**309. *Coffea liberica***
Leaves less than 12 cm broad. ....**318-318a. *Pavetta, Ixora***

8. Main lateral veins fairly indistinct and rather irregularly shaped. Reticulate venation prominent. Stipules c. 0.3 cm long. ....**310. *Craterispermum***
Combination of characters not as above. 9

9. Venation prominent. Leaves rather thin and papery. Stipules with narrow, strap-shaped tips (see Plate 25). ....**311. *Vangueria*; 327b-d. *Rytigynia***
Combination of characters not as above. 10

10. Main lateral veins markedly raised and very prominent on undersurface of leaf, c. 15-20 on each side. Leaves often c. 14 x 5 cm or less. 11
Main lateral veins not so prominent and numerous or, if so, then leaf larger. 13

11. Tree in lake-belt forests. ....**308. *Pauridiantha viridiflora***
Tree at higher altitudes (above c. 1300 m). 12

12. Branches knobbly due to retention of the lowermost parts of the petioles at each node. ....**321. *Psychotria***
Branches not markedly knobbly. ....**320. *Galiniera***

13 Wood bright yellow. Stipules large and rounded, c. 2 cm across, falling off early. ....**312. *Morinda***
Wood not bright yellow. Stipules not as above. 14

14. Mature leaves over 15 cm long and/or 7 cm wide. 15
Mature leaves less than 15 x 7 cm. 28

15. Shoot growth peculiar. At each node the shoot terminates in a flower or leaf (so that the leaves may appear to be in whorls of 3) and growth continues by one or both axillary shoots. ....**323. *Rothmannia whitfieldii***
Shoot growth normal. 16

16. Base of leaf obtuse to rounded. 17
Base of leaf acute to cuneate. 22

17. Tree found above c. 1400 m. 18
Tree found below 1400 m. 20

18. Branchlets knobbly, due to the retention of the lowermost parts of the petioles at each node. Main leaf veins reddish. ....**321. *Psychotria***
Branchlets not markedly knobbly. Veins not reddish. 19

19. Leaves normally hairy on lower surface. ....**320. *Galiniera***
Leaves more or less glabrous below, except sometimes for tufts of hairs in vein axils. ....**317. *Oxyanthus speciosus***

20. Young shoots and/or undersurface of midrib usually markedly hairy. ...**313. *Bertiera***
Young shoots and midrib more or less glabrous. 21

21. Trunk straight, branchlets curving down. ....**314. *Coffea canephora***
Trunk crooked, at least at base. Branchlets either irregular or horizontal. ....**317. *Oxyanthus speciosus***

22. Trunk tending to be straight, with branches more or less at right angles (but sometimes slightly ascending or drooping down). 23
Habit not as above. 25

23. Young shoots and leaves glabrous or almost so (except sometimes for hairs in the vein axils below). 24
Young leaves and/or shoots hairy. ....**320. *Galiniera***

24. Main lateral veins comparatively numerous (c. 11-15 on each side of the midrib) and ascending at a comparatively less steep angle than below (see Plate 24). ....**314. *Coffea canephora*; 317. *Oxyanthus speciosus***
Main lateral veins comparatively few (c. 6-9 on each side of the midrib) and ascending at a comparatively steep angle (see Plate 24). ....**315. *Belonophora***

25. Midrib and main lateral veins reddish in colour. Branchlets knobbly due to the persistence of the petiole bases at each node. ....**321. *Psychotria***
Midrib and main veins not normally reddish. Branchlets not markedly knobbly. ...26

26. Stipules comparatively short, abruptly pointed (see Plate 25). ....**319. *Tarenna***
Stipules comparatively long, not abruptly pointed (except sometimes right at the end). 27

27. Slash pink. Trunk often bearing short woody outgrowths. ....**316. *Leptactina***
Slash not pink. Trunk without outgrowths, though often irregular. ....**317. *Oxyanthus speciosus***

28. Shoot growth peculiar. At each node, the shoot terminates in a flower or leaf (so that the leaves may seem to be borne in threes) and growth continues by one or both of the axillary shoots. ....**322-323. *Rothmannia***
Shoot growth normal. 29

29. Main lateral veins 7 or fewer on each side of the midrib. 30
Main lateral veins normally more than 7 on each side of the midrib. 33

30. Leaves markedly hairy below, particularly on the veins. Veins c. 4-6 on each side of the midrib, very prominent on undersurface and ascending at a fairly steep angle. ....**324. *Heinsenia***
Leaves glabrous on veins below or, if more or less hairy, then veins not as above; several other Rubiaceae, including *Aidia* (326) may key out here. 31

31. Small understorey trees with reddish young leaves. Leaves relatively small (often c. 9 x 3.25 cm or less). Venation often relatively obscure. ....**327. *Coffea eugenioides*; 327b-d. *Rytigynia***
Trees in open forest or on forest edges. Young leaves not reddish. Leaves often larger than above. 32

32. Leaf base comparatively long cuneate (see Plate 25). ......**318. *Pavetta*; 318a. *Ixora***
Leaf base comparatively short cuneate or acute. ....**325. *Psydrax***

33. Main leaf veins reddish. Branchlets knobbly due to the retention of the bases of the petioles at each node. ....**321. *Psychotria***
Main leaf veins not reddish. Branchlets not, or slightly, knobbly. 34

34. Main lateral veins normally more than 10 on each side of the midrib. 35
Main lateral veins normally fewer than 11 on each side of the midrib (*Galiniera* may also key out here). 37

35. Trunk straight with branches at right angles. ....**320. *Galiniera***
Trunk crooked, at least at base. 36

36. Stipules comparatively short, abruptly pointed (see Plate 25). ....**319. *Tarenna***
Stipules comparatively long, not abruptly pointed (except sometimes right at the end). ....**317. *Oxyanthus speciosus***

37. Leaves comparatively small (often c. 9 x 3.25 cm or less), reddish when young, often with venation comparatively obscure. ....**327. *Coffea eugenioides***
Combination of characters not as above. Leaves often larger, not reddish when young, venation conspicuous. 38

38. Leaves comparatively long cuneate at base (see Plate 25). ....**319. *Tarenna***
Leaves acute to cuneate at base, but not as markedly cuneate as above. . **327. *Aidia***

Fleroya stipulosa (DC.) Y.F. Leroy (303) Rubiaceae
SYNONYMS: *Hallea stipulosa* (DC.) Y.F. Leroy; *Mitragyna stipulosa* (DC.) O. Kuntze
Munyamaizi (am); Nzingu (ga, tn); Muzingu (ga); Obul (lo); Oo (md); Omuho (no); Abura (tn).
30 m. Trunk usually crooked, bearing a rounded to fairly spreading crown of large leaves. Knee roots (pneumatophores) often present. Bark grey, smooth or rough. Slash soft and thick, fibrous, light brown, pinkish or red, turning darker. Leaves simple, opposite, c. 20-30 x 15-20 cm (much larger on young plants), with c. 7-11 main lateral veins on each side of the midrib, apex rounded to obtuse, base cordate to shortly cuneate. Petiole c. 3 cm long. Stipules long and broad, c. 4 x 2.5 cm, stiff (rather than papery), with 10-20 veins, vein reticulum obscure, glabrous.
OCCURRENCE: U1, 2 and 4. In swamp forest. Abundant in Mengo.
CONSERVATION STATUS: Global: NE (IUCN), LC (TOU); National VU (WCS).

Fleroya rubrostipulata (K. Schum.) Y.F. Deng (304) Rubiaceae
SYNONYMS: *Hallea rubrostipulata* (K. Schum.) J.-F. Leroy; *Mitragyna rubrostipulata* (K. Schum.) Havil.
Munyamaizi (am); Muzingu (ga, to); Nzingu (ga); Engomera, Omuziko (ki), Muziko (na); Mutoro, Mutororo (to).
15 m. Trunk crooked, branches large and few, with large leaves in an often rounded crown. Bark light brown, fairly thick and rough, with vertical or irregular fissures. Slash thick, very soft, fibrous, pink or yellow to light green above, pink below, usually with white lines, turning darker. Leaves simple, opposite, c. 20 x 14 cm (much larger on young trees), with c. 7-11 main lateral veins on each side of the midrib, apex very shortly acuminate, base subcordate, rounded or shortly cuneate, hairy on veins beneath, with conspicuous tufts of hairs in vein axils. Petiole c. 3 cm long. Stipules long and wide, c. 7 x 3.5 cm (but very variable in size), thin and papery, with 6-10 veins branching to form a reticulum, hairy.
OCCURRENCE: U1, 2 and 4. Abundant in swamps in lake-belt forests, Kigezi, Ankole and Kibale Forest. Extending to higher altitudes (2000 m) than *Fleroya stipulosa*.

Nauclea diderrichii (De Wild. & T. Durand) Merr. (305) Rubiaceae
Kibuki-lingi, Kilingi (am); Opepe (tn).

Plate 24. Rubiaceae (304-323): see also Plate 25

304. *Fleroya rubrostipulata* 307. *Pauridiantha callicarpoides* 314. *Coffea canephora* 315. *Belonophora coffeoides* subsp. *hypoglauca* 316. *Leptactina arborescens* 317. *Oxyanthus speciosus* 323. *Rothmannia whitfieldii*

Actual sizes: leaves, fruits and stipule x 2.

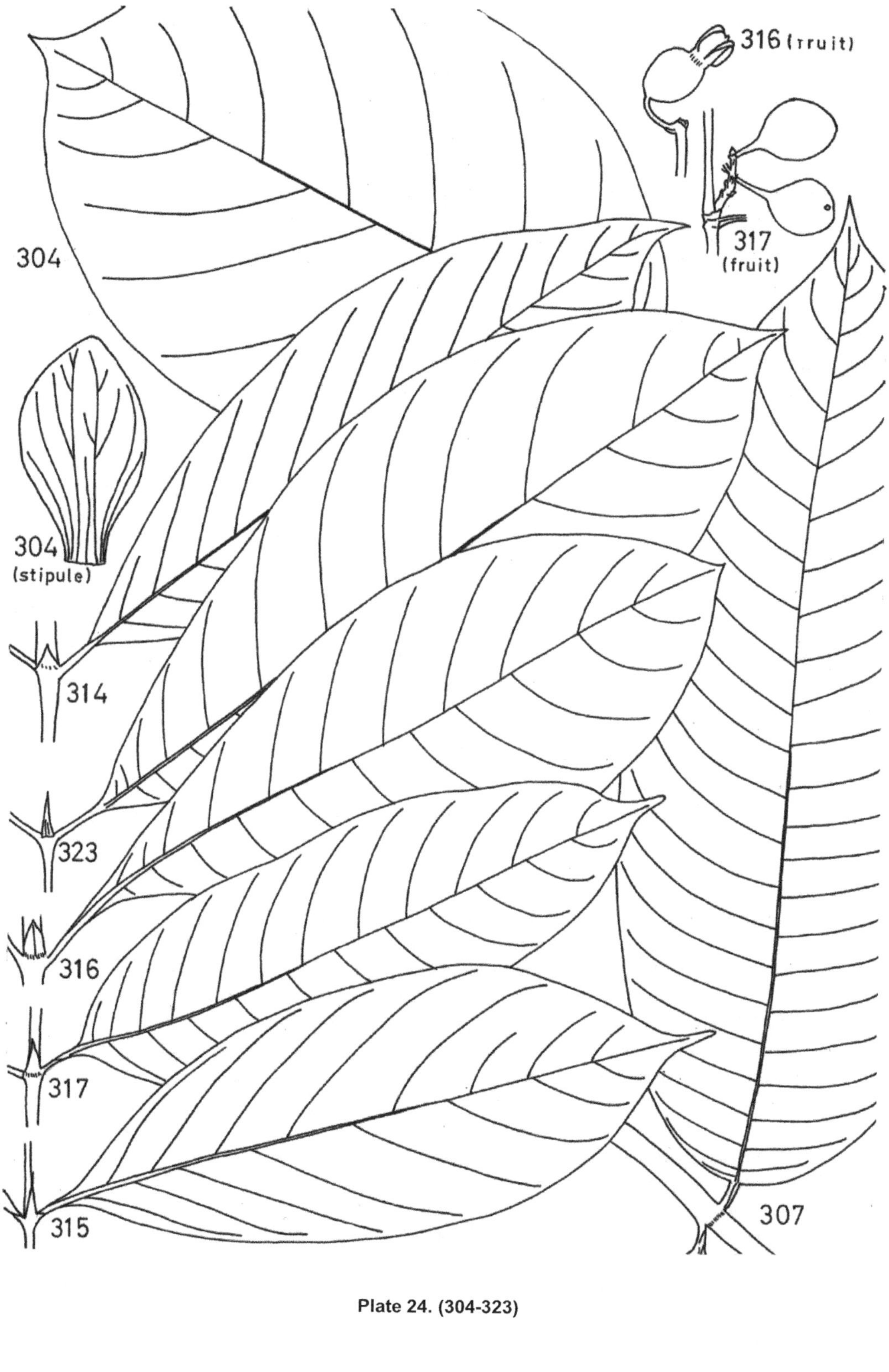

Plate 24. (304-323)

35 m. Trunk straight, rather thin, usually cylindrical (but sometimes fluted). Bark light brown with small longitudinal fissures. Slash fibrous, white or brownish, turning darker. Leaves simple, opposite, c. 14 x 8 cm (larger on young trees), with c. 5 main lateral veins on each side of the midrib, apex rounded to obtuse, base shortly cuneate. Petiole c. 15 cm long. Stipules large c. 2.25 x 1.5-2 cm.
OCCURRENCE: U2. Only recorded from Bwamba.
CONSERVATION STATUS: Global VU (IUCN), LC (TOU); National VU (WCS).

Oxyanthus unilocularis Hiern (306) Rubiaceae
Tree or understorey shrub to 6 m. Branchlets hollow. Leaves simple, opposite, very large (c. 33 x 22 cm), with c. 9-15 main lateral veins on each side of the midrib, apex acute, base usually subcordate and asymmetric, hairy on the veins beneath, more or less sessile or petiole to 1 cm long. Stipules c. 2.5 x 1.5 cm, pointed. Corolla tube very long, c. 15-23 cm.
OCCURRENCE: U1, 2 and 4. In more open habitats. Occasional.

Pauridiantha callicarpoides (Hiern) Bremek. (307) Rubiaceae
15 m. Trunk straight, with branches at right angles. Slash soft, white, usually with yellow dots, sometimes turning darker. Leaves simple, opposite, large (c. 30 x 10 cm), with c. 20-25 main lateral veins on each side of the midrib, apex acuminate, base cordate to rounded. Petiole c. 1 cm long. Stipules c. 2 x 1.4 cm.
OCCURRENCE: U2. In wet highland forests, 1300-1650 m. Common in the Kayonza area.

Pauridiantha viridiflora (Hiern) Hepper (308) Rubiaceae
Understorey tree to 8 m. Trunk irregular, with a spreading dense crown. Bark of moderate thickness, light brown, rough in places, with vertical fissures. Phellogen yellowish. Slash fibrous, light yellow. Leaves simple, opposite, c. 14 x 5 cm, with c. 15-20 main lateral veins on each side of the midrib, these veins raised and very prominent on the lower surface, apex acuminate, base acute, glabrous to hairy on the midrib and main veins below. Petiole c. 1.25 cm long. Stipules c. 1 x 0.7 cm, broad for most of their length, with a narrow tip. Berries spherical, c. 0.6 cm diameter, numerous in loose inflorescences.
OCCURRENCE: U4. Mainly in lake-shore forests. On forest edges.
NOTES: Other tree species of *Pauridiantha* occur. *P. paucinervis* (Hiern) Bremek. (Syn.: *P. holstii* (K. Schum.) Bremek.) is a shrub or small tree to 6 m found in Kalinzu and Bwindi forests (U2). *P. dewevrei* (De Wild. & T. Durand) Bremek. is a shrub or small tree to 8 m, only recorded from U2.

Coffea liberica Hiern (309) Rubiaceae
Mwanyi (am, to); Shari coffee (en); Mumwanyi (ga).
8 m (rarely to 15 m). Understorey tree with a short wavy trunk. Bark grey. Slash yellow. Leaves simple, opposite, large, c. 30-40 x 15-20 cm, with c. 9-14 main lateral veins on each side of the midrib, apex shortly acuminate, base cuneate, more or less glabrous. Petiole c. 2 cm long. Stipules c. 0.5 x 0.8 cm. Flowers white, borne in the leaf axils. Berry red, 1.25-2 cm long.
OCCURRENCE: U1 and 2. Only known from Bwamba, Itwara, Zoka and Kilak (Acholi) forests.

Craterispermum schweinfurthii Hiern (310) Rubiaceae
SYNONYM: *Craterispermum laurinum* (Poir.) Benth. (*sensu* ITU & UFT)
Mpoomerezi (ga);Munura (na); Enura (to).
6 m. Understorey tree with a crooked trunk and dense crown. Bark thin, light brown, with prominent ring marks. Phellogen green. Slash yellow and/or white. Young stems more or less glabrous. Leaves simple, opposite, c. 15 x 6.5 cm (sometimes considerably smaller), with prominent reticulate venation and comparatively indistinct main lateral veins, widest in the upper part, apex very shortly acuminate, base cuneate. Petiole c. 1.5 cm long. Stipules short and broad, c. 0.3 x 0.5 cm, the stipules on opposite sides of the stem being fused to one another.

OCCURRENCE: U1, 2 and 4. Abundant in many forests, particularly lake-shore forests and Kalinzu Forest.
NOTE: The Luganda, Runyankore and Rutooro names refer to the sweetness of the bark when chewed.

Vangueria apiculata K. Schum. (311) Rubiaceae
Amalere (at); Mutugunda (ga); Etoukoroi (ka); Omuyagare (ki); Kasogo (ko); Amalera (la); Kidangerere, Shikomosi (ms); Lutegankofu (so).
10 m. Deciduous tree or shrub. Leaves simple, opposite, rather thin and papery, c. 13 x 6.5 cm (but variable in size), with c. 7-11 main lateral veins on each side of the midrib, vein reticulum very prominent, apex acuminate, base more or less rounded (occasionally cuneate), glabrous. Petiole c. 0.7 cm long. Stipules long and thin, c. 0.8 x 0.3 cm, upper part strap-shaped. Fruit yellow-brown.
OCCURRENCE: U1-4. Widely distributed. Mainly in open places (on forest edges and under open forest). Also in agricultural land.
CULTIVATION AND PROPAGATION: Moderately fast-growing. Can grow on a wide range of soils, including depleted land. Collect fruits from the tree or from the ground beneath. Remove the seeds manually. Nick the hard seed coat before sowing.
NOTES: The fruit (*ettugunda* in Luganda) is eaten. Another species of *Vangueria*, *V. madagascariensis* J.F. Gmelin (Syn.: *V. acutiloba* Robyns) (illustrated on Plate 25), sometimes occurs on forest edges and in evergreen and riverine forests. The leaves are usually somewhat larger than *V. apiculata* (up to 22 x 12 cm) and not so conspicuously acuminate.

Morinda lucida Benth. (312) Rubiaceae
Mukiringi (am); Mubajjansayi (ga); Muganzura, Mularankoba (na); Omukalabafu (no); Musinganjovu (sa); Mulyambwa (so).
12 m. Spreading tree with a crooked trunk and dark-coloured crown. Bark smooth, grey. Slash yellow. Wood bright yellow (very characteristic). Leaves simple, opposite, c. 13 x 6 cm (but variable in size), with c. 7-10 main lateral veins on each side of the midrib, apex very shortly acuminate, base rounded to slightly cuneate, glabrous (except sometimes for base of midrib). Petiole c. 1 cm long. Stipules large and rounded, c. 2 cm broad, falling off early. Fruits fused together.
OCCURRENCE: U2-4. Common on forest edges. A light-demanding species.
NOTES: Another species, *Morinda titanophylla* E.M.A. Petit, occurs in Kigezi and Kasyoha-Kitomi forests. It is normally a bush to 3 m tall, but is said to sometimes reach 8 m. The leaves are large (c. 36 x 12 cm) and long attenuate at base.

Bertiera racemosa (G. Don) K. Schum. (313) Rubiaceae
5 m. Understorey tree or shrub. Branchlets 4-angled. Leaves c. 22 x 10 cm, with c. 9-12 main lateral veins on each side of the midrib, apex acute or very shortly acuminate, base obtuse to rounded, hairy on the midrib and main veins below. Petiole c. 0.8 cm long. Stipules c. 1.8 x 0.7 cm.
OCCURRENCE: U4. Lake-shore forests. Uncommon.
NOTE: Two other species of *Bertiera* found in Ugandan forests are *B. capitata* De Wild. (lake-shore forests) and *B. globiceps* K. Schum. (lshasha Gorge).

Coffea canephora A. Froehner (314) Rubiaceae
Mwanyi (am, to); Wild robusta coffee (en); Mumwanyi (ga); Omwanyi (no).
6 m. Understorey tree with a dense crown. Branchlets curving down. Leaves simple, opposite, c. 22 x 8 cm, with c. 11-15 main lateral veins on each side of the midrib, apex acuminate, base rounded to cuneate, more or less glabrous, except sometimes for tufts of hairs in the vein axils

below. Stipules variable in shape, c. 0.4 x 1 cm to 1.0 x 0.5 cm, triangular, pointed. Berry red, c. 1.25 cm long.
OCCURRENCE: U1-4. Abundant in Zoka Forest and parts of Kibale Forest.
NOTE: Robusta coffee is a form of this species.

Belonophora coffeoides Hook. f. subsp. _hypoglauca_ (Welw. ex Hiern) S.E. Dawson & Cheek (315) Rubiaceae
SYNONYM: *Belonophora hypoglauca* (Welw. ex Hiern) A. Chev.
Nabbumba (ga).
12 m. Understorey tree with a straight thin trunk and branches either at right angles or slightly ascending. Bark light brown, quite thick, with vertical fissures. Slash fibrous, white to light yellow, sometimes turning darker. Young shoots and leaves glabrous or almost so. Leaves simple, opposite, c. 22 x 8.5 cm, with c. 6-9 main lateral veins on each side of the midrib, widest in upper half, apex acuminate, base acute to cuneate. Petiole c. 1 cm long. Stipules c. 12 x 0.3 cm, triangular, gradually tapering to a sharp point. Flowers sessile.
OCCURRENCE: U1, 2 and 4. Common in parts of Mabira.

Leptactina arborescens (Welw. ex Benth. & Hook. f.) De Block (316)
Rubiaceae
SYNONYM: *Dictyandra arborescens* Hook. f.
10 m. Understorey tree with an irregular, knobbly, trunk and a heavy spreading crown. The trunk characteristically bears blunt woody 'spines'. Bark brown, of moderate thickness, fairly rough, with small vertical fissures, flaking. Slash pink with short orange lines. Young shoots flattened, more or less glabrous. Leaves simple, opposite, c. 20 x 9 cm, with c. 10-13 main lateral veins on each side of the midrib, widest in upper half, apex very shortly acuminate, base long cuneate, glabrous except sometimes for tufts of hairs in the vein axils below. Petiole c. 1.5 cm long. Stipules c. 1 x 0.7 cm, rounded or triangular, with a short sharp point. Fruit c. 2 x 1.5 cm, capped by the large persistent sepals.
OCCURRENCE: U2 and 4. An abundant understorey tree in lake-belt forests.

Oxyanthus speciosus DC. (317) Rubiaceae
10 m. Understorey tree. Trunk often not cylindrical, fluted and knobbly (at least in lower part), larger branches vertical with branchlets at right angles. Bark light brown, fairly thick and rough, vertically fissured. Slash fibrous, light yellow to orange (occasionally white), turning brown. Young shoots flattened, more or less glabrous. Leaves simple, opposite, c. 15 x 6.5 cm (but rather variable in size), with c. 11-14 main lateral veins on each side of the midrib, glabrous except for tufts of hairs in vein axils below, apex acuminate, base shortly cuneate. Petiole c. 1 cm long. Stipules c. 0.8 x 0.5 cm, triangular.
OCCURRENCE: U2-4. An abundant understorey tree in lake-belt forests.
NOTE: Other species of *Oxyanthus* occur. *O. formosus* Planch grows to 5 m.

Plate 25. Rubiaceae (308-327); see also Plate 24

308. *Pauridiantha viridiflora* 310. *Craterispermum schweinfurthii*
311. *Vangueria apiculata* 312. *Morinda lucida* 318. *Pavetta molundensis*
319. *Tarenna pavettoides* 320. *Galiniera saxifraga* 321. *Psychotria mahonii*
322. *Rothmannia urcelliformis* 325. *Psydrax parviflora* 327. *Coffea eugenioides*

Actual sizes: leaves and fruit x 2.

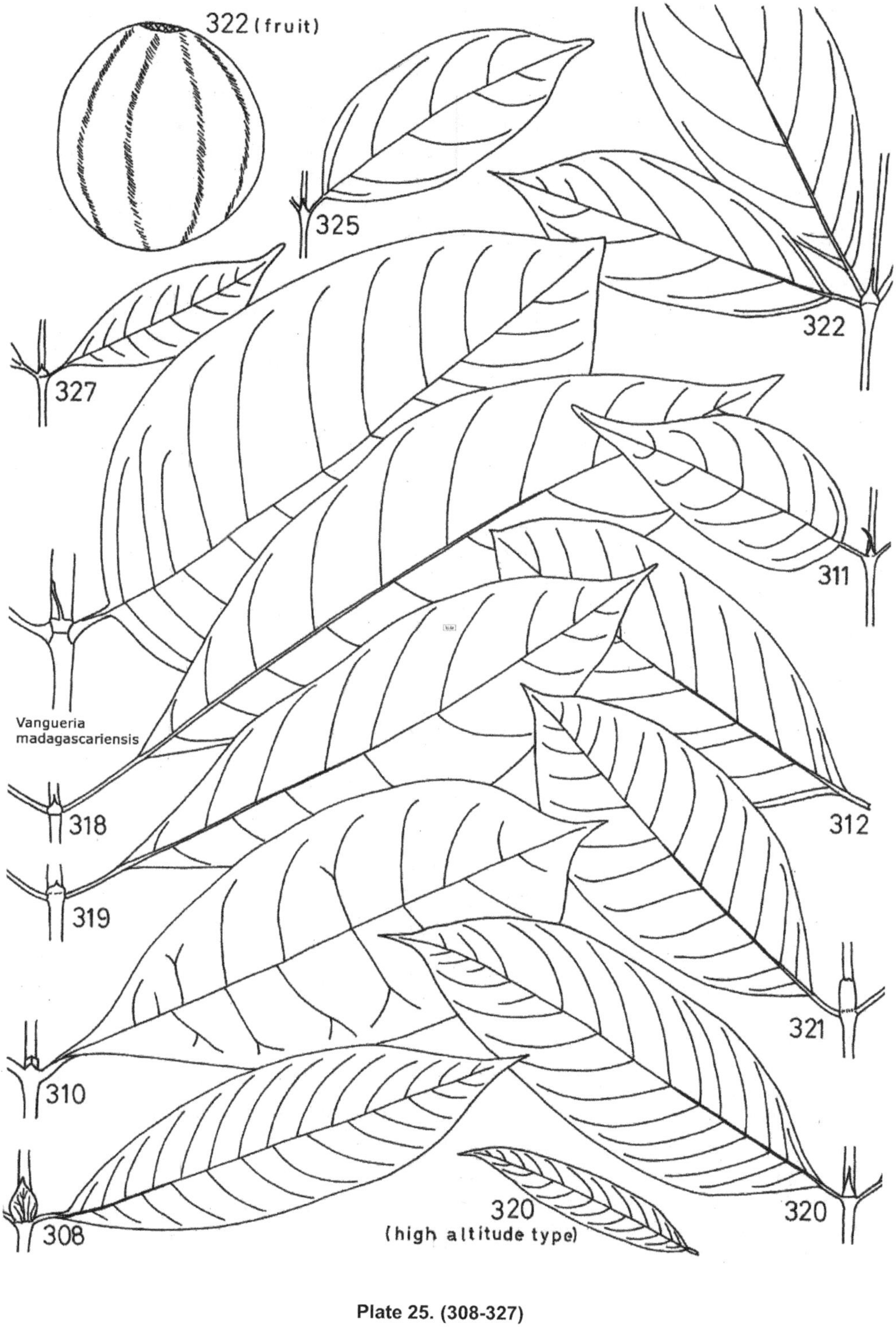

Plate 25. (308-327)

Pavetta molundensis **K. Krause (318)** Rubiaceae

SYNONYM: *Pavetta insignis* Bremek.

7 m. Usually a shrub, occasionally a tree. Leaves simple, opposite, shiny, c. 20 x 7.5 cm (but variable in size, sometimes up to 30 cm long, sometimes only 10 cm), with c. 6-10 main lateral veins on each side of the midrib, apex acuminate, base cuneate, glabrous or hairy on the midrib and main veins below, often with scattered small black dots. Petiole c. 1.5 cm long (but variable in size). Stipules c. 0.5 cm long, with an abrupt sharp tip, the stipules on opposite sides clearly fused to one another and forming a well-marked ring. Flowers white, tinged with red on outside and yellow in the throat, sweet-smelling.

OCCURRENCE: U2 and 4. On forest edges and in open places in forests. Common.

NOTES: Various other species of *Pavetta* are forest shrubs or small trees, often with attractive inflorescences. *Pavetta urundensis* Bremek. is a shrub to small tree nearly impossible to distinguish from *P. molundensis* based on easily visible vegetative characters, but often having bacterial nodules on the nerves (these are lacking in *P. molundensis*). Its conservation status is: Global VU (IUCN); National NE. Other shrubs to small trees are *P. acrochlora* Bremek., *P. oliveriana* Hiern, *P. ruwenzoriensis* S. Moore and *P. ternifolia* (Oliv.) Hiern.

Ixora seretii **De Wild. (318a)** Rubiaceae

Small tree or shrub to 4.5 m. Bark more or less smooth. Young branches glabrous. Leaves glabrous, 8-19.5 cm long, 3.3-6 cm wide, acute to subacuminate at apex, obtuse or cuneate at base, nerves not impressed above, smaller nerves apparent beneath. Domatia and bacterial nodes absent (they are usually present in *Pavetta*). Petiole 0.4-1 cm long.

OCCURRENCE: U2. Only known from Semuliki National Park in Uganda; also in eastern D.R. Congo. A range-restricted species of narrow endemism. Reported from forest and riverine forest.

Tarenna pavettoides **(Harv.) Sim (319)** Rubiaceae

Kuizuzu (ms).

10 m. Trunk crooked. Branches not obviously at right angles. Bark light brown, quite thin, rough, flaking or granular in places. Slash yellow, sometimes becoming reddish towards the bark. Leaves simple, opposite, thin and papery, c. 17 x 5.5 cm (but variable in size), with c. 8-12 (-16) main lateral veins on each side of the midrib, glabrous or hairy on veins below. Petiole c. 1.5 cm long. Stipules c. 0.4 cm long, not (or not very obviously) fused to one another (as in *Pavetta*) and more triangular in shape, with a short-pointed tip.

OCCURRENCE: U2-4. Usually a savanna species.

NOTES: This species closely resembles *Pavetta molundensis* K. Krause. The stipules provide the best characters for separation. A related species, *Coptosperma graveolens* (S. Moore) Degreef (Syn.: *Tarenna graveolens* (S. Moore) Bremek.), occurs in savanna.

Galiniera saxifraga **(Hochst.) Bridson (320)** Rubiaceae

SYNONYM: *Galiniera coffeoides* Delile

Omugushagwenkombe, Omuryanyonyi (ki); Mulyangote (ko); Labatwa (ms); Mulyansule (to).

15 m. Branches at right angles, either horizontal or drooping. Bark smooth, pink-grey. Slash light brown with orange markings. Leaves simple, opposite, c. 13 x 4.5 cm (but very variable in size), with c. 11-20 main lateral veins on each side of the midrib, apex shortly acuminate, base obtuse to cuneate, shiny above, usually markedly hairy on main veins and midrib below, veins sometimes with pits in axils below. Petiole c. 1 cm long (but variable in size). Stipules c. 0.8 x 0.6 cm (but variable in size), apex pointed. Fruit red, c. 1 cm long.

OCCURRENCE: U1-3. In montane forest, c. 1700-2500 m. Abundant in Kigezi and in Kalinzu Forest.

NOTE: Higher altitude specimens tend to have small, narrow, leaves (c. 7 x 1.5 cm) and much shorter petioles and stipules than specimens at lower altitudes.

***Psychotria mahonii* C.H. Wright (321)** Rubiaceae

15 m. Understorey tree with an irregular trunk. Bark moderately rough, tending to split into squares, scaling. Underside of bark red. Slash granular, pink. Branchlets knobbly due to the persistence of the bases of the petioles at each node. Leaves simple, opposite, c. 13 x 6 cm (but variable in size), with c. 8-18 main lateral veins on each side of the midrib, usually with conspicuous pits and sometimes tufts of hairs in vein axils below, apex acuminate, base cuneate to obtuse, hairs sometimes present on midrib and main veins below. The midrib and main veins are reddish in colour. Petiole c. 0.5-2.5 cm long. Stipules c. 1 x 0.5 cm, broad in upper part, tapering at end to a rounded or acute apex.
OCCURRENCE: U2 and 4. Common on Rwenzori and in Kigezi, 1900-2600 m, also growing in Kalinzu Forest and reported to occur in Mengo, Ankole and Bwamba.
NOTES: Various other species of *Psychotria* occur, including *P. riparia* (K. Schum. & K. Krause) E.M.A. Petit on the Ssese Islands and *P. bagshawei* E.M.A. Petit, which grows in riverine forest and secondary forest. *Psychotria bagshawei* has knobbly branchlets and leaves 11 x 3.5 cm in size.

***Rothmannia urcelliformis* (Hiern) Robyns (322)** Rubiaceae

SYNONYM: *Randia urcelliformis* (Hiern) Eggeling
Munyaburo (to).
10 m. Understorey tree with a thin, often leaning, trunk. Bark brown, fairly thin and smooth, with small vertical fissures. Slash yellow. Branches tending to be at right angles. This and other species of *Rothmannia* have unusual branching, making the genus easy to recognize: the shoot terminates in a flower or leaf at each node (so that the leaves may appear to be borne in threes) and growth continues by one or both of the axillary shoots. Leaves simple, opposite, c. 11 x 3-5 cm, with c. 5-8 main lateral veins on each side of the midrib, apex acuminate, base cuneate, glabrous except sometimes for tufts of hairs in vein axils below. Petiole c. 0.7 cm long. Stipules c. 0.5 x 0.4 cm, sharp pointed. Flowers borne upright on the branches, large, c. 6-8 long. Fruit c. 6 cm long, 2.5 cm diameter.
OCCURRENCE: U1-4. Common understorey species, often in drier forest types. Common on Mt Kadam.
NOTE: *Rothmannia longiflora* Salisb. is a rather similar-looking species. It has very long flowers (c. 16 cm) and usually conspicuous pits in the axils of the veins beneath.

***Rothmannia whitfieldii* (Lindl.) Dandy (323)** Rubiaceae

5 m. Understorey tree. Shoot growth as with *Rothmannia urcelliformis*. Leaves c. 21 x 9.5 cm, c. 9-12 main lateral veins on each side of the midrib, apex shortly acuminate, base cuneate, glabrous except sometimes with a few hairs on the midrib and main veins below. Petiole c. 1.2 cm long, thick. Stipules c. 1 x 0.4 cm, gradually tapering to a sharp point.
OCCURRENCE: U1, 2 and 4. Not common.

***Heinsenia diervilleoides* K. Schum. (324)** Rubiaceae

10 m. Understorey tree. Leaves simple, opposite, c. 11 x 3 cm, apex acute, base long attenuate, with c. 4-6 main lateral veins on each side of the midrib, these veins very prominent below and ascending at a fairly steep angle, leaves hairy below, particularly on the veins. Petiole c. 0.5 cm long. Stipules c. 0.3 x 0.3 cm, sharp pointed.
OCCURRENCE: U2 and 3. Kalinzu Forest and Mt Elgon.

***Psydrax parviflora* (Afzel.) Bridson (325)** Rubiaceae

SYNONYM: *Canthium vulgare* (K. Schum.) Bullock
Kamwanyimwanyi (ga).

12m. Shrub or tree with a spreading, rather dark-coloured, crown. Branches at right angles. Bark fairly thick, rough, rather dark-coloured. Slash light pink, yellow/orange to brown, turning darker, with a strong smell of almonds. Leaves simple, opposite, c. 10 x 4 cm (but variable in size), with c. 3-5 main lateral veins on each side of the midrib, sometimes with pits in vein axils below, apex acuminate, base shortly cuneate, glabrous. The midrib and main veins are usually yellowish. Petiole c. 0.6 cm long. Stipules c. 0.5 x 0.3 cm, pointed. Flowers small, sweet-scented.
OCCURRENCE: U1-4. Common on forest edges.
NOTES: Many other species of *Psydrax* occur. *Psydrax schimperiana* (A. Rich.) Bridson (Syn.: *Canthium schimperianum* A. Rich.) is a tree to 12 m found in dry forests and savanna. *Psydrax acutiflora* (Hiern) Bridson (Syn.: *Canthium lacus-victoriae* Bullock) is a shrub of lake-shore forests.

***Aidia micrantha* (K. Schum.) Bullock ex F. White (326)** Rubiaceae
10 m. Understorey tree with a straight trunk and long narrow crown. Branchlets drooping. Bark thin and smooth, brown. Slash thin, whitish, with some yellow granules. Leaves simple, opposite, c. 12 x 4.5 cm with c. 4-9 main lateral veins on each side of the midrib, apex acuminate, base acute to cuneate, glabrous except sometimes in the vein axils below. Petiole c. 0.6 cm long, Stipules c. 0.5 x 0.3 cm, pointed. Flowers deep reddish-purple outside, apricot coloured within.
OCCURRENCE: U2 and 4. Widespread, but uncommon except in the vicinity of lshasha Gorge.

***Coffea eugenioides* S. Moore (327)** Rubiaceae
Mwanyi (na, to); Mumwanyi (ga); Omwanyi (no); Nkinga (to).
4 m. Shrub or small understorey tree. Trunk wavy. Crown small, dark-coloured. Bark light brown, quite rough, flaking. Slash yellow-white. Young leaves red. Leaves simple, opposite, c. 8.5 x 3 cm (sometimes rather smaller), with c. 4-10 main lateral veins on each side of the midrib, these veins not always easily distinguishable from the secondary lateral veins, apex acuminate, base acute, glabrous. Petiole c. 0.5 cm long. Stipules c. 0.3 x 0.3 cm, sharp-pointed.
OCCURRENCE: U1, 2 and 4 (rare in 4). Abundant in Kibale and Maramagambo forests and parts of Kalinzu.
NOTES: Fruits edible. *Calycosiphonia spathicalyx* (K. Schum.) Robbr. (Syn.: *Coffea spathicalyx* K. Schum.) is a similar-looking plant, recorded from Bwamba and Zoka forests and possibly present in Kayonza. Its leaf has a very long, thin acumen.

***Chassalia subochreata* (De Wild.) Robyns (327a)** Rubiaceae
Small shrub or tree to 9 m, with slender, branched, glabrous stems. Leaves oblanceolate to narrowly elliptic, up to 18 x 5.5 cm, apex narrowly acuminate, base narrowly cuneate, thin, glabrous. Petiole to 3.5 cm long.
OCCURRENCE: U2. Bwindi and Kalinzu forests. In evergreen forest.

***Rytigynia acuminatissima* (K. Schum.) Robyns (327b)** Rubiaceae
Small tree.
U3. Montane forest on Mt Elgon.
CONSERVATION STATUS: Global CR (IUCN), LC (TOU); National NE.

***Rytigynia kigeziensis* Verdc. (327c)** Rubiaceae
Small tree.
OCCURRENCE: U2. Bwindi Forest. In evergreen forest, bamboo forest and, less often, wooded grassland.

CONSERVATION STATUS: Global VU (IUCN), LC (TOU); National NE.
NOTE: Provides an important local herbal medicine.

Rytigynia ruwenzoriensis (De Wild.) Robyns (327d) Rubiaceae
Small tree.
OCCURRENCE: U2. Montane forest, in evergreen and bamboo forest and on forest edges.
CONSERVATION STATUS: Global VU (IUCN), LC (TOU); National NE.

Tricalysia bagshawei S. Moore subsp. _bagshawei_ (327e) Rubiaceae
Small tree.
OCCURRENCE: U1, 3 and 4. In forest, riverine and lakeside forest, evergreen thicket and secondary bushland.
NOTE: The genus *Tricalysia* appears to be closest to *Coffea*. Stipules sheathing in *Tricalysia*; very shortly united above the axils in *Coffea*.

Xymalos monospora (Harv.) Warb. (328) Monimiaceae
Umuhotora (fu); Ttokekkulu (ga); Omuhotora (ki); Chebanatit, Lusuari, Namalindi (ms); Mubarara (na); Lemon wood (tn).
15 m. Understorey tree with a crooked trunk and irregular, spreading, crown. Bark thick, rough, vertically fissured, flaking, light brown. Slash granular, light yellow to white, with darker streaks. Leaves opposite to sub-opposite, thick and glabrous, c. 15 x 5.5 cm, generally with a few large teeth, gland-dotted. The crushed leaves smell of lemons. Petiole c. 2 cm long.
OCCURRENCE: U1-4. Widely distributed, but only common at 1400-2500 m (e.g. in Kibale and Bwindi forests and on Mt Elgon) and in lake-shore forests in Masaka.
NOTE: The wood is attractive and durable.

Elaeodendron buchananii (Loes.) Loes. (329) Celastraceae
SYNONYM: *Cassine buchananii* Loes.
Sunwa (ku); Munyakakabale (to).
30 m, but usually only to 20 m. Trunk straight, cylindrical to irregularly shaped. Crown spreading. Buttresses sometimes present. Bark grey to brown, either fairly thin and smooth with conspicuous lenticels or rough and thick and longitudinally fissured. Inner surface of bark bright orange. Slash brittle, pink to red, sometimes with white lines. Leaves usually opposite (occasionally alternate), c. 10 x 3.5 cm (but variable in size), margin typically crenate (but variable). Petiole c. 1 cm long.
OCCURRENCE: U1-4. Widely distributed, found up to 2300 m. Probably uncommon, except in Kibale Forest (where it reaches the canopy).
NOTE: Plant poisonous to domestic animals.

Catha edulis (Vahl) Forssk. (330) Celastraceae
Khat tea, Somali tea (en); Ameirungi (ki); Tumeyondet (ku); Kitandwe, Lutandwe (ms); Mutabungwa (na).
20 m. Trunk straight and thin with a small crown. Bark smooth, grey, with shallow vertical fissures and numerous lenticels. Slash white with red streaks. Leaves opposite (rarely alternate on sterile shoots), c. 9 x 3 cm, with many regularly arranged crenations, base cuneate. Petiole c. 0.7 cm long.
OCCURRENCE: U1-3. A colonizing species, 1500-2500 m. One of the most abundant trees on the lower northern slopes of Mt Kadam.
NOTES: The leaves are usually narrower and the crenations more numerous than with *Elaeodendron buchananii*, but it can sometimes be difficult to distinguish between the two species in the vegetative state. The leaves contain a drug that causes loss of appetite and inhibits sleep.

Buddleja polystachya **Fresen. (331)** Scrophulariaceae
Butterfly bush (en); Lugohago (ms).
Shrub to 5m. Leaves simple, opposite, lanceolate. c. 10 x 2.5 cm, covered below with pale brown or whitish hairs. Flowers orange.
OCCURRENCE: U1 and 3. On Elgon and the Karamoja mountains, 2200-2750 m. In open forests.

OLEACEAE

With the exception of *Schrebera alata* (379), which has pinnate leaves, the Oleaceae have simple, opposite, entire leaves. The slash is usually white or yellow, often with fibrous bands and/or stone cells. The flowers have 4 petals and 2 carpels. The fruit is a capsule in *Schrebera*, but indehiscent in the other genera.

Key to Oleaceae.

1. Leaves pinnate. ....**379. *Schrebera alata***
   Leaves simple. 2
2. Bark dark brown, fibrous. Lower surface of leaves yellow to brown (due to a covering of scales). Leaves relatively narrow (c. 2 cm broad). ....**335. *Olea europaea***
   Not as above. 3
3. Petiole 2.25 cm long or longer. 4
   Petiole shorter than 2.25 cm. 5
4. Slash turning rapidly darker. ....**332. *Schrebera arborea***
   Slash not turning rapidly darker. ....**333. *Olea capensis* subsp. *welwitschii***
5. Leaves comparatively broad in proportion to length, often c. 8-10 x 4.5 cm (see Plate 26). 6
   Leaves comparatively narrow in proportion to length, often c. 13-19 x 5 cm. 7
6. Slash discolouring rapidly. ....**332. *Schrebera arborea***
   Slash not discolouring rapidly. ....**334. *Olea capensis* subsp. *hochstetteri***
7. Leaves often over 16 cm long. ....**336. *Chionanthus africanus***
   Leaves often less than 16 cm long. ....**337. *Chinonanthus mildbraedii***

Schrebera arborea **Chev. (332)** Oleaceae
Muhona (am); Ndera (ga, tn); Munalibo (ko); Nabulamu, Nawulamu (sa).
Deciduous tree to 20 m. Trunk long and thin (but only rarely completely straight), sometimes branched low down. Small buttresses sometimes present. Bark thin and smooth, light brown to orange, with prominent lenticels, flaking in pieces c. 2-5 cm across. Slash white and brown, granular or fibrous, turning very rapidly darker. Leaves simple, opposite, c. 8 x 4.5 cm. Petiole c. 2 cm long. Fruit a 2-valved woody capsule, 4-6 cm long.

Plate 26. Oleaceae, Myrtaceae and others (328-342)

328. *Xymalos monospora* 329. *Elaeodendron buchananii* 330. *Catha edulis*
332. *Schrebera arborea* 333. *Olea capensis* subsp. *welwitschii*
334. *Olea capensis* subsp. *hochstetteri* 335. *Olea europaea* 336. *Chionanthus africanus*
337. *Chionanthus mildbraedii* 338. *Olinia rochetiana* 339. *Syzygium guineense*
340. *Syzygium cordatum* 341. *Eugenia bukobensis* 342. *Memecylon myrianthum*

Actual sizes: leaves x 2; trunk base x 80; tree profiles x 800.

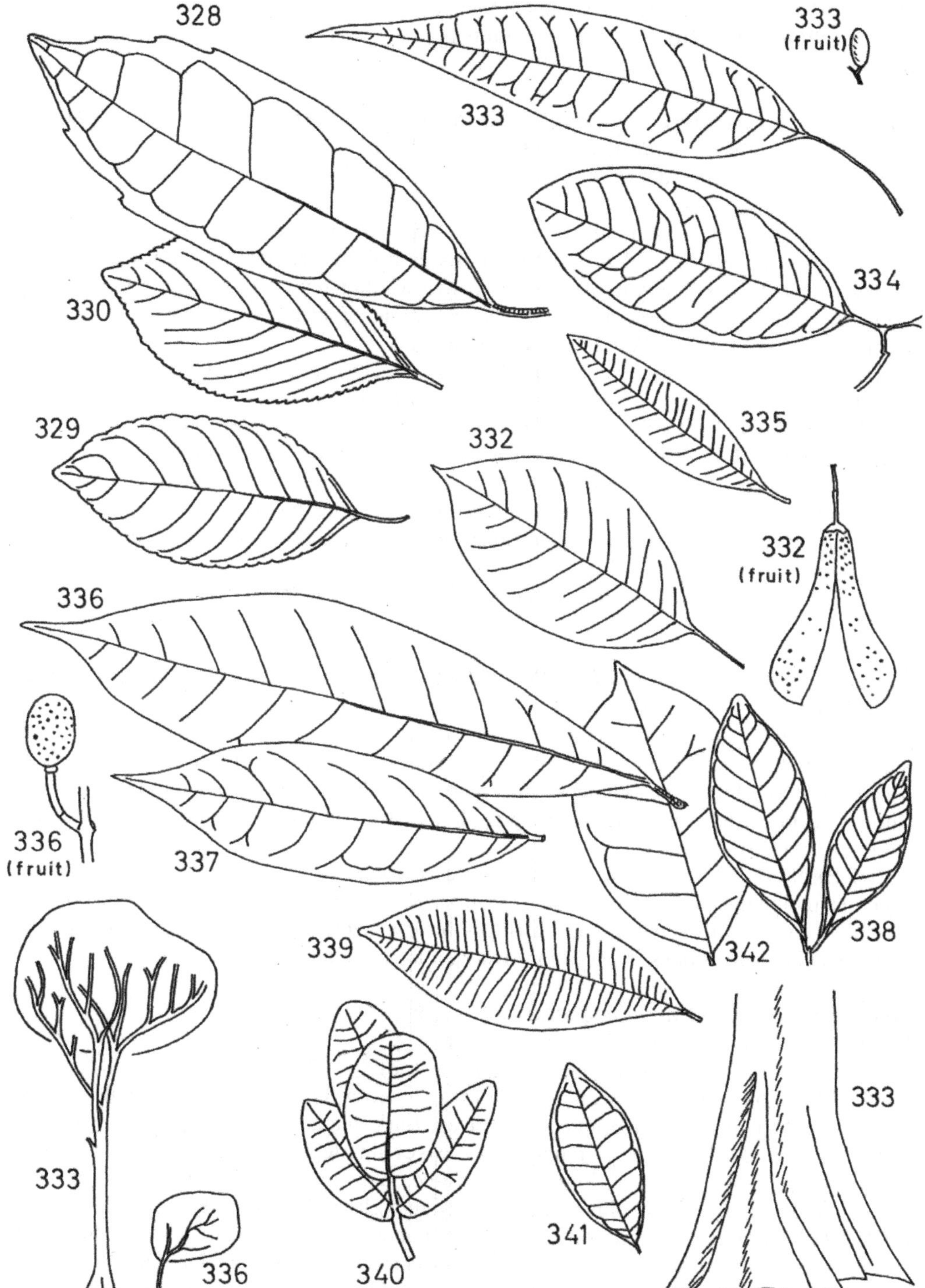

Plate 26. (328-342)

OCCURRENCE: U1, 2 and 4. Occasional in Mengo, Masaka and Bunyoro, but rare elsewhere.
CULTIVATION AND PROPAGATION: Probably best grown under shade in mixed stands. Propagation as for *Schrebera alata* (379).

Olea capensis L. subsp. _welwitschii_ (Knobl.) Friis & Green (333) Oleaceae
SYNONYM: *Olea welwitschii* (Knobl.) Gilg & Schellenb.
Omugandu (ki); Pekeriaondet (ku); Gibengeyi (ms); Musoke, Musoko (na, to); Elgon olive, Loliondo (tn); Musodo (to).
20 m. Large tree, trunk cylindrical, straight or wavy, bearing large ascending branches and a large, but fairly open, crown. Fairly large buttresses usually present. Bark thick and rough, deeply vertically fissured, brown. Slash white to yellow, often with orange streaks, both fibrous and granular, with a strong pleasant smell. Leaves simple, opposite, rather long and thin, c. 13 x 4 cm, with a long acumen. Petiole c. 2.5-4 cm long.
OCCURRENCE: U2-4. Often on forest edges. Up to 2150 m altitude. Common in Central Kibale Forest and (formerly) on the western slopes of Mt Elgon.
CONSERVATION STATUS: Global NE (IUCN), LC (TOU); National VU (WCS).
CULTIVATION AND PROPAGATION: Moderately fast-growing. Requires shade for the first few years, thereafter full light. Collect fruits from the ground beneath mother trees and remove pulp by soaking in water. Dry the seeds and either plant soon or store at a low temperature. Germination rate low and irregular. Retain the seedlings in the nursery for a year or more, only transplanting seedlings that are strong.
NOTE: The tree is well known for its handsome wood, which smells of olives when freshly cut.

Olea capensis L. subsp. _hochstetteri_ (Baker) Friis & P.S. Green (334) Oleaceae
SYNONYM: *Olea hoschstetteri* Baker
East African olive (en); Masgat (ku); Musharagi (tn).
25 m. Tree with a cylindrical trunk and steeply ascending branches. Buttresses absent. Bark smooth, grey to whitish, becoming slightly vertically channelled on older trees. Slash off-white, with brown to orange markings, granular. Leaves simple, opposite, c. 10 x 4.5 cm. Petiole c. 0.5-1.25 cm long.
OCCURRENCE: U1-3. Ankole (Kalinzu Forest), Kigezi, Tooro and Mt Elgon and the Karamoja mountains. Generally between 1500 and 2500 m altitude (to 2900 m on drier mountains). A tree of dry sites, e.g. on ridges on Rwenzori and in Kigezi and on the drier Northern and Eastern slopes of Mt Elgon (where it is very abundant).

Olea europaea L. (335) Oleaceae
SYNONYM: *Olea africana* Mill.
Brown olive, Wild olive (en); Yemit (ku); Murama (na).
25 m on Mt Elgon, but more commonly to 10 m. Sometimes multi-stemmed. Trunk crooked, fluted, bearing a much-branched, spreading, crown. Bark rough, dark brown, vertically fibrous. Slash white with small brown or orange lines. Leaves simple, opposite, rather stiff, c. 7.5 x 2 cm, narrowing gradually towards both ends, covered below with small yellow or brown scales (these are individually visible only using a magnifying glass). Petiole c. 0.5 cm long.
OCCURRENCE: U1-4. Abundant at 2000-3000 m in dry montane forest on Mt Elgon and the Karamoja mountains. Also abundant in crater forests in Queen Elizabeth National Park and in thickets in Masaka. Apparently absent from Kigezi and the Ugandan side of Rwenzori.
NOTES: This species (as a wild plant) has a huge geographical range, being found from South Africa to China. The cultivated olive tree originated from this species.

Chionanthus africanus **(Knobl.) Stearn (336)** Oleaceae

SYNONYM: *Linociera johnsonii* Baker
Mukutulankizi (ga).
15 m. Understorey tree with a crooked trunk, branching from near base and with a fairly spreading crown. Bark thin and smooth, greenish, light brown to whitish, with very prominent brown lenticels, with small vertical fissures on some trees. Phellogen sometimes green. Slash white, with orange/yellow markings, turning red-brown. Young leaves red. Leaves simple, opposite, c. 19 x 5 cm. Petiole thick and scaly, c. 1.75 cm long.
OCCURRENCE: U2 and 4. A widely distributed and often common tree.

Chionanthus mildbraedii **(Gilg & Schellenb.) Stearn (337)** Oleaceae

SYNONYM: *Linociera latipetala* M. Taylor
12 m. Shape as for *Chionanthus africanus*. Bark smooth, grey, with a prominent green phellogen. Slash yellow. Leaves simple, opposite, thick, c. 15 x 5 cm (sometimes rather smaller), apex acuminate. Petiole c. 0.5 cm long.
OCCURRENCE: U2 and 4. Uncommon. Recorded from Mengo, Ankole and Tooro.
NOTES: The leaf and petiole are shorter than with *Chionanthus africanus*.

Olinia rochetiana **A. Juss. (338)** Penaeaceae

SYNONYM: *Olinia usambarensis* Gilg
Omubaba (ki); Nerekio (ku).
15 m, but often less. Canopy tree with a straight trunk or (sometimes) a small, much-branched, shrub. Bark red-brown, rough, flaking in pieces c. 1.25 cm wide. Slash white to pale pink (on larger trees), with a characteristic, unpleasant, burnt smell. Branchlets 4-angled. Leaves simple, opposite, red when young, c. 7 x 3 cm (sometimes rather larger), with a red midrib and entire margin (toothed on young plants). Vein reticulum prominent. Petiole c. 0.3 cm long, grooved. Flowers small, borne in strong-smelling bunches. Drupes red, attractive.
OCCURRENCE: U1-3. In montane forest, 1600-2900 m. Sometimes in climax forests on drier sites, but more often in secondary forest or on forest edges.

MYRTACEAE

A family of woody plants, the flowers often having very conspicuous and numerous brightly coloured stamens. The species contain essential oils. Additional to indigenous species of *Syzygium* (339-340) and *Eugenia* (341), the Myrtaceae includes a number of prominent introduced trees, including *Eucalyptus*, the bottlebrush tree (*Callistemon*), guava (*Psidium guajava* L.) and Java plum (*Syzygium cumini* (L.) Skeels), a commonly planted fruit tree.

Key to *Syzygium*.

1. Petiole more than 6 mm long. 2
   Petiole very short (often to 2 mm long (occasionally to 5 mm) ....**340. *S. cordatum***
2. Leaves often more than 9 cm long. ....**339. *S. guineense***
   Leaves often less than 9 cm long. ....**339a. *S. congolense***

Syzygium guineense **(Willd.) DC. (339)** Myrtaceae

Lugwaya (ac); Oya (al); Kalunginsanvu (ga); Anigo, Kuzu (gb); Ekuyam (ka); Omugoote (ki); Lemaiyua, Reberwo (ku); Kano (lo); Chiemo, Sizanzass, Wandiviri (ms); Mosimangwa (na); Mutuli (sa).

30 m. Trunk occasionally straight, but more usually irregular, with a dark-coloured crown. Large flutes and/or buttresses sometimes present. Bark thin, light brown. Slash brittle, light brown, dark brown to red-brown (probably always with some red shade). Young leaves reddish, old leaves yellowish. Leaves simple, opposite, c. 10 x 4 cm, with numerous lateral veins, thick, slightly fragrant when crushed. Petiole more than 6 mm long.
OCCURRENCE: U1-4. Widely distributed, occurring up to 2300 m. Common in lake-shore forests and in wet montane forest.
CULTIVATION AND PROPAGATION: Moderately fast-growing, grows best in moist places. Probably best grown as isolated trees or in pure stands at wide spacing. Light-requiring and intolerant of competition. Collect ripe berries from mother trees or else from the ground beneath soon after they have fallen. Remove the pulp and dry the seeds for 2-3 hours. Sow as soon as possible.
NOTE: A well-known medicinal plant in Uganda.

Syzygium congolense Vermoesen (339a) Myrtaceae
Tree to 30 m, sometimes buttressed, with a much-branched rounded habit and a tendency to be flat-topped. Bark cream, silvery-grey to reddish brown, thin and smooth. Slash dark brown, reddish or white, watery. Leaves simple, opposite, oblong, oblong-elliptic or (mostly) obovate to obovate-oblong, often less than 9 cm long (but up to up to 16 x 6.3 cm in size), rounded, apex abruptly apiculate to shortly acuminate, base cuneate, more or less revolute at the margin, often subcoriaceous, midrib strongly impressed above. Petiole more than 6 mm. Young foliage reddish to ruby red.
OCCURRENCE: U2 and U4. Appears to be a much rare tree than *Syzygium guineense*.
NOTE: The leaf is typically less than 9 cm long (compared with typically more than 9 cm long in *S. guineense*).

Syzygium cordatum Krauss (340) Myrtaceae
Kalunginsanvu (ga); Omukoondo (ki); Lemaiyua, Lomoiyo, Reberwo (ku); Munyabariko (na).
10 m. Much-branched tree. Bark dark brown, thick, fissured. Slash soft, fibrous, dark red. Leaves simple, opposite, c. 5 x 3 cm, more or less sessile and clasping the stem, gland-dotted, fragrant when crushed. Inflorescence dense, with fragrant white flowers.
OCCURRENCE: U1-4. A common tree in swamps, particularly in and near papyrus. Also by rivers. The dominant tree in swamp forest in Kigezi at about 2000 m.

Eugenia bukobensis Engl. (341) Myrtaceae
Small tree to 5 m. Leaves aromatic when crushed. Young stems hairy (contrasting with *Syzygium* in which they are glabrous). Leaves simple, opposite, c. 5.5 x 2.5 cm (sometimes a bit bigger). Petiole c. 0.3 cm long.
OCCURRENCE: U2-4. On forest edges.

Memecylon myrianthum Gilg (342) Melastomataceae
10 m. Understorey tree. Bark grey. Slash white. Leaves simple, opposite, c. 9 x 5 cm, with very indistinct pinnate venation. Berries blue.
OCCURRENCE: U2 and 4. Widely distributed. Abundant only in lake-shore forests.

Balanites wilsoniana Dawe & Sprague (343) Zygophyllaceae
Kikirri (am); Naliggwalimu (ga); Omukunga (no); Lukoyo, Rukoyo (to).
35 m. Deciduous tree with a crooked to fairly straight trunk bearing large spreading branches. Trunk deeply fluted, sometimes with compound thorns in the hollows. The flutes are irregular and reach far up the trunk. Bark moderately thick, light brown, flaking or fissuring in places, but general effect smooth. Slash hard, yellow. Branchlets often armed. Leaves opposite, with 2 leaflets. Leaflets c. 8.5 x 5 cm, apex acuminate. Petiolules c. 1 cm long. Fruit 6-10 cm long, 5-ribbed, with an unpleasant smell.

OCCURRENCE: U2 and 4. Mainly on dry sites. Common in places in Central Kibale Forest. Tends to be gregarious.
NOTES: *Balanites* is the only genus among Ugandan trees to regularly have bifoliolate leaves. The fruit is a favourite food of the elephant, which is thought to be mainly responsible for distributing the tree.

Erythrina excelsa **Baker (344)** Fabaceae (Faboideae)

Mubajjangabo (ga); Mulungula, Mulungulu (ko).
30 m. Trunk straight and cylindrical, with a deciduous crown. Trunk armed with woody, conical, horizontally-elongated, bosses. Bark thin, smooth, brown, with prominent lenticels. Slash yellow, fibrous, turning darker, smelling vaguely of peas. Leaves trifoliolate, alternate. Leaflets c. 12 x 7.5 cm (sometimes much larger on young plants). Flowers orange to red.
OCCURRENCE: U1-4. Widely distributed, nearly always in swamp forest.
NOTES: The wood was traditionally used in Buganda for making shields, as the Luganda name implies (*-bajja* 'carve' *engabo* 'shield'). *Erythrina abyssinica* Lam. ex DC. is the common savanna and farmland species.

Erythrina droogmansiana **De Wild. & T. Durand (345)** Fabaceae (Faboideae)

SYNONYMS: *Erythrina* sp. C and *Erythrina* sp. D (both of FTEA)
20 m. Similar to *Erythrina excelsa* in most characters, but with larger leaflets (leaflets on saplings very large), longer pods with very large seed bulges, and only 1-3 seeds per pod. Flowers bright red.
OCCURRENCE: U2. Only known with certainty from Budongo Forest. Possibly also in Bwamba. On well-drained sites.
CONSERVATION STATUS: Global DD (IUCN), LC (TOU); National NE.

Balsamocitrus dawei **Stapf (346)** Rutaceae

Omuboro (ki).
20 m. Deciduous. Bark smooth, fairly dark-coloured. Slash yellow. Spines present, at least on some shoots. Leaves trifoliolate, alternate. Leaflets c. 11 x 5 cm (but variable in size), gland-dotted. Petiolules c. 4 cm long. Fruit elongate-spherical, c. 10 cm long, with a very hard wall and numerous seeds embedded in a fibrous red pulp.
OCCURRENCE: U1, 2 and 4. Widely distributed, but nowhere common.
NOTES: The fruits are eaten by elephants. The bark and leaves are said to have aphrodisiac properties.

Vepris nobilis **(Delile) Mziray (347)** Rutaceae

SYNONYM: *Teclea nobilis* Delile
Achacha (ac); Oya (al); Mubio (am); Ejoroi, Ekude (at); Nzo (ga); Nakomole (gw); Ekodep (ka); Omuzo (ki); Gurio (ku); Achacho (la); Lutati (ms); Mugangwe (nl); Omuzo (no); Mudati (sa); Muzu (so); Muzo (to).
10 m, but usually less. Understorey tree with a crooked trunk, often branching from near base. Crown fairly spreading, dark-coloured. Bark very thin, smooth, often with ring marks, brown to light brown. Phellogen usually present, green. Slash variable in texture, yellow, turning darker. Leaves 1 to 3-foliolate, alternate. Leaflets c. 14 x 4 cm, gland-dotted. Petiolules short, to 0.75 cm long.
OCCURRENCE: U1-4. Abundant in many forests, particularly on drier sites, up to 2500 m.
CULTIVATION AND PROPAGATION: Slow-growing. Can be planted in pure or mixed stands. Collect the fruits when ripe and extract the seeds manually. Sow as soon as possible. Alternatively, collect wildings from near mother trees.
NOTES: The wood is very hard and heavy and is used for making walking sticks and bark cloth mallets. *Vepris nobilis* is by far the commonest species of *Vepris* in Uganda.

Vepris trichocarpa (Engl.) Mziray (347a) Rutaceae
Synonym: *Teclea trichocarpa* (Engl.) Engl.
9 m. Tree or shrub. Very similar to *Teclea nobilis*, except that the inflorescence is a panicle in *V. nobilis* and a raceme in *T. trichocarpa*.
OCCURRENCE: U2-4. Recorded from Ankole, Busoga, Mengo and Tooro.

Vepris grandifolia (Engl.) Mziray (348) Rutaceae
Synonym: *Teclea grandifolia* Engl.
Nzo (ga); Omuzo (no).
Similar to *Vepris nobilis*, but distinguished by leaflets either lacking gland dots or with only one or two large dots on each.
OCCURRENCE: U1, 2 and 4. Recorded from Mengo, Ankole, Bunyoro and Busoga (Mutai). Not common.

Vepris eggelingii (Kokwaro) Mziray (349) Rutaceae
Understorey tree similar to *Vepris nobilis*, but easily distinguished by having a dense cover of stiff, spreading, grey or yellow, hairs on young stems, petioles and midribs.
OCCURRENCE: U2 and 4. Only known from Mabira and Itwara forests. Rare.
CONSERVATION STATUS: Global NE (IUCN), VU (TOU); National NE.

ALLOPHYLUS

The genus may be recognized by having trifoliolate leaves, and leaflets with non-entire (toothed to crenate) margins and lacking gland dots. Many specimens of *Allophylus*, as encountered in the field, appear to be intermediate between the species given below.

Allophylus abyssinicus (Hochst.) Radlk. (351) Sapindaceae
Gulindi, Kirindi (ms); Bionwa (ku).
20 m. Trunk straight, often fluted. Crown fairly small. Bark smooth, grey. Slash off-white to light pink-brown, turning darker, smelling of pepper. Leaves trifoliolate, alternate. Leaflets c. 15 x 6.5 cm, crenate or serrate, glabrous except for tufts of hairs in the axils of the veins beneath.
OCCURRENCE: U1-4. The common species of *Allophylus* on Mt Elgon and the Karamoja mountains, 1400-2500 m.
NOTE: Differs from *Allophylus ferrugineus* in having glabrous shoots and leaves (except for the tufts of hairs mentioned above).

Allophylus ferrugineus Taub. (352) Sapindaceae
SYNONYM: *Allophylus macrobotrys* Gilg
Omushusha (ki); Mutete (na).
20 m, but usually less. Usually a small, spreading, understorey, tree. Short spines occasionally present on trunk (especially on trees growing in swamps). Bark thin, fairly smooth. Phellogen green. Slash fibrous, white, sometimes turning darker. Young shoots hairy. Leaves trifoliolate, alternate. Leaflets c. 10 x 6 cm, hairy to glabrous, toothed to crenate.

Plate 27. Rutaceae, _Allophylus_ and others (343-366); see also Plate 28

343. *Balanites wilsoniana* 344. *Erythrina excelsa* 346. *Balsamocitrus dawei*
347. *Vepris nobilis* 352. *Allophylus ferrugineus* 355. *Ritchiea albersii*
366. *Musanga cecropioides*

Actual sizes: leaves and fruits x 2; trunk base x 80; tree profiles x 800.

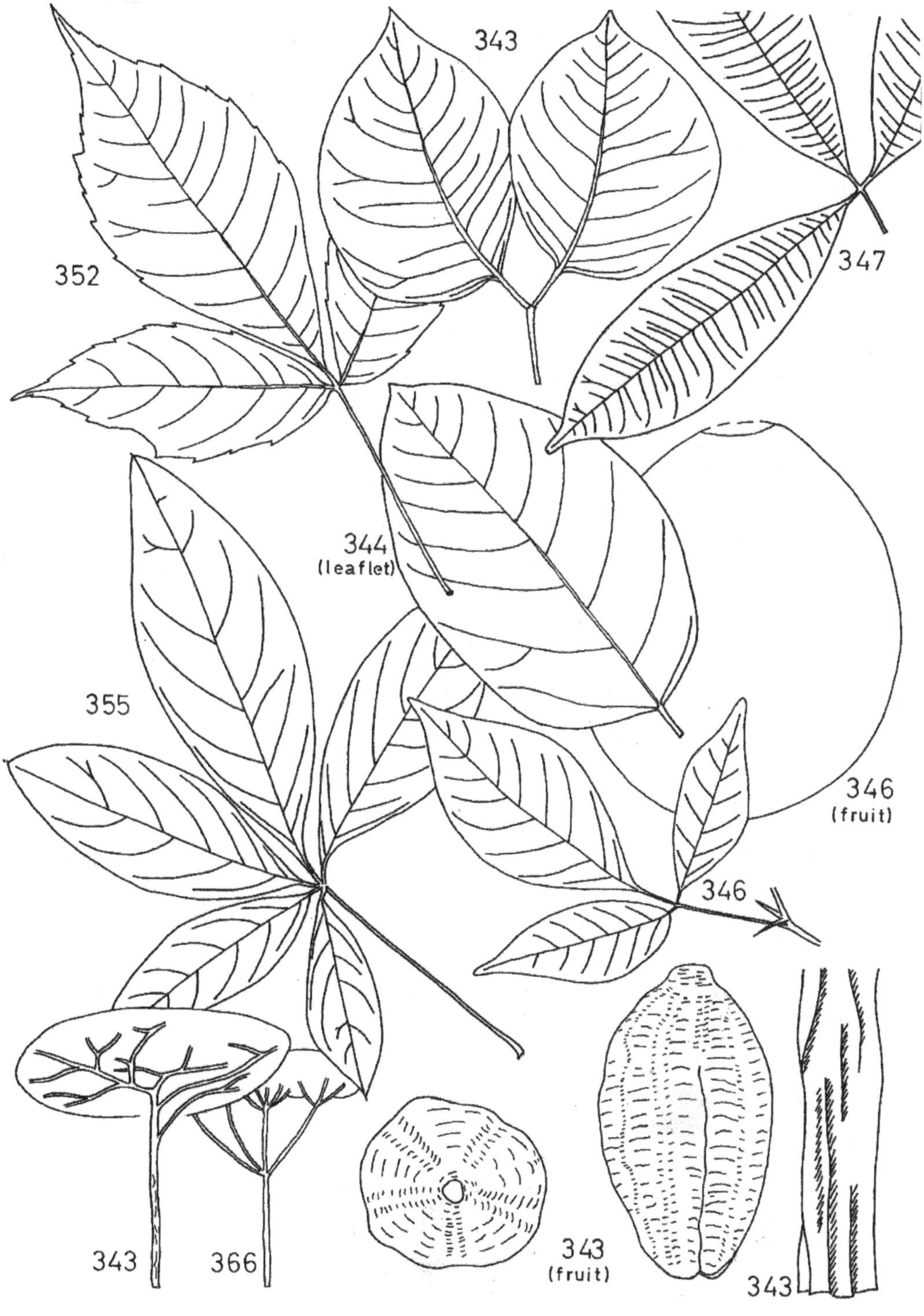

Plate 27. (343-366)

OCCURRENCE: U2-4. The most abundant species of *Allophylus* in south and south-west Uganda. Altitudinal range in Kigezi c. 2000-2500 m. On dry land and in swamp forest.
NOTE: Sometimes assuming the habit of *Allophylus abyssinicus*, but distinguishable by the hairy young shoots.

Allophylus dummeri **Baker f. (353)** Sapindaceae

10 m. Spreading understorey tree. Bark thin, light brown. Phellogen green. Slash fibrous, white to light yellow, sometimes turning darker. Leaves trifoliolate, alternate. Leaflets large, c. 20 x 11 cm. Fruits red when mature, fragrant.
OCCURRENCE: U2 and 4. Widespread in lower altitude primary and secondary forests. Abundant in Budongo Forest.
NOTE: Distinguished from other species of *Allophylus* by its large leaves.

Euadenia eminens **Hook. f. (354)** Capparaceae

5 m. Small tree or shrub. Leaves trifoliolate, alternate. Petioles c. 7 cm long. Leaflets c. 10 x 4 cm, margin entire, the lateral leaflets markedly asymmetric. Petiolules c. 0.5 cm long. Fruit 15-20 cm long and 1.25 cm diameter.
OCCURRENCE: U2 and 4. Widespread, but uncommon. Usually on swampy ground.

Ritchiea albersii **Gilg (355)** Capparaceae

Omuhenvu (ki); Munwabisani (to).
10 m. Understorey tree with a crooked trunk, irregularly branched. Crown spreading. Bark thin and fairly smooth, with conspicuous raised lenticels, brown (sometimes with layers of darker and lighter brown). Slash of even texture, white to yellow, sometimes turning darker. Leaves usually with 3 leaflets (but varying from 1 to 5-foliolate on the same tree), alternate. Leaflets c. 11 x 4.5 cm, apex mucronate. Petiole c. 6 cm long. Petiolule c. 0.5 cm long. Flowers large, green. Fruit c. 3.5 cm long.
OCCURRENCE: U1-4. Widespread, ascending to 2340 m. In open forest types.

Cussonia holstii **Engl. (356)** Araliaceae

Shikurati (ms).
15 m. Tree shaped like a pawpaw tree, with a straight trunk and few branches. Leaves large, alternate, digitate, with 3-7 leaflets, clustered at the ends of the branches. Leaflets c. 13 x 6 cm, crenate, acuminate. Petiole c. 20 cm long. Petiolule c. 6 cm long.
OCCURRENCE: U1-3. Mainly on dry sites.
NOTE: Used as a live fence around Kisoro.

Cussonia spicata **Thunb. (357)** Araliaceae

Cabbage wood (en); Mokyobelyo (ku).
Tree of similar shape to *Cussonia holstii*, but with compound (rather than simple) leaflets.
OCCURRENCE: U1 and 3. On Mt Elgon and the Karamoja mountains, 1500-2300 m.

SCHEFFLERA

Schefflera is unusual in that the plants often begin life as epiphytes or climbers, only later becoming self-supporting. This is a habit shared with some species of *Ficus*.

Key to *Schefflera*.

1. Lateral veins numerous (see Plate 28). Leaflet base rounded to cordate. On Rwenzori. ....**361. *Schefflera myriantha***
   Lateral veins fewer than above. 2

2. Leaflets cordate to distinctly rounded at base. 3
   Leaflets more or less cuneate at base (the very base of the leaflet may be rounded, but the general shape is cuneate). 4

3. Leaflets entire or nearly so. In western Uganda and lake-belt forests. ....**360. *Schefflera barteri***
   Leaflets crenate to toothed. Most common in eastern Uganda. ....**358. *Schefflera abyssinica***

4. In montane forests in eastern Uganda. ....**359. *Schefflera volkensii***
   In western Uganda and lake-belt forests. ....**360. *Schefflera barteri***

Schefflera abyssinica **(A. Rich.) Harms. (358)** Araliaceae

Omuwamira (ki); Litwalet (ku); Geyeyo, Lududu, Musolodi (ms).
25 m. Trunk very irregular, with a large spreading crown. Bark dark-coloured, thick and rough. Leaves alternate, digitate. Leaflets 5-7, c. 15 x 7 cm. Petiole c. 25 cm long. Petiolule c. 5 cm long. Old leaflets turn yellow.
OCCURRENCE: U1-3. Mt Elgon, the Karamoja mountains and the Imatongs, 2000- 2600 m. Also recorded from the Bufumbira Volcanoes.

Schefflera volkensii **(Engl.) Harms (359)** Araliaceae

Kwalet, Kwelet (ku); Chichipeno, Mubondwe (ma).
25 m. Trunk very irregular, with a large spreading crown. Leaflets 4-6, c. 19 x 6.5 cm. Petiole long. Petiolule c. 6 cm long.
OCCURRENCE: U1and 3. Mt Elgon and the Karamoja mountains, 1800-2750 m. Common on Mt Elgon.

Schefflera barteri **Harms (360)** Araliaceae

10 m. Small tree with a spreading crown. Leaflets 4-11, c. 13 x 6 cm (but sometimes much bigger). Petiole c. 25 cm long. Petiolule c. 4 cm long.
OCCURRENCE: U2 and 4.
Two varieties are recognized, var. *urostachya* (Harms) Tennant, a common tree in swamps at lower altitudes (occurrence U2 and 4), and var. *barteri* at higher altitudes in western Uganda (occurrence U2).

Schefflera myriantha **(Baker) Drake (361)** Araliaceae

SYNONYM: *Schefflera polysciadia* Harms
Small tree with a weak trunk and spreading crown. Leaflets c. 16 x 7.5 cm. Lateral veins on leaf more numerous than with other species of *Schefflera*.
OCCURRENCE: U2. Only known from Rwenzori.

Ricinodendron heudelotii **(Baill.) Heckel (362)** Euphorbiaceae

Kisongo (am) Omusodo (no, tn); Cork wood, Erimado, Musodo (tn).
40 m. Deciduous tree with a very straight and cylindrical trunk. Branches tending to be whorled and at right angles on younger trees, but not so on older specimens. Buttresses and flutes absent. Bark grey, thin, smooth, becoming scaly with age, with vertical lines of lenticels. Slash soft, granular, pink to red, often with white dots and/or orange granules, sometimes exuding red exudate. Leaves alternate, digitate, with 3-6 sessile (or near-sessile) leaflets. Leaflets c. 18 x 7 cm. Fruit 2 or 3-lobed, c. 3 cm wide.
OCCURRENCE: U1, 2 and 4. A light-demanding species, scattered throughout lower altitude forests. Abundant in secondary forest in Budongo Forest.

CULTIVATION AND PROPAGATION: Fast-growing. Prefers full light. Collect fruits from ground beneath mother trees. The rotten fruits can be boiled to remove the pulp and extract the seeds. The seeds can then be boiled again until the coat cracks, after which they should be dried. Sow the seeds as soon as possible.
NOTES: The wood is light, soft and can be used for carving. *Ricinodendron* is unusual among Euphorbiaceae in having digitate leaves. See before tree 94 for an overview of the family.

Bombax buonopozense P. Beauv. (363) Malvaceae
Bulanka, Kitutube (am); Wild kapok tree, Wild silk cotton tree (en); Mulungula (ko); Omulimbi (no).
40 m. Large deciduous tree, with a straight cylindrical trunk and branches in whorls. Trunk armed with conical woody bosses. Bark smooth, with large lenticels in vertical columns. Slash fibrous, white to greenish white (to red?). Leaves digitate, with 5-9 leaflets. Leaflets c. 14 x 4 cm, usually acuminate. Flowers large, attractive, red to orange.
OCCURRENCE: U2 and 4. Mainly in swamp forest. Recorded from Tooro, Bunyoro, Mengo and Kigezi; also in Maramagambo and Mabira forests, where it is very rare.
NOTES: The baobab tree *Adansonia digitata* L. (not recorded from Uganda) is a member of this family, as too is the kapok tree (also called silk cotton tree) *Ceiba pentandra* (L.) Gaertn. The kapok tree was introduced into Uganda to be a source of a type of cotton (the silky fibres associated with its seeds).

Myrianthus arboreus P. Beauv. (364) Urticaceae
Giant yellow mulberry (en); Mugunga (ga); Omusinyanuro (no).
10 m. Understorey tree with a short trunk, with branches from near base. Crown spreading, untidy. Stilt roots often present. Bark thin, smooth, brown. Slash usually off-white with brown lines (but occasionally red), turning darker. Leaves digitate, red when young, with 5-7 leaflets. Leaflets c. 25 x 9 cm (but sometimes much larger), coarsely toothed. Fruit hard, spherical, 3-8 cm diameter, with a moderate number of closely-packed seeds.
OCCURRENCE: U2 and 4. Mengo, Masaka, Mubende, Bunyoro. Mainly, but not always, in swamp forest.
CULTIVATION AND PROPAGATION: Prefers damp sites. Extract the seeds from the fruit pulp and dry. Soak seeds for a day before sowing.
NOTE: The fruits are edible.

Myrianthus holstii Engl. (365) Urticaceae
Kibanda, Kibende (am); Omwufa (ki); Kiruhura (na); Mwebende, Mwibende (to).
10 m, occasionally to 20 m. Similar to *Myrianthus arboreus* in general appearance, but differing in the following ways. Bark usually smooth (but sometimes flaking). Slash brittle, pink, red to red-brown, turning darker. Leaves digitate. Leaflets usually less than 25 cm long. Fruits usually smaller than those of *M. arboreus*.
OCCURRENCE: U2. Ankole, Kigezi and Tooro. A higher altitude species than *Myrianthus arboreus*, found mainly above 1300 m. Mainly on raised sites (rather than in damp places).

Plate 28. _Schefflera_, Urticaceae and others (357-366); see also Plate 27

357. *Cussonia spicata* 358. *Schefflera abyssinica* (leaflet)
359. *Schefflera volkensii* (leaflet) 361. *Schefflera myriantha* (leaflet)
363. *Bombax buonopozense* 365. *Myrianthus holstii* 366. *Musanga cecropioides*

Actual sizes: leaves and fruit x 2 (except 365 x 4).

Plate 28. (357-366)

Musanga cecropioides **Tedlie (366)** Urticaceae

Kigere, Kikumbu (am); Umbrella tree (en); Kaliba, Nnamagulu (ga).
30 m. Trunk straight and cylindrical, bearing a spreading, umbrella-shaped crown, only one leaf thick, Stilt roots often present, sometimes entirely supporting the tree. Short spines sometimes present on trunk. Bark very thin, smooth, whitish to brown, with ring marks, often with corky outgrowths. Phellogen green. Slash white to red. The inner part of the slash and the wood turn darker. Leaves palmately compound, with 11-25 segments. Leaf segments large, to 45 x 15 cm. Petiole to 60 cm long. Stipules very large, furry, with the texture of mammal skin.
OCCURRENCE: U2 and 4. Mengo, Bwamba and Budongo Forest. In secondary and swamp forest, sometimes abundant. A light-requiring species.
NOTE: The Luganda name *kaliba* means 'small animal skin'.

Musanga leo-errerae **Hauman & J. Léon. (367)** Urticaceae

Umbrella tree (en).
Similar in appearance to *Musanga cecropioides*, but trunk slenderer and with only 8-11 segments on each leaf.
OCCURRENCE: U2. In Ankole and Kigezi, above c. 1300 m. Abundant in Kalinzu and Bwindi forests.

Vitex ferruginea **Schumach. & Thonn. (368)** Lamiaceae

SYNONYM: *Vitex amboniensis* Gűrke (*sensu* ITU: 442 and UFT: 368)
Rwata (no).
15 m. Understorey tree. Slash very thin, off-white, turning rapidly green-brown. Branchlets clothed with long yellow hairs. Leaves opposite, digitate, with 5-7 leaflets. Petiole c. 10 cm long. Leaflets c. 9 x 3 cm, acuminate.
OCCURRENCE: U1, 2 and 4. Mainly on shallow soils.
NOTE: Another species of *Vitex*, *V. keniensis* Turrill, occurs in forests in Kenya and has been planted in Uganda.

BIGNONIACEAE

A family easily recognized by its opposite (or whorled) and pinnate leaves, a combination of characters only shared by *Fagaropsis angolensis* (372) and *Schrebera alata* (379). The flowers are showy and the fruits large.

Key to Bignoniaceae.

1. Large 'stipules' present at base of leaves. ....**369. *Markhamia lutea***
   Large 'stipules' absent. 2
2. Leaflets very hairy below. ....**370. *Spathodea campanulata***
   Leaflets glabrous or only moderately hairy below. ....**371. *Kigelia Africana***

Markhamia lutea **(Benth.) K. Schum. (369)** Bignoniaceae

SYNONYM: *Markhamia platycalyx* (Baker) Sprague
Misola (al); Mukana, Ndora (am); Musambya (ga, to); Lusambya (ga); Abonigo (gb); Omusavu (ki); Swaya (ku); Lusola (ms); Muchambye, Rusambya (na); Solwa (nl); Ilisiola, Musoolya (sa).

25 m. Tree with crooked trunk and fairly small, untidy, crown. Trunk sometimes fluted. Bark thin to thick, fairly rough, fibrous, fissured or flaking. Slash soft, off-white to light yellow, sometimes with brown markings, turning brown. Leaves opposite, imparipinnate, with 7-11 leaflets. Leaflets c. 15 x 5 cm. Each leaf has a pair of large, stipule-like, outgrowths at its base. Flowers conspicuous, large, yellow, trumpet-shaped. Fruit very long (to 1 m), thin, containing many winged seeds.

OCCURRENCE: U1-4. Common on forest edges; also within forests where the canopy is open, or where there has been a large gap. Particularly abundant in Mubende District and Central Kibale Forest. Frequently planted or retained in farmland.

CULTIVATION AND PROPAGATION: Fast-growing. Grows well on a variety of soils (but not in wet places). Drought tolerant. Unsuitable for planting close to crops, as it has an extensive root system. Readily coppices and is suitable for production of firewood and poles. Remove the seeds from the ripe fruits (when greyish in colour) and plant as soon as possible. The seedlings have very long taproots, so transplant early or prune the taproots.

NOTES: The flowers are eaten by monkeys. The wood is resistant to termite attack and much used for construction and furniture.

Spathodea campanulata P. Beauv. subsp. _nilotica_ (Seem.) Bidgood (370)
Bignoniaceae

Lapengwata (ac); Kikussu (am); Flame of the forest, Flame tree, Tulip tree (en); Kifabakazi (ga); Ekifurafura (ki); Chemungwa (ku); Opal (la); Kichubi, Kijubu (ms); Munyara (na); Mwatanshare (na); Mungobe (nl); Omunyaara (no); Mudungudungu (sa); Kinyhalisa (so).

20 m. Trunk crooked. Crown irregular. Bark fairly thin and smooth, becoming rough and flaking on older trees, brown (sometimes nearly white on outside), with prominent lenticels. Slash off-white, sometimes with brown markings, turning brown. Leaves opposite, imparipinnate, with c. 9-13 leaflets. Leaflets c. 9 x 5 cm, markedly hairy below. Flowers red, large and attractive, trumpet-shaped. Fruits ascending, paired, 15-25 cm long, containing numerous winged seeds.

OCCURRENCE: U1-4. Forest edges and farmland.

CULTIVATION AND PROPAGATION: Fast-growing. Tolerant of a range of soils. Can grow on degraded land. Drought tolerant. Preferably collect the fruits on the tree when they are beginning to open. Extract the seeds by hand. Sow as soon as possible. Unlike *Markhamia*, rarely found in forest interiors. Commonly planted.

NOTE: Medicinal.

Kigelia africana (Lam.) Benth. (371) Bignoniaceae

Sausage tree (en); Mussa (ga); Omuwifuzo (ki); Mwimbiri, Nyajungu (ko); Omuikya (no); Naibere (so).

10 m. Trunk short and crooked, branching from near base, with a spreading crown. Bark thin and smooth, grey. Phellogen green. Slash thin, off-white to light brown, with brown markings, slowly turning darker. Leaves opposite, imparipinnate, with c. 11-13 leaflets. Leaflets c. 14 x 5 cm, usually, but not always, toothed. Flowers orange, trumpet-shaped, borne in long, pendulous inflorescences. Fruit cylindrical, c. 40 cm or more long.

OCCURRENCE: U1-4. Mainly in swamp forest or by water, but also on raised sites in forest in W. Elgon, where it is common up to 2150 m. It is also found in savanna.

NOTES: Bark used medicinally. The forest subspecies is subsp. *moosa* (Sprague) Bidgood & Verdc., which has a different overall appearance and differently coloured flowers from specimens of *Kigelia africana* found in savanna.

RUTACEAE

Leaves often pinnate (372-378) or trifoliolate (346-349), more rarely simple (in *Aeglopsis eggelingii* (167), which can also have bi- or trifoliolate leaves). The leaves are gland-dotted, aromatic when crushed and lack stipules. Citrus trees belong to this family.

***Fagaropsis angolensis* (Engl.) Dale (372)** Rutaceae

Muyinja (ga, tn); Omukaka (ki); Kabegwi (ku); Mafu (tn); Mumara (to).
15 m. Trunk usually straight and cylindrical, with a spreading crown. Buttresses and flutes absent. Bark light brown, moderately thin to moderately thick, quite rough, flaking in pieces c. 0.3-5 cm across. Underside of bark characteristic, bright orange. Slash brittle, sometimes more or less granular, yellow, sometimes with orange stone cells, aromatic (at least when freshly cut). Leaves opposite, imparipinnate, with 5-11 leaflets. Leaflets variable in shape, often c. 7.5 x 2.75 cm, apex acuminate, with a single row of more or less prominent gland-dots near the margin.
OCCURRENCE: U2-4. Widely distributed, mainly on forest edges and in big gaps. Common in Central Kibale Forest and in places on Mt Elgon. Below 2200 m.
CONSERVATION STATUS: Global NE (IUCN), LC (TOU); National VU (WCS).
CULTIVATION AND PROPAGATION: Moderately fast-growing. Grow in mixed stands. Collect fruits when ripe (purple in colour). Sow seeds as soon as possible.
NOTE: The wood has a very attractive appearance.

Key to *Zanthoxylum*.

1. Leaflet base asymmetric. Leaflets usually over 15 cm long. **373. *Zanthoxylum gilletii***
   Leaflet base symmetric. Leaflets usually under 15 cm long. 2
2. Leaf rachis unarmed. Leaflets often over 3.5 cm broad. **374. *Zanthoxylum rubescens***
   Leaf rachis armed. 3
3. Young branches and rachises completely glabrous. ......**375. *Zanthoxylum leprieurii***
   Young branches and rachises with at least some hairs. **376. *Zanthoxylum mildbraedii***

***Zanthoxylum gilletii* (De Wild.) P.G. Waterman (373)** Rutaceae

SYNONYM: *Fagara macrophylla* (Oliv.) Engl. of ITU: 364. UFT: 373
Kitutube (am); Munyenye (ga); Nyakabonde (ko); Sagawat (ku); Shukuma (ms); Mulemankobe (na); Entalyerungu (no); East African satinwood, Olon (tn); Mutatembwa (to).

Plate 29. Bignoniaceae, Rutaceae and others (369-377)

369. *Markhamia lutea* 370. *Spathodea campanulata* 371. *Kigelia africana* (leaflet)
372. *Fagaropsis angolensis* 373. *Zanthoxylum gilletii*
374. *Zanthoxylum rubescens* (leaflet) 375. *Zanthoxylum leprieurii* (leaflet)
377. *Clausena anisata* 378. *Citropsis articulate*

Actual sizes: leaves, leaflets and flower x 2; trunk bases x 80; tree profiles x 800.

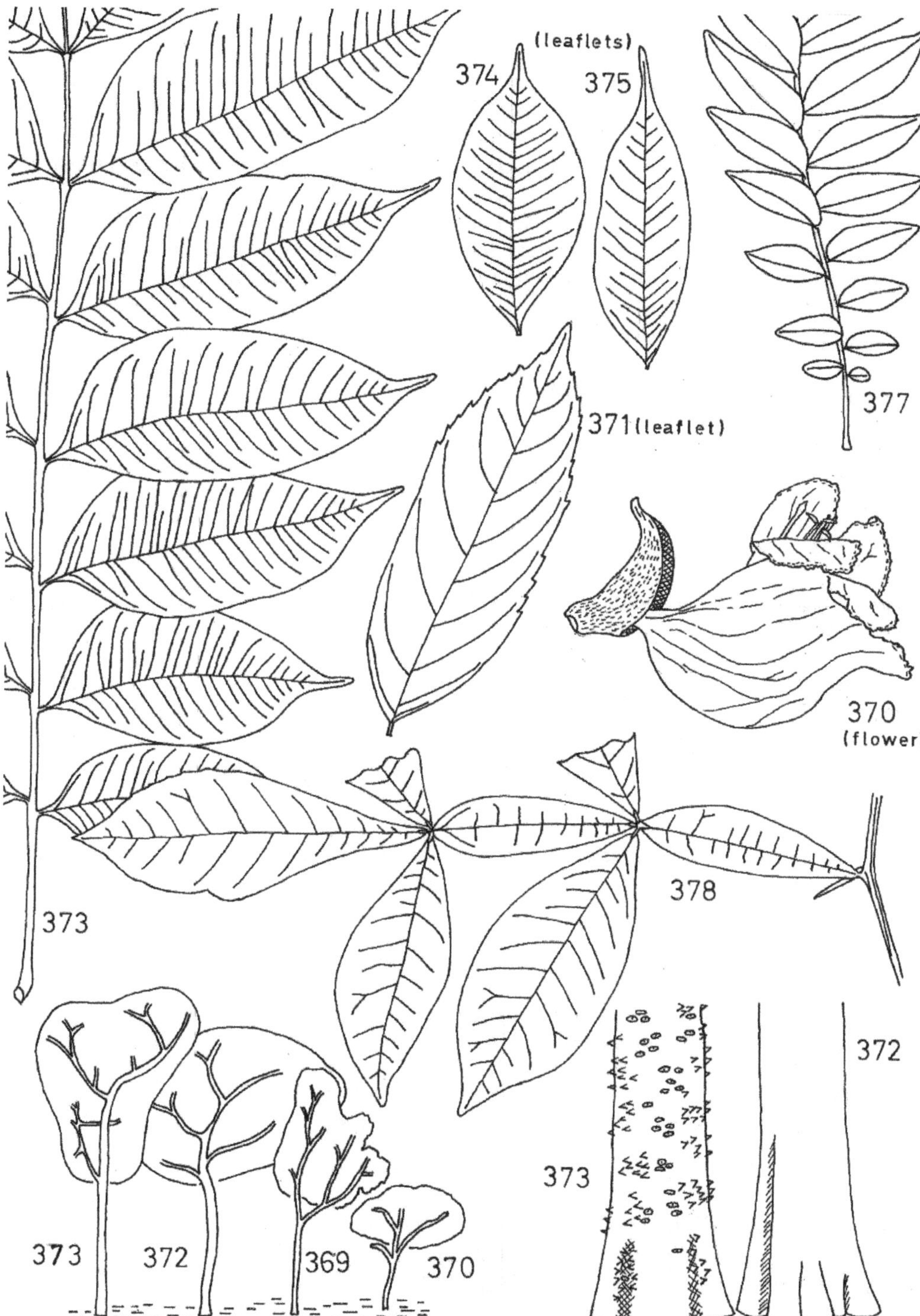

Plate 29. (369-377)

35 m. Large tree with a thick cylindrical trunk and fairly spreading, deciduous, crown. Trunk armed with conical woody bosses at base (these may be inconspicuous on very large trees). Bark thin, brown, with small vertical fissures, fairly smooth. Slash yellow to white, sometimes with orange stone cells, sometimes turning darker, fragrant. Young stems armed. Leaves alternate, imparipinnate, 25-100 cm long, with 5-27 leaflets. Leaflets alternate, c. 14 x 4.5 cm (sometimes much larger), very unequal-sided at base, with prominent gland-dots. Fruits red, with black seeds.
OCCURRENCE: U1-4. Widely distributed. Found up to 2000 m. Abundant in Kalinzu Forest.
CULTIVATION AND PROPAGATION: Moderately fast-growing. Grow in mixed stands. Collect fruits while they are still closed, either from mother trees or from the ground beneath. Leave fruits under shade until they split. Seed dormancy can be difficult to break. Can be propagated from stem cuttings, especially if coppice shoots are used.

Zanthoxylum rubescens Hook. f. (374) Rutaceae
SYNONYM: *Fagara rubescens* (Hook. f.) Engl.
Munyenye (ga); Nyakabonde (ko); Entalyerungu (no).
8 m. Leaflets opposite to alternate, c. 13 x 5.5 cm (usually larger than those of *Zanthoxylum leprieurii* and *Zanthoxylum mildbraedii* and broader in proportion to length), base symmetrical, gland-dotted. Rachis unarmed (unlike *Zanthoxylum leprieurii*).
OCCURRENCE: U1, 2 and 4. Sometimes in swamps.

Zanthoxylum leprieurii Guill. & Perr. (375) Rutaceae
SYNONYM: *Fagara leprieurii* (Guill. & Perr.) Engl.
Munyenye (ga, tn); Ntaleyeddungu (ga); Entalyerungu (no).
15 m. Understorey tree with a wide trunk, branched from low down. Crown spreading, deciduous. Trunk armed with conical woody bosses. Bark thin, brown, flaking, but general effect smooth. Slash of even texture, yellow, turning darker (brown). Leaves alternate, imparipinnate, with c. 11-17 leaflets. Leaflets opposite to sub-opposite, c. 7 x 2.5 cm, with prominent gland-dots, apex long acuminate, base symmetrical.
OCCURRENCE: U1, 2 and 4. Abundant in Tooro and Mengo; common in Bunyoro.
NOTE: A smaller tree than *Zanthoxylum gilletii*, from which it may be further distinguished by its opposite to sub-opposite leaflets (rather than alternate) and smaller leaflets with symmetrical bases.

Zanthoxylum mildbraedii (Engl.) P.G. Waterman (376) Rutaceae
Omuremankobe, Omushaaga (ki); Mulemankobe (na).
30 m. Similar to *Zanthoxylum leprieurii* in leaflet size, shape and arrangement, but rachises and young branches with at least some hairs (those of *Z. leprieurii* are completely glabrous). Also, the inflorescence is corymbose (more or less flat-topped, with the branches of the inflorescence starting at different levels on the rachis, but all reaching to about the same level), rather than paniculate (inflorescence not flat-topped; branches of inflorescence not all reaching to about the same level).
OCCURRENCE: U2 and 4. Mengo, Ankole, Kigezi and Tooro.
CONSERVATION STATUS: Global VU (IUCN), LC (TOU); National NE.

Clausena anisata (Willd.) Benth. (377) Rutaceae
Musaniko (ga, so); Mfunyandudu (gw); Omutana, Omutanwa (ki); Muhaya, Muhayi (ko); Mutana, Mutanwa (na); Musali (nl); Omutonwa (no); Luselasimba, Musivasimba (sa); Musaniko, Musokolindo, Muwonera, Muwonesa, Nsaniko (so); Mutonwa (to)
8 m. Unarmed tree. Leaves alternate, pinnate. Leaflets 11-37, c. 6 x 2.25 cm, with gland-dots, strongly aromatic when crushed.
OCCURRENCE: U1-4. Widely distributed. Often abundant on forest edges; also in secondary forest.

Citropsis articulata **(Spreng.) Swingle & Kellerman (378)** Rutaceae
Katimboro (to).
Small shrub or tree to 5 m. Leaves alternate, pinnate. Easily distinguished by its winged rachis and gland-dotted leaves (see Plate 29). Fruit c. 2.25 cm across, aromatic.
OCCURRENCE: U2 and 4. Widespread, on damper sites.
CONSERVATION STATUS: Global NE (IUCN), LC (TOU); National VU (WCS).

Schrebera alata **(Hochst.) Welw. (379)** Oleaceae
Gagawa (ku).
25 m. Deciduous tree. Bark smooth, dark grey. Slash off-white to yellow. Leaves opposite, imparipinnate, with c. 5 leaflets, petiole and rachis winged. Leaflets c. 7 x 3.5 cm. Fruit a 2-valved woody capsule, 2-3 cm long.
OCCURRENCE: U1 and 3. Mt Elgon and the Karamoja mountains, 1900-2200 m.
CULTIVATION AND PROPAGATION: Moderately fast-growing. Can be planted in pure or mixed stands. If possible, collect the unopened fruits from mother trees just before they ripen. Dry the fruits in the sun and separate the seeds from the capsules once opened. Remove insect-damaged seeds. Sow as soon as possible.

Bersama abyssinica **Fresen. (380)** Melianthaceae
Omukaka (ki); Sigirwo (ku); Gisombe, Shikisombe (ms); Muhingura, Mukore (na); Muhungura, Njajungu (to).
15 m. Trunk crooked, with an irregular crown. Bark rough, flaking. Slash brown, with whitish streaks. Leaves alternate, imparipinnate, to 60 cm long, often with a winged rachis (especially at higher altitudes), with 7-27 leaflets which are opposite to sub-opposite. Leaflets c. 12 x 3.5 cm, usually toothed. There are large intrapetiolar stipules. Capsule with 4-5 valves, reddish on outside, splitting to reveal red seeds with yellow arils.
OCCURRENCE: U1-4. Widely distributed. Abundant on wetter mountains, 2000-2500 m.
CULTIVATION AND PROPAGATION: Fast-growing when older, slow-growing when young. Tolerates a variety of soils. Prefers open places. Can be planted alone or in mixed stands. Collect unopened capsules on the tree and store until they split open. Pick out the seeds and remove the arils. Germination irregular (mostly 3-10 weeks). Provide seedlings with shade and water for at least 6 months before transplanting. Seedlings are often abundant under mother trees.

Hagenia abyssinica **(Bruce) J.F. Gmel. (381)** Rosaceae
Leo (ac); Omugyesi, Omujeesi (ki); Sigurwa (ku); Kisichetwa, Museregego, Nafuru (ms).
20 m. Trunk wide, crooked, often branched from near base and gnarled. Crown spreading. Bark red-brown, rough, flaking raggedly. Branchlets covered with yellow hairs. Leaf scars very conspicuous. Leaves tufted at ends of branches, alternate, imparipinnate, with 11-13 leaflets, with irregular outgrowths on rachis. Leaflets c. 10 x 3 cm, with numerous teeth. Dioecious. Inflorescences very large, pendulous, reddish.
OCCURRENCE: U1-3. Common at 2300-3500 m on most mountains. Descends to 2000 m. Abundant in the *Hagenia-Rapanea* Zone on Mt Elgon, Rwenzori and the Bufumbira Volcanoes. Also, very abundant among bamboo on the western slopes of Mt Elgon.
CULTIVATION AND PROPAGATION: Very fast-growing. Fire-resistant. Suitable for land reclamation, at least down to 1500 m altitude. Can be planted as single specimens or in pure or mixed stands, but dislikes competition. Can be lopped and pollarded for poles and firewood. Seeds can be stored in sealed containers for up to 12 months.
NOTE: The dried female inflorescence is used in Kigezi to rid goats of worms.

***Harrisonia abyssinica* Oliv. (382)** Rutaceae

13 m. Trunk irregular, branched from near base, often multi-stemmed, often with sucker shoots. Crown spreading. Spines often present on small shoots and sucker shoots, sometimes on old wood. Bark light brown, quite thick, rough, flaking. Slash fibrous, pink, often with white lines. Leaves imparipinnate, with c. 9-19 leaflets (on forest specimens). Rachis winged. Leaflets c. 4 x 1.5 cm (sometimes rather larger), usually crenate or toothed.
OCCURRENCE: U2-4. Widely distributed. Common in South and Central Kibale Forest and occasional in Mabira Forest.

***Quassia undulata* (Guill. & Perr.) D. Dietr. (383)** Simaroubaceae

SYNONYM: *Hannoa longipes* (Sprague) G.C.C. Gilbert (*sensu* UFT)
50 m. Large tree with a straight cylindrical trunk and fairly rounded crown. Buttresses absent. Bark probably light-coloured and fairly smooth. Slash soft and thick, white, with fibres. Leaves alternate, imparipinnate, with c. 9-13 leaflets. Leaflets c. 11 x 4.5 cm. Petiolules c. 1.5 cm. Venation rather obscure on upper surface of leaflets.
OCCURRENCE: U2. Kigezi, Tooro and Ankole (Kalinzu Forest). Uncommon, except in Kalinzu.

***Polyscias fulva* (Hiern) Harms (384)** Araliaceae

Ssettaala (ga); Mungu (na); Omungo (ki); Kyango, Kyungu, Mukungu (ko); Lamadi (ku); Gafuti, Gufuri, Gufuru (ms); Mujugantara (to).
25 m. Trunk straight (sometimes curved), cylindrical, ending abruptly and dividing into a number of main branches (without the central trunk continuing), these branches ascending, spreading out and dividing in the same manner. Crown umbrella-shaped, one leaf thick. Bark thin, light brown, smooth, with vertical lines of lenticels, flaking on older trees. Slash soft, light yellow to white, sometimes with orange spots, turning darker and becoming greenish near the wood. Leaves alternate, imparipinnate (rarely paripinnate), with c. 12-15 leaflets. Leaflets c. 11 x 4.25 cm (sometimes much larger). Petiolules c. 0-0.5 cm long.
OCCURRENCE: U1-4. A light-requiring species. Abundant on forest edges and in larger gaps, ascending to 2200 m on wetter mountains. Absent from Bunyoro.
CULTIVATION AND PROPAGATION: Fast-growing. Suitable for restoring depleted land or providing a light shade for slower-growing species. Can be grown in pure stands. Collect fruits when ripe (purple) on the ground beneath mother trees, remove the seeds from the pulp

***Pseudospondias microcarpa* (A. Rich.) Engl. (385)** Anacardiaceae

Kiboru, Mbolu (am); Muziru (ga, so); Mungu (ko); Muhohote (nl); Bagambanimpyata (no, to).
30 m, but occasionally to 45 m. Trunk irregular, often gnarled, sometimes fluted, with large spreading branches often from near base. The branches characteristically bear many epiphytes. Buttresses usually present. Bark on older trees thick, brown, with horizontal and vertical fissures, flaking. Bark on young trees thin, smooth. Slash white to pink (perhaps to dark red), slowly exuding small drops of red exudate. Green phellogen present on young trees. Leaves alternate, imparipinnate, with 3-11 leaflets. Leaflets c. 13 x 5.5 cm (but variable in size and shape), base unequal-sided. Fruit c. 2 cm long, blue-black when ripe.

Plate 30. Simaroubaceae, Anacardiaceae and others (379-389)

379. *Schrebera alata* 380. *Bersama abyssinica* 381. *Hagenia abyssinica*
382. *Harrisonia abyssinica* 384. *Polyscias fulva*
385. *Pseudospondias microcarpa* 387. *Lannea welwitschii*
388. *Antrocaryon micraster* (leaflet) 389. *Canarium schweinfurthii* (leaflet)

Actual sizes: leaves, leaflets and fruit x 2; tree profiles x 800.

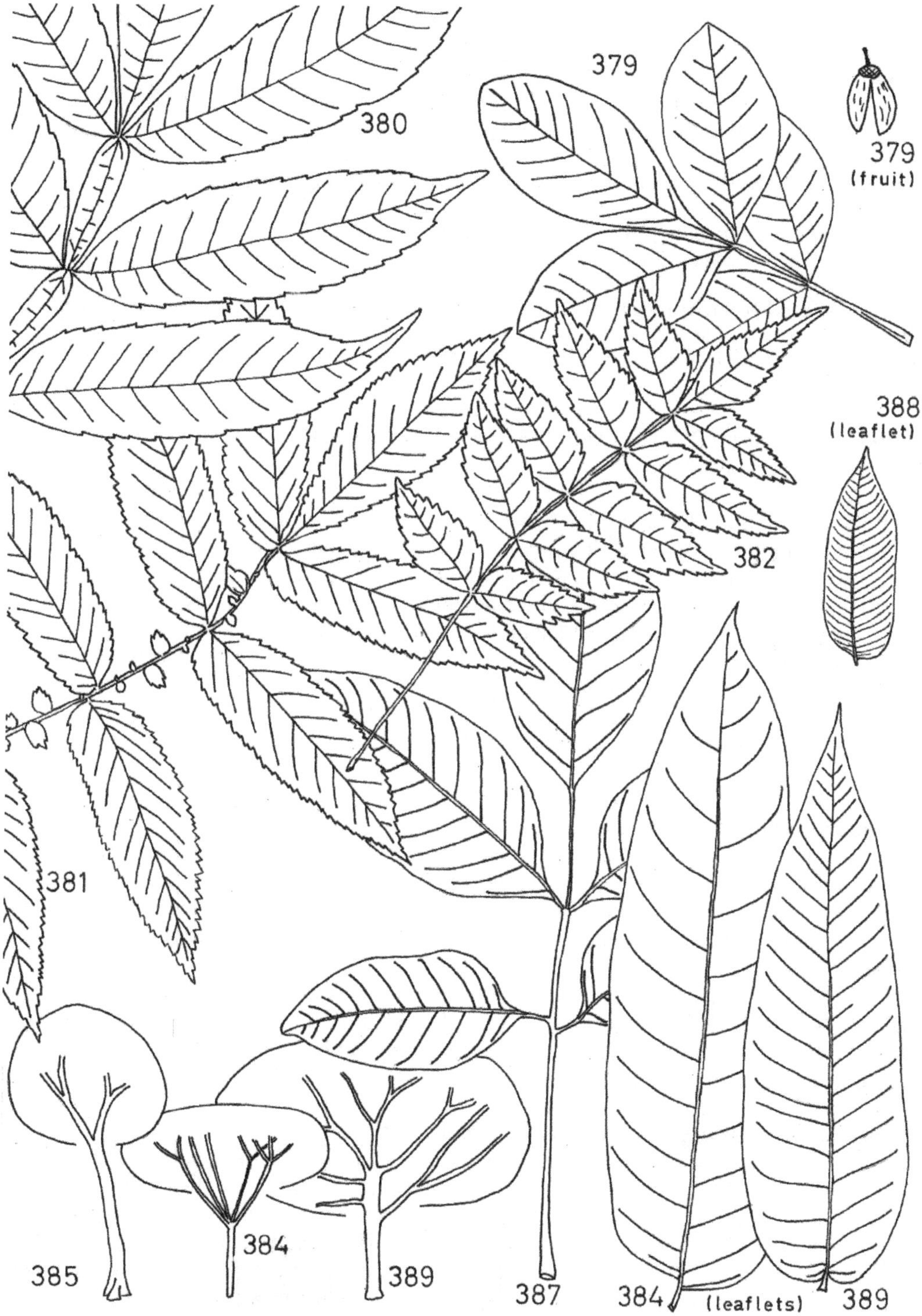

Plate 30. (379-389)

OCCURRENCE: U1-4. Widespread and often abundant, especially in swamp forest and on its margins, and along rivers. Occasionally found on raised sites.
NOTE: The fruit (*enziru* in Luganda) is edible.

Sorindeia juglandifolia (A. Rich.) Oliv. (385a) Anacardiaceae
Shrub or tree to 10 m.
OCCURRENCE: U4. Recorded from Masaka. Moist forest and swamp forest, c. 1200 m.

Trichoscypha lucens Oliv. (386) Anacardiaceae
SYNONYM: *Trichoscypha submontana* Van der Veken
Mushaya (na).
25 m. Bark dark-coloured, rough. Slash red, with fairly abundant red exudate. Leaves alternate, imparipinnate, with 8-14 leaflets. Leaflets c. 10 x 3.5 cm. Fruit c. 2.5 cm long, purple when ripe.
OCCURRENCE: U2. Quite common in parts of Kalinzu Forest. Not known from elsewhere. On raised sites.

Lannea welwitschii (Hiern) Engl. (387) Anacardiaceae
Kingalangala, Mukowa (ga).
30 m. Trunk straight and cylindrical, with large branches and a spreading crown. Buttresses and flutes absent. Bark thin and smooth, brown (to almost white on outside), flaking in pieces c. 5 cm across, with conspicuous lenticels. Slash mostly fibrous, pink to red, sometimes with white lines, with lines of orange granules (which are particularly abundant near the bark). Leaves clustered at ends of branches, alternate, imparipinnate, with c. 5-9 leaflets. Leaflets c. 13 x 5.5 cm, with c. 9-15 prominent lateral veins on each side of the midrib, tufts of hairs in vein axils below, apex acuminate, base unequal-sided. Petiolule of terminal leaflet c. 2.5-4.5 cm long. Fruit c. 0.8 cm long, purple when ripe.
OCCURRENCE: U2 and 4. Widely distributed, but nowhere common.
NOTE: It can be confused with Meliaceae and is included in the Meliaceae key (before tree 390).

Antrocaryon micraster A. Chev. & Guill. (388) Anacardiaceae
35 m. Deciduous tree with a cylindrical trunk. Buttresses absent. Bark dark grey, thick, fairly smooth to rough, with vertical fissures, scaling on older trees. Slash pink to red, with vertical white lines, scented. Leaves imparipinnate, with c. 11-12 leaflets. Leaflets c. 7 x 2.5 cm with numerous, prominent, lateral veins, unequal-sided at base. Fruit depressed-globose, c. 5 cm diameter, strong smelling.
OCCURRENCE: U1, 2 and 4. Recorded from Mabira, Budongo and Zoka forests.
CONSERVATION STATUS: Global VU (IUCN), LC (TOU); National CR (WCS).
NOTE: Best distinguished by the size and venation of the leaflets. It can be confused with Meliaceae and is included in the Meliaceae key (before tree 390).

Canarium schweinfurthii Engl. (389) Burseraceae
Buhura, Byoro (am); Incense tree (en); Muwafu (ga); Omubani (ki); Mubani (na, to); Omubani (no), Omusanki (no); Mubafu (so); African canarium (tn).
40 m. Trunk straight and cylindrical, bearing large branches. Deciduous. Crown spreading, umbrella-shaped. The branches of young trees are in whorls, at right angles, and curve upwards. Leaves clustered at ends of branches. Buttresses absent. Bark very thick and rough, red-brown (to almost white on the outside), flaking in pieces c. 5-30 x 2-10 cm in size. Slash fibrous, pink, red to red-brown, not turning rapidly darker, smelling of incense. Leaves imparipinnate, with c. 13-31 leaflets. Leaflets c. 15 x 4 cm (but much larger on young trees), hairy below (at least on the main veins), apex acuminate. Fruit purple, c. 2.5 cm long.

OCCURRENCE: U1-4. Widely distributed. Uncommon in forest interiors, except in lake-shore forests. Very rare in forests in Ankole and Kigezi. Abundant in farmland in Mengo and Busoga.
CULTIVATION AND PROPAGATION: Moderate to slow-growing. Can be planted as specimen trees. The stone containing the seeds can be stored for a long time. Crack open the stone and separate the seeds. Soak seeds in water for 24 hours before sowing. Germination rate irregular and rather low. Transplant seedlings after about 6 months in the nursery.
NOTES: The resin (*obubaane* in Luganda), which exudes from the slash after some time, is burnt for incense in traditional shrines and Catholic churches. The fruits are eaten, mainly by children. The Luganda name for the fruit is *empafu* and for the seed *enje*.

MELIACEAE

A family well-represented in Ugandan forests by trees of all sizes. The leaves are pinnate (except *Turraea*), alternate (borne in a spiral), usually on rather thick stems. The slash is usually some shade of red and is sometimes strongly scented. Fruits of many different kinds. Seeds often with arils. *Lannea welwitschii* (387), *Antrocaryon micraster* (388) and *Canarium schweinfurthii* (389) may sometimes be mistaken for Meliaceae and are included in the key below.

Key to Meliaceae and superficially similar-looking trees

1. Leaves simple. ....**226-229a. *Turraea***
   Leaves compound. 2
2. Slash yellow. Leaflets 20-36 on a leaf. ....**408. *Turraeanthus***
   Slash not yellow. Leaflets fewer than 25. 3
3. Slash brown, strongly scented. Leaflets usually 8. ....**405. *Leplaea cedrata***
   Dominant colour of slash either whitish or a shade of red. 4
4. Slash with a white or off-white exudate (which can come slowly).
   ....**392-393. *Trichilia dregeana. Trichilia martineaui***
   White or off-white latex absent. 5
5. Leaflets comparatively small (often c. 7.5 x 3 cm), shiny above, much paler and glaucous (bluish or greyish) beneath. A forest-edge species, usually less than 10 m tall, only rarely found within forests and then growing to 25 m. ...**390. *Ekebergia***
   Leaflets not both small and glaucous. 6
6. Understorey or second storey trees, to 25 m, usually with more or less crooked trunks. Foliage usually fairly easily accessible. Leaves normally imparipinnate. 7
   Canopy or emergent species, with straight trunks and foliage only high up; leaves paripinnate or imparipinnate. Note: young individuals of these species are clearly not understorey trees, having straight thin stems with prominent leaf scars and leaves only at their tops. 13
7. Tree growing above c. 1300 m (including Kalinzu, Kayonza and Bwindi forests). ...8
   Tree growing below c. 1300 m. 11
8. Slash strongly scented. 9
   Slash not strongly scented. 10

9. Bark thick and fissured. Leaflets with more than 16 main lateral veins on each side of the midrib. ....**398. *Leplaea mayombensis***
Bark quite thin and smooth. Leaflets with fewer than 16 main lateral veins on each side of the midrib. ....**395-396. *Trichilia rubescens*, *Lepidotrichilia***

10. Rachis and petiole together normally over 40 cm long. ....**397. *Carapa***
Rachis and petiole together normally under 40 cm long. ....**390. *Ekebergia***

11. Understorey tree to 25 m, with a crooked trunk and large leaves (often over 50 cm long) which are imparipinnate and red when young. Slash not scented. Abundant in North Kibale and Kalinzu forests and in parts of Kigezi, but rare elsewhere. ....**397. *Carapa***
Not as above. Note: *Lovoa* (406-407) may occasionally key out here. 12

12. Youngest part of the stem conspicuously covered with stellate yellow hairs. ....**396. *Lepidotrichilia***
Youngest part of the stem not conspicuously covered with yellow hairs. Hairs on plant simple, not stellate. Large lenticels sometimes conspicuous on young stems. ....**394-395. *Trichilia prieuriana*. *T. rubescens***

13. Leaflets small (often c. 7 x 2.5 cm), with numerous, closely spaced, lateral veins (see illustration). ....**388. *Antrocaryon***
Not as above. 14

14. Bark splitting into fairly well-defined squares or rectangles. 15
Bark not splitting into fairly well-defined squares or rectangles. 17

15. Tree growing mainly above 1500 m altitude. ....**390. *Ekebergia***
Tree found below 1500 m (except perhaps on Rwenzori, where *Entandrophragma utile* is reported to ascend to 1860 m). 16

16. Slash strongly scented. ....**389. *Canarium*; 406. *Lovoa trichilioides***
Slash not strongly scented. ....**399. *Entandrophragma utile***

17. Slash strongly scented. 18
Slash not strongly scented. 22

18. Slash smelling of incense. Leaves imparipinnate. ....**389. *Canarium***
Slash smelling spicy. Leaves paripinnate. 19

19. Bark thick, flaking to leave shallow concave depressions, c. 2-5 cm diameter. Lenticels not prominently raised. ....**403. *Khaya anthotheca***
Bark not as above or, if bark flaking to leave depressions, then lenticels prominently raised. 20

20. Bark flaking to leave lines and rings of prominent raised lenticels. ....**405. *Leplaea cedrata***
Lenticels, if prominent, not as above. 21

21. Buttresses large. ....**400. *Entandrophragma cylindricum***
Buttresses absent. ....**406. *Lovoa trichilioides***

22. Tree growing above 1500 m. 23
Tree growing below 1500 m. 24

23. Leaves imparipinnate. ....**390. *Ekebergia***
Leaves paripinnate. ....**402. *Entandrophragma excelsum***

24. Orange stone cells present in slash. Buttresses usually absent. ....**387. *Lannea***
Orange stone cells absent from slash. Buttresses present or absent. 25

25. Slash with numerous, concentric layers of different shades. **407. *Lovoa swynnertonii***
Slash not as above. ....**401-402. *Entandrophragma angolense, E. excelsum***

21. Buttresses large. ....**400. *Entandrophragma cylindricum***
Buttresses absent. ....**406. *Lovoa trichilioides***

22. Tree growing above 1500 m. 23
Tree growing below 1500 m. 24

23. Leaves imparipinnate. ....**390. *Ekebergia***
Leaves paripinnate. ....**402. *Entandrophragma excelsum***

24. Orange stone cells present in slash. Buttresses usually absent. ....**387. *Lannea***
Orange stone cells absent from slash. Buttresses present or absent. 25

25. Slash with numerous, concentric layers of different shades. **407. *Lovoa swynnertonii***
Slash not as above. ....**401-402. *Entandrophragma angolense, E. excelsum***

Ekebergia capensis **Sparm. (390)** Meliaceae

Synonym: Ekebergia senegalensis A. Juss.
Kitwalabafu (ga); Omufumba (ki); Bumet (ku); Gusira, Musalamumali (ms).
30 m, but often shorter. Trunk straight or crooked, usually branched low down (sometimes with very big low branches). When found inside dense forest (rare), it has a long, straight trunk. Crown rounded or spreading. Buttresses absent or poorly developed, lower part of trunk sometimes fluted on large trees. Bark grey, deeply fissured, flaking into square pieces c. 5 x 5 cm in size. Slash granular, dark red (sometimes with white lines), white on young trees. Leaves imparipinnate, with c. 5-17 leaflets. Leaflets c. 7.5 x 3.5 cm (sometimes rather larger, but variable in size), shiny above, much paler and rather glaucous (bluish or greyish) beneath, usually with a few black spots near the midrib, asymmetric at base. Fruit a drupe, c. 2 cm long.
OCCURRENCE: U1-4. Usually on forest edges at lower altitudes. Within forest at higher altitudes (1800-2500 m). Mainly on drier sites.
CULTIVATION AND PROPAGATION: Moderately fast-growing. Plant as isolated specimens or in pure or mixed stands. Preferably collect the fruits on the tree when mature (red in colour) and remove the flesh to release the seeds. Sow seeds as soon as possible.

Trichilia dregeana **Sond. (392)** Meliaceae

Ssekkoba, Ssesambya (ga); Marawatawula (so).
40 m. Trunk straight and cylindrical, with big branches and a rather rounded crown. Buttresses absent or small. Bark fairly thin and smooth, with prominent lenticels, brown. Slash red and white, sometimes with some yellow colour, turning darker, slowly exuding small drops of off-white latex from near the wood. Leaves imparipinnate, with c. 5-13 leaflets. Leaflets often c. 11 x 4 cm (but sometimes much larger), with c. 7-16 main lateral veins on each side of the midrib. Fruit a red/pink, hairy capsule, with black seeds with red arils.
OCCURRENCE: U1-4. Widely distributed. Common in Central Kibale Forest.

Trichilia martineaui **Aubrév. & Pellegr. (393)** Meliaceae

Musuga (ga).
35 m. Trunk straight. Buttresses absent. Bark light brown, fairly thin and smooth (to greyish-black and rough), flaking on older trees. Slash pink, with darker (red) lines, foul-smelling, exuding drops of off-white latex (the only other species of *Trichilia*, apart from *T. dregeana*, to do so). Leaves imparipinnate, with c. 8-15 leaflets. Leaflets c. 10 x 2.5 cm (i.e. narrower than those of *Trichilia dregeana*), acuminate.

OCCURRENCE: U2 and 4. Uncommon. Recorded from Budongo, Bugoma, Mabira and Kayonza forests. Probably also in other forests.

Trichilia prieuriana A. Juss. (394) Meliaceae

Ssesambya (ga); Ojo (la); Musanhilapindi (nl); Omuralike (no); Mwangati (sa).
25 m (usually considerably less). Spreading understorey tree, with a crooked (rarely fairly straight) trunk and dense crown. Trunk characteristically deeply fluted. Bark light brown, vertically fissured, rather fibrous. Slash fibrous, pink to pink and yellow, white on young trees, layered, plate-like, sometimes with a strong smell. Leaves imparipinnate, with c. 5-11 leaflets. Leaflets c. 15 x 6 cm, red when young. Fruit a capsule, containing black seeds with red arils.
OCCURRENCE: U1-4. Widespread understorey tree, also in savanna. It is very abundant in Mabira and other Mengo forests. Not recorded from Kayonza and Bwindi forests.

Trichilia rubescens Oliv. (395) Meliaceae

Omugaba (no).
15 m. Spreading understorey species. Trunk crooked, cylindrical or fluted. Bark thin, smooth, greenish to light brown, sometimes flaking. Slash fibrous, pink to red, turning darker, strongly scented. Leaves imparipinnate, with c. 9-17 leaflets. Leaflets c. 18 x 6 cm. Fruit a red capsule, containing black seeds with red arils (like *Trichilia prieuriana*).
OCCURRENCE: U1, 2 and 4. Abundant in Mengo and Bunyoro forests, with a tendency to grow in somewhat wetter places than *Trichilia prieuriana*. Present in Kalinzu Forest.

Lepidotrichilia volkensii (Gürke) Leroy (396) Meliaceae

Omukavu, Omuzo (ki); Bionwa (ku); Chiwiwi (ms).
10 m. Small, spreading, much-branched, understorey tree. Bark fairly thin. Slash pink to red, sometimes with white lines, scented. Leaves imparipinnate, with c. 5-11 leaflets. Leaflets c. 13 x 5 cm (but variable in size). The petiole, rachis, young stems and (usually) lower surface of the leaves are markedly hairy. Hairs stellate.
OCCURRENCE: U1-4. Mainly at 1200-2750 m (occasionally down to 1000 m). Often found with mountain bamboo.
NOTE: This is the only African member of the Meliaceae to have stellate hairs.

Carapa grandiflora Sprague (397) Meliaceae

SYNONYM: *Carapa procera* Sprague
Omuruguya (ki); Mutongana (na); Uganda crabnut, Uganda crabwood (tn); Muhumbulia, Mujogo (to).
25 m, but usually less. Understorey tree. Trunk crooked, usually branched from low down. Crown spreading, dense. Bark thin, usually smooth, greenish to brown, becoming vertically fissured and flaking with age. Slash fibrous, pink to red, sometimes with white lines, not fragrant. Leaves very large, sometimes over 1 m long, conspicuously red when young, imparipinnate, with c. 6-18 leaflets. Leaflets c. 25 x 9 cm, with c. 10 main lateral veins on each side of the midrib. Fruit large, c. 15 cm diameter, with 5 leathery valves and up to 10 large seeds.

Plate 31. Meliaceae (390-398): see also Plate 32

390-391. *Ekebergia capensis* 392. *Trichilia dregeana* 393. *Trichilia martineaui*
394. *Trichilia prieuriana* 395. *Trichilia rubescens* 396. *Lepidotrichilia volkensii*
397. *Carapa grandiflora* 398. *Leplaea mayombensis*

Actual sizes: leaves, leaflets, fruit and seed x 2.

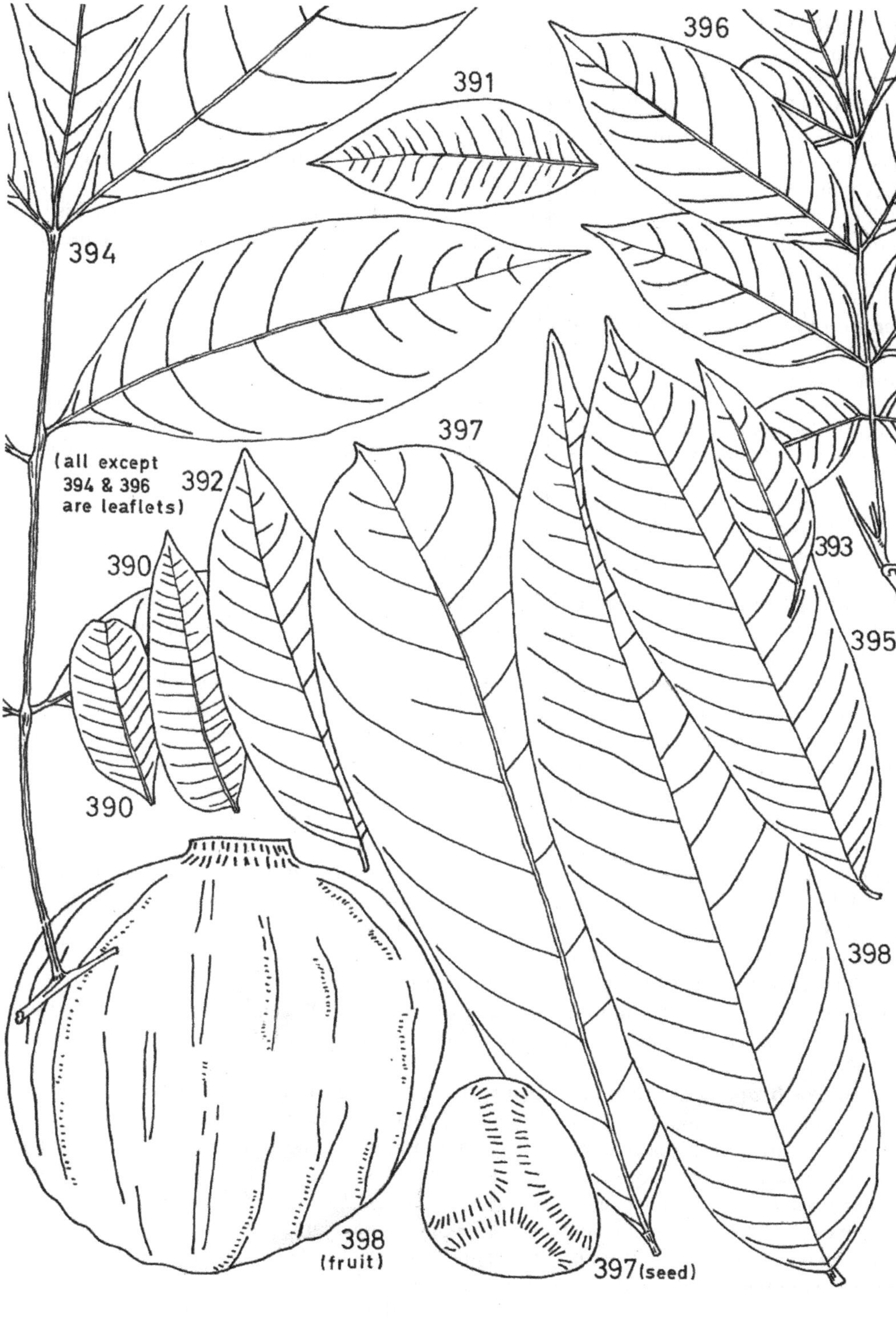

Plate 31. (390-398)

OCCURRENCE: U2 and 4. Abundant at 1200-1800 m, particularly in North Kibale, Kalinzu, Kayonza and North Bwindi forests. It grows in the Sango Bay forests in Masaka and in Mityana Forest, Mengo (rare in the later).
CULTIVATION AND PROPAGATION: Fast-growing. Perhaps best grown in mixed stands. Collect fruits beneath mother trees and remove the large seeds by hand. Sow seeds as soon as possible. Seedlings can be ready for transplanting within 6 months. Young plants may grow weakly in first months after transplanting, but do well once established.
NOTES: The large leaves, which are bright red when young, and the large fruits make this tree distinctive. The timber is decorative. The seeds are used in Kigezi for the manufacture of a type of butter.

Leplaea mayombensis (Pellegr.) Staner (398) Meliaceae

SYNONYM: *Guarea mayombensis* Pellegr.
Omucuraga (ki).
Similar in general appearance to *Carapa*, with which it often grows. Trunk crooked, with spreading branches from low down and a dark-coloured crown. Bark light brown, thick, with deep vertical fissures in places. Slash fibrous, very strongly scented, pink, turning darker. The leaflets are longer and thinner than those of *Carapa* (see Plate 31) and have c. 17-26 main lateral veins on each side of the midrib. The fruits contain only 1 huge seed in each of the 1-5 loculi.
OCCURRENCE: U2. Recorded from Bwindi and Kayonza forests.
CONSERVATION STATUS: Global VU (IUCN), LC (TOU); National EN (WCS).

Entandrophragma utile (Dawe & Sprague) Sprague (399) Meliaceae

Mahogany (en); Muyovu (ga); Omufumbi, Omuyovu (no); Budongo heavy mahogany, Feather sapele, Utile (tn).
50 m. Very large, deciduous tree with a wide, straight, cylindrical trunk and a large, spreading, open crown. The relatively numerous and rather narrow leaflets give the crowns of tall trees a feathery appearance. Buttresses present, usually quite large, but not extending far up the trunk. Bark brown, thick and rough, fissuring into rectangles. Slash fibrous, red, often with white lines, not scented. Leaves crowded at ends of branches. Leaves paripinnate, with c. 16-22 leaflets. Leaflets c. 11 x 3 cm (i.e. about 3 to 4 times as long as broad), with c. 12-21 main lateral veins on each side of the midrib. Petiolule c. 0.3 cm long. Capsule c. 20 cm long, club-shaped, splitting into five woody valves, containing winged seeds which are c. 6-10 cm long.
OCCURRENCE: U1, 2 and 4. Abundant in Budongo and common in Mabira forests, but rare elsewhere. Normally below 1400 m, but said to occur up to 1830 m on Rwenzori.
CONSERVATION STATUS: Global VU (IUCN), LC (TOU); National EN (WCS).
CULTIVATION AND PROPAGATION: Similar to *Entandrophragma angolense* (401).
NOTES: This is one of the most commercially important of Uganda's mahoganies. The tree has a superficial resemblance to *Mildbraediodendron excelsum* (423).

Entandrophragma cylindricum (Sprague) Sprague (400) Meliaceae

Mahogany (en); Muyovu (ga, tn, to); Omuyovi (ki); Omuyovu (no); Sapele (tn).
55 m. Very large deciduous tree, with a very long, straight, cylindrical trunk and rather rounded crown. From a distance the trunk appears to be twisted (like eucalyptus and *Strychnos*). Buttresses large (but smaller than those of *Khaya anthotheca*). Bark quite thin, brown, with prominent lenticels, smooth on young trees, flaking in quite large pieces up to 60 cm across on older trees. Slash red, often with white markings, turning darker and red-brown, scented. Leaves paripinnate, with c. 10-16 leaflets. Leaflets c. 11 x 3.5 cm (i.e. about 3 to 4 times as long as broad), with c. 6-12 main lateral veins on each side of the midrib. Capsule c. 6-10 cm long, with five valves which are less tough than those of *Entandrophragma utile* and *E. angolense*.

OCCURRENCE: U2 and 4. Abundant in Bunyoro forests. Occasional in Mengo. It occurs up to 1500 m in Bwindi Forest.
CONSERVATION STATUS: Global VU (IUCN), LC (TOU); National EN (WCS).
CULTIVATION AND PROPAGATION: Similar to *E. angolense* (401), but believed to be slower growing than other species of *Entandrophragma.*
NOTES: A very valuable timber tree, found also in W. Africa. It is reported to be extensively used for the manufacture of plywood and production of decorative veneer. A candidate for being the tallest of all indigenous African trees, for which another candidate is *Cephalosphaera usambarensis* Warb. growing in the forests of the East Usambara Mountains in Tanzania.

Entandrophragma angolense (Welw.) C. DC. (401) Meliaceae
Mahogany (en); Mukusu (ga, tn, to); Muyovu (ga, to); Kikura (ko); Omukusu, Omuyovu (no); Budongo mahogany, Gedu nohor (tn).
50 m. Very large deciduous tree with a somewhat wavy trunk and rather dense crown. Leaves clustered at ends of branches. Buttresses present, usually not large. Surface roots usually well developed. Bark light-coloured, quite thin and smooth, with prominent lenticels, flaking in pieces c. 10-20 cm across to leave concave scars. Slash red to dull red, sometimes with whitish streaks, not or only slightly scented. Leaves paripinnate, with c. 10-16 leaflets. Leaflets c. 13 x 5.5 cm (i.e. 2 to 3 times as long as broad), with c. 7-10 main lateral veins on each side of the midrib. Capsule c. 18 cm long, with 5 valves.
OCCURRENCE: U1-4. Frequent in most south Mengo forests. Occasional in Budongo and Mabira forests. Growing below 1220 m.
CONSERVATION STATUS: Global VU (IUCN), LC (TOU); National EN (WCS).
CULTIVATION AND PROPAGATION: Moderately fast-growing (but fast-growing on good sites). Should be planted under the shade of pioneer species, which should be removed after c. 10-15 years. Best grown in mixed stands. Collect unopened seed pods (if available) under mother trees or else the seeds. Sow seeds as soon as possible. Seedlings should be transplanted when strong (8-12 months). Seedlings develop long taproots, so long pots should be used.
NOTE: Produces a good timber of high market value.

Entandrophragma excelsum (Dawe & Sprague) Sprague (402) Meliaceae
Mahogany (en); Omuyovi (ki); Kikula (ko); Gumurumba, Murumba (ms); Muyovu (na, to); Omuhungura (no).
40 m. Deciduous tree with a straight trunk and large crown. Leaves clustered at ends of branches. Buttresses large. Bark smooth on young stems, flaking on older stems (as with *Entandrophragma angolense*). Slash red with whitish streaks (similar to *Entandrophragma angolense*), but a bit darker. Leaves paripinnate, with c. 14 (8-20) leaflets. Leaflets c. 17 x 7 cm (i.e. 2 to 3 times as long as broad), with c. 10-11 main lateral veins on each side of the midrib. Capsule c. 20-30 cm long, with 5 valves.
OCCURRENCE: U2-4. Frequent in Kigezi, in Kalinzu and Itwara forests, and on Mt Elgon. Also in Uni and Nakiga forests (west Mengo). Altitudinal range 1280-2150 m (extending to higher altitudes than other species of *Entandrophragma*).
CULTIVATION AND PROPAGATION: Similar to *E. angolense.*
NOTE: This species resembles *Entandrophragma angolense.*

Khaya anthotheca (Welw.) C. DC. (403) Meliaceae
Kirumbo (am); Mahogany (en); Omunyama (no); Munyama (to); African mahogany, Uganda mahogany (tn).

50 m. Very large deciduous tree with a long, wavy or straight, trunk and a massive, spreading crown. Buttresses very large. Bark brown, quite thick, fairly smooth, flaking in small circular pieces to leave concave depressions, c. 2-5 cm in diameter. Slash brittle, red, strongly scented (occasionally only slightly). Leaves paripinnate, with c. 6-10 leaflets. Leaflets c. 12 x 5.5 cm, with c. 6-9 main lateral veins on each side of the midrib. Fruit a spherical capsule with 4 valves, c. 6 cm diameter.
OCCURRENCE: U2. Tooro and Bunyoro. The commonest mahogany in Budongo Forest, where it is very abundant. Rare in Bugoma Forest.
CONSERVATION STATUS: Global VU (IUCN), LC (TOU); National EN (WCS).
CULTIVATION AND PROPAGATION: Moderately fast-growing. Plant in partial shade under fast-growing pioneer species, which should be removed once the plants are well established. Collect capsules from mother trees if possible; if not, from the ground beneath. Discard insect-damaged seeds and sow as soon as possible.
NOTES: The tree produces an attractive timber, similar in strength to 'true' mahogany (*Swietenia macrophylla* King from Central and South America). Saplings of this and other mahoganies may be confused with those of *Canarium* and *Polyscias*.

Khaya grandifoliola C. DC. (404) Meliaceae
Mahogany (en).
Large tree, very similar to *Khaya anthotheca*, but with larger leaflets (c. 18 x 9 cm) and c. 10-14 main lateral veins on each side of the midrib. Fruit 5-valved.
OCCURRENCE: U1-2. In forest outliers near Budongo Forest and in riverside vegetation in north-west Uganda.
CONSERVATION STATUS: Global VU (IUCN), LC (TOU); National EN (WCS).

Leplaea cedrata **(A. Chev.) E.J.M. Koenen & J.J. de Wilde (405)** Meliaceae
SYNONYM: *Guarea cedrata* (A. Chev.) Pellegr.
Scented guarea (tn).
45 m. Deciduous tree with a cylindrical trunk and a dense, rounded crown. Buttresses present on some trees (perhaps only in Budongo Forest), these not extending far up the trunk, but spreading out from the trees for some distance. Bark quite thin, light-coloured, smooth, flaking in moderate-sized pieces, leaving characteristic raised lines and rings of prominent lenticels. Slash brown or pink, with orange stone cells, rather granular, turning darker, strongly scented. Leaves paripinnate, with c. 8 leaflets. Leaflets c. 19 x 5 cm. Petiolule c. 0.5 cm long. Fruit an orange capsule with leathery valves, opening to reveal black seeds with orange arils.
OCCURRENCE: U2 and 4. Widely distributed through Mengo and Bunyoro, but mature trees uncommon. Young trees are very common in Budongo Forest and in parts of Mabira Forest.
CONSERVATION STATUS: Global VU (IUCN), LC (TOU); National EN (WCS).
NOTE: Not easily distinguishable from *Entandrophragma angolense* in the field.

Lovoa trichilioides **Harms (406)** Meliaceae
Nkoba (ga, tn); Enkoba (no); Uganda walnut (tn); Mukusu (to).

Plate 32. Meliaceae (392-407); see also Plate 31

392. *Trichilia dregeana* 394. *Trichilia prieureana* 399. *Entandrophragma utile*
400. *Entandrophragma cylindricum* 401. *Entandrophragma angolense*
402. *Entandrophragma excelsum* 403. *Khaya anthotheca* 405. *Leplaea cedrata*
406. *Lovoa trichilioides* 407. *Lovoa swynnertonii*

Actual sizes: leaflets, fruits and seeds x 2; trunk bases x 80; tree profiles x 800.

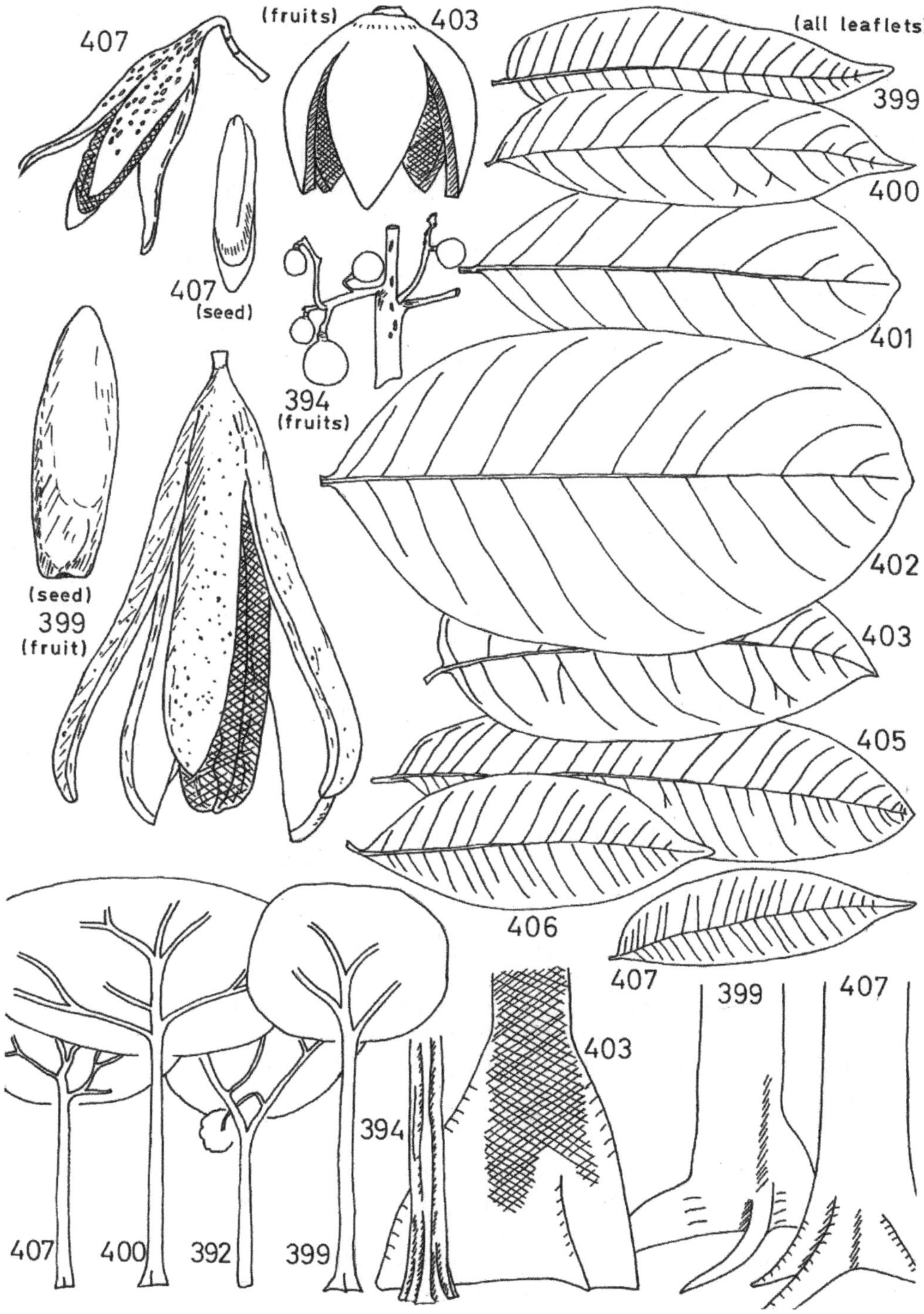

Plate 32. (392-407)

40 m. Tree with a straight or (more often) wavy trunk and a small, rounded, dark-coloured crown (occasionally the crown may be fairly spreading). Buttresses absent. Trunk sometimes flaring at base. Bark thin and smooth, brown (sometimes very dark-coloured), with vertical lines of lenticels, tending to fissure on older trees (sometimes into rectangles). Slash fibrous, red, sometimes with paler lines, strongly scented. Leaves paripinnate, with c. 6-12 leaflets. Leaflets c. 10 x 3.5 cm, with c. 14-24 main lateral veins on each side of the midrib. Flowers white, produced in large numbers. Fruit a 4-valved capsule, c. 5 cm long.
OCCURRENCE: U2 and 4. Very common in lake-shore forests, but very rare elsewhere. Often associated with *Piptadeniastrum africanum*.
CONSERVATION STATUS: Global LC (IUCN, TOU); EN (WCS).
CULTIVATION AND PROPAGATION: Initially slow-growing, later speeding up. Grow in pure or mixed stands using fast-growing shrubs or trees to provide shade for the first 5-10 years. Collect seeds beneath mother trees, remove those that are insect-damaged and plant as soon as possible.
NOTE: Produces a good quality timber.

Lovoa swynnertonii Baker f. (407) Meliaceae
Nabulagala (ga); Enkoba (ki); Mukusu (na, to).
40 m. Trunk straight and cylindrical, branching high up. Crown spreading. Buttresses absent to large. Surface roots often well developed. Bark brown, fairly thin and smooth (sometimes locally thick), with prominent lenticels, flaking in pieces c. 2-30 cm across. Slash fibrous, red, red-brown, or red and white, characteristically with very many, thin, concentric layers of different colours, not fragrant. Leaves paripinnate, with (6-)12-16 leaflets. Leaflets c. 11 x 3.5 cm, with c. 15-30 main lateral veins on each side of the midrib. Capsule 4-valved, c. 5 cm long.
OCCURRENCE: U2 and 4. Common in Kibale Forest, but rare elsewhere.
CONSERVATION STATUS: Global NT (IUCN), LC (TOU); National EN (WCS).
NOTE: The tree may be distinguished from *Lovoa trichilioides* by the scentless slash and the (usually) greater number of leaflets.

Turraeanthus africanus (C. DC.) Pellegr. (408) Meliaceae
Mbahira (to).
20 m. Poorly-shaped understorey tree, with a short trunk and a dense, rounded crown. Bark grey, rough, scaling in small pieces. Slash yellow. Leaves large, paripinnate, with c. 20-36 leaflets. Leaflets c. 13 x 3.5 cm, with well-marked lateral veins on lower surface, apex abruptly acuminate. Fruit spherical, c. 2.5 cm diameter.
OCCURRENCE: U2. Budongo, Itwara, Kagombe, Kalinzu-Maramagambo, Kasyoha-Kitomi, Semliki and Kibale forests. In riparian and poorly drained forests.
CONSERVATION STATUS: Global VU (IUCN), LC (TOU); National VU (WCS).

SAPINDACEAE

Understorey trees (except *Majidea* and sometimes *Blighia*), bark usually thin and smooth, slash often yellow to orange (sometimes with traces of red, particularly towards the bark). The leaves are paripinnate (except in *Allophylus*), but may appear to be imparipinnate due to abortion of one of the terminal leaflets. Stipules absent. Flowers small. Fruits often conspicuous, either dehiscent or indehiscent, often trilocular.

Key to Sapindaceae.

1. Leaves trifoliolate. ....**351-353. *Allophylus***
Leaves pinnate, sometimes with only two leaflets. 2

2. Leaves usually 6 or fewer on a leaf. 3
Leaves usually more than 6 on a leaf. 8

3. Leaflets 4 or 6, the basal pair (of at least some of the leaves) rounded and placed so near the stem as to simulate stipules. ....**418. *Glenniea***
Not as above. 4

4. Leaflets usually 6 or 8, comparatively broad in proportion to length and comparatively small (often c. 8 x 3.5 cm). ....**416. *Zanha***
Not as above. 5

5. Smallest stems not (or only slightly) furrowed, being more or less circular in cross-section. 6
Smallest stems furrowed (not circular) in cross-section. 7

6. Young stems (those with leaves) comparatively thin (often less than 0.3 cm diameter) and generally with small (but fairly conspicuous) light-coloured lenticels. ....**417. *Pancovia***
Young stems comparatively thick (often over 0.3 cm diameter), lacking conspicuous light-coloured lenticels. ....**420. *Lepisanthes***

7. Petiolules not markedly swollen. ....**419. *Blighia unijugata***
Petiolules markedly swollen. ....**420. *Lepisanthes***

8. Underside of leaflets with a dense covering of red-brown hairs. An uncommon understorey tree. ....**410. *Deinbollia fulvotomentella***
Not as above. 9

9. Main lateral veins very numerous, appearing yellow to yellow-brown against the green lamina, c. 15 on each side of the midrib on larger leaflets. Leaflets normally toothed in upper half. A small tree. ....**412. *Lychnodiscus***
Not as above. If veins prominent and yellow to yellow-brown, then leaflets with fewer than c. 13 main lateral veins on each side of the midrib and not toothed. 10

10. Leaflets relatively small (c. 7 x 3 to 11 x 5 cm), a little over twice as long as broad. Slash yellow. ....**416. *Zanha***
Leaflets not shaped as above or, if so, then slash not yellow. 11

11. Leaflets usually over 4.5 cm wide. 12
Leaflets usually less than 4.5 cm wide. 13

12. Leaf rachis and petiole together usually over 25 cm long. An uncommon tree to 7 m. ....**411. *Deinbollia kilimandscharica***
Leaf rachis and petiole together less than 25 cm long. A frequent tree. ....**415. *Blighia welwitschii***

13. Leaflets up to 26 on a leaf. Slash generally yellowish. ....**409. *Majidea***
Leaflets usually fewer than 13 on a leaf. Slash pink to red-brown. ....**413-414. *Lecaniodiscus***

Majidea fosteri **(Sprague) Radlk. (409)** Sapindaceae

Muwunda (ga).

35 m. Tree with a straight trunk and a spreading, deciduous crown. Buttresses absent to well developed. Bark yellowish, thin and smooth, with ring marks and small vertical fissures, sometimes flaking in large pieces. Slash granular, pale yellowish to yellow-brown, often with white vertical streaks, smelling of disinfectant (thymol). Leaves paripinnate, with c. 8-24 leaflets. Leaflets c. 8.5 x 2.25 cm (but rather variable in shape), base unequal-sided, margin slightly toothed or crenate. Fruit a capsule, c. 3.5 cm long, bright red on the inner wall, containing three seeds.

OCCURRENCE: U1, 2 and 4.

Deinbollia fulvotomentella **Baker f. (410)** Sapindaceae

7 m. Understorey tree. Branchlets hollow. Leaves paripinnate, up to 90 cm long, with c. 14-16 leaflets. Leaflets c. 15 x 5.5 cm, more or less glabrous on upper surface, with a dense covering of red-brown hairs below.

OCCURRENCE: U2 and 4. Mengo, Ssese and Kasyoha-Kitomi Forest.

Deinbollia kilimandscharica **Taub. (411)** Sapindaceae

Small tree to 7 m. Leaves paripinnate, with c. 8-12 leaflets. Leaflets c. 13 x 5 cm (but sometimes much larger), glabrous.

OCCURRENCE: U2-4. Recorded from Mengo, Mt Elgon and South Maramagambo Forest.

Lychnodiscus cerospermus **Radlk. (412)** Sapindaceae

15 m. Understorey tree, with a weak, often leaning, trunk. Bark thin and smooth, greenish, with ring marks and prominent, dark brown, lenticels. Phellogen green to reddish. Slash of a rather indeterminate colour (red-brown, pink-brown, pink-yellow to orange). Leaves paripinnate, with c. 8-14 leaflets. Leaflets c. 15 x 5 cm (sometimes a bit smaller), with c. 12-20 main lateral veins on each side of the midrib, these veins being well defined, parallel to one another and appearing rather yellowish as seen from the undersurface, margin normally toothed in upper half. Fruit a lobed capsule, hairy on the outside, reddish when ripe, with a pink inner wall, containing 3 bright red seeds.

OCCURRENCE: U1, 2 and 4. Sometimes abundant, e.g. in Mpanga and Central Kibale forests.

NOTE: Only one variety of this species, var. *cerospermus*, occurs in Uganda.

Lecaniodiscus fraxinifolius **Baker (413)** Sapindaceae

15 m (occasionally to 20 m). Very badly shaped tree, with numerous spindly shoots growing from near the base of the trunk. Bark grey. Slash red-brown to pink. Leaves paripinnate, with c. 6-10 leaflets. Leaflets c. 8 x 2.5 cm.

OCCURRENCE: U1, 3 and 4. East Mabira Forest, Busoga and stream valleys at the base of Mts Morongole, Kadam and Napak in Karamoja.

Lecaniodiscus cupanioides **Benth. (414)** Sapindaceae

Tayi (am).

Plate 33. Sapindaceae (409-420)

409. *Majidea fosteri* 412. *Lychnodiscus cerospermus* 415. *Blighia welwitschii*
417. *Pancovia turbinata* 418. *Glenniea africana* 419. *Blighia unijugata*
420. *Lepisanthes senegalensis*

Actual sizes: leaves, leaflets and fruits x 2; tree profile x 800.

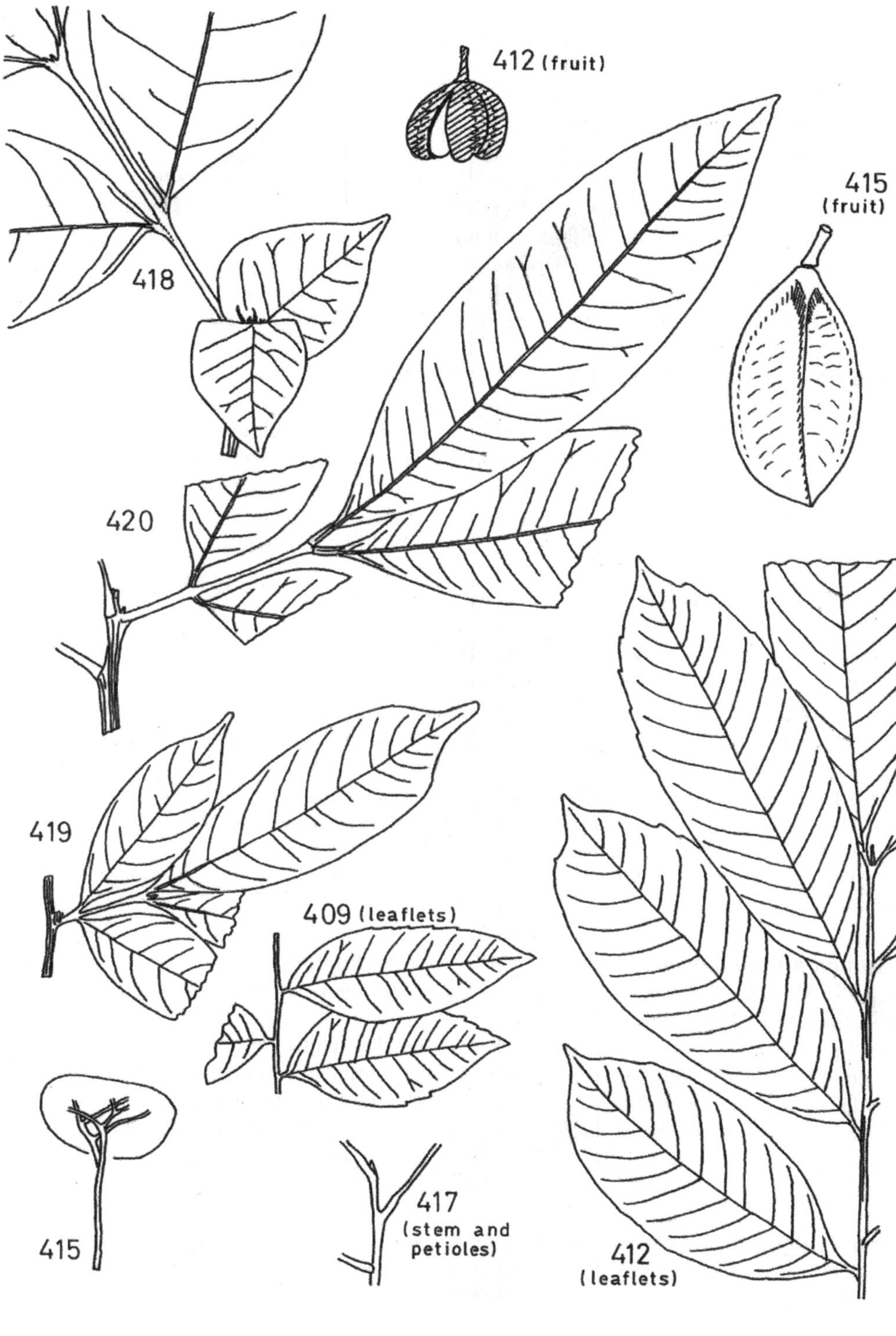

Plate 33. (409-420)

15 m. Spreading understorey tree. Bark grey. Slash red-brown. Leaves paripinnate, with 8-12 leaflets.
OCCURRENCE: U2. Recorded from Budongo, Semliki, Itwara, Kalinzu, Kasyoha-Kitomi and Kibale forests.

Blighia welwitschii **(Hiern) Radlk. (415)** Sapindaceae

Mukuzannyana (ga).
25 m. Second storey tree, with a wavy trunk and a moderately spreading crown casting a heavy shade. Bark brown, fairly thin and smooth, with very small vertical fissures, usually with ring marks. Slash yellow to orange, often becoming reddish towards the bark, either granular or both granular and fibrous. Leaves paripinnate, with c. 6-8 leaflets. Leaflets c. 18 x 7 cm (sometimes smaller). Fruit c. 6 cm long, dull red when ripe, three-sided, with sharp edges.
OCCURRENCE: U2 and 4. Mengo, Bunyoro and Tooro (Kibale Forest).

Zanha golungensis **Hiern (416)** Sapindaceae

Ekwalakwala (at); Muyiki (ga); Mukaka (na).
20 m. Deciduous understorey tree, with a wavy, cylindrical, trunk and a fairly spreading to rounded crown. Bark thin, red-brown to light brown, usually with conspicuous lenticels, flaking in patches to give a mottled appearance of lighter and darker colours. Slash fibrous, yellow, smelling of antiseptic (thymol). Leaves paripinnate, with c. 6-8 leaflets. Leaflets alternate to sub-opposite (rarely opposite), rather broad in comparison to length, often c. 8 x 3.5 cm, base symmetrical.
OCCURRENCE: U1-4. Widespread and sometimes common.

Pancovia turbinata **Radlk. (417)** Sapindaceae

SYNONYM: *Pancovia* sp. near *turbinata* (*sensu* ITU & UFT)
20 m, but usually much less. Understorey tree, branching from near base, casting a heavy shade. Trunk characteristically gnarled. Bark very thin and smooth, dark green to brown. Phellogen often reddish. Slash yellow to brown, becoming red near the bark. Leaves paripinnate, with c. 4-6 leaflets. Terminal leaflets the largest, c. 14 x 5 cm. Petiolules thick and grooved, c. 0.5 long. The young stems are quite thin (often less than 0.3 cm wide) and more or less circular in cross-section. Young leaves purple-grey.
OCCURRENCE: U2 and 4. Mengo, Kalinzu and Kibale forests, also Ishasha Gorge. Common in Central Kibale Forest.

Glenniea africana **(Radlk.) Leenh. (418)** Sapindaceae

SYNONYM: *Melanodiscus sp. nov.?* (*sensu* ITU); *Melanodiscus* sp. (*sensu* UFT)
Omwatibale (no); Mwatibale (to).
25 m. Spreading understorey tree, with crooked trunk, branching from low down and with low foliage. Bark thin and smooth, green to brown, with prominent lenticels and ring marks. Slash granular, yellow (or white and orange), turning white after some time. Leaves paripinnate, with 2-6 (-8) leaflets. Basal pair of leaflets rounded and so near the stem as to simulate stipules. Leaflets c.16 x 6 cm (but variable in size).
OCCURRENCE: U1-4. Very common in Budongo and Mabira forests.

Blighia unijugata **Baker (419)** Sapindaceae

Mukuzannyana, Nkuzannyana (ga); Muhohote (sa); Mukuzadhyna, Musandikira (so); Mwatihale (to).

15 m (occasionally to 30 m). Usually an understorey tree, with a crooked, often gnarled, trunk, branching from near the base. Occasionally an upperstorey tree. Crown usually dense and casting a heavy shade, resembling that of a mango. Bark thin and smooth, with ring marks, light-coloured. Slash variable, white to orange-red, sometimes with orange streaks, both fibrous and granular. Leaves paripinnate, with c. 2-6 leaflets, the upper pair the largest. Leaflets c. 15 x 5 cm (but sometimes considerably smaller), the midrib and main veins appearing yellowish on the lower surface. Petiolules absent to 0.7 cm long. Young stems noticeably grooved. Fruit a red capsule, c. 3.5 cm long, three-sided, with sharp corners.
OCCURRENCE: U2-4. Abundant in some areas, e.g. Mengo and Mubende. Mainly on forest edges and in secondary forest.

Lepisanthes senegalensis **(Poir.) Leenh. (420)** Sapindaceae
SYNONYM: *Aphania senegalensis* (Poir.) Radlk.
Mukaka (na); Kobwa (to).
15 m. Understorey tree with a crooked trunk and a heavy, spreading crown. Bark light brown, thin and fairly smooth, flaking. Slash yellow, orange to orange-brown, with a strong smell. Leaves paripinnate, with 2-6 leaflets, the terminal pair being the largest. Leaflets c. 15 x 5 cm. Petiolules thick and grooved. Young stems furrowed. Fruit a red juicy drupe, eaten by monkeys.
OCCURRENCE: U1-4. Widespread and locally common.

FABACEAE AND CONNARACEAE

Fabaceae (formerly known as Leguminosae) is a large family, which includes beans and peas, as well as numerous trees. The leaves are stipulate, alternate and compound (but *Baphia* and *Baphiopsis* have compound unifoliolate leaves that appear simple). The fruits are typically pods (like beans), which are winged in *Tetrapleura* and indehiscent and spherical in *Dialium* and *Midbraediodendron*. The roots of legumes typically have nodules containing nitrogen-fixing bacteria. There are three subfamilies (formerly regarded as families, e.g. in ITU):

Subfamily Mimosoideae. All forest trees in Uganda with bipinnate leaves, except *Erythrophleum* (Caesalpinioideae), are in this subfamily. The flowers are radially symmetric, grouped together and often have many stamens (usually the most conspicuous part of the flower). *Acacia*, a common savanna genus, and *Entada*, which includes woodland trees and large forest climbers with huge pods, are included here. *Acacia pennata* (L.) Maslin is a common forest climber.

Subfamily Caesalpinioideae. The leaves are pinnate, except in *Erythrophleum* (bipinnate). The flower is bilaterally symmetric, the uppermost petal lying inside the laterals.

Subfamily Faboideae. The leaves are pinnate, except in *Baphia* (apparently simple leaves) and *Erythrina* (trifoliolate). The flowers are bilaterally symmetric, the uppermost petal lying outside the laterals.

The Connaraceae is a family of shrubs, small trees and large climbers related to Fabaceae, but differing in lacking stipules. The genera *Agelaea* and *Connarus* and the species *Rourea thomsonii* (Baker) Jongkind are shrubs that can produce fast-growing leaders which can turn into lianes.

Key to Fabaceae and Connaraceae

1. CHOOSE FROM ONE OF THESE 4 OPTIONS
   Leaves apparently simple. Small trees. ....**230-231. *Baphia*; 232. *Baphiopsis***
   Leaves trifoliolate. Spines present, persistent as conical woody bosses on the trunk. ....**344-345. *Erythrina***
   Leaves pinnate. 2
   Leaves bipinnate 9

2. CHOOSE FROM ONE OF THESE 3 OPTIONS
   Leaflets more than 24 on a leaf. ....**422. *Cnestis*; 423. *Mildbraediodendron***
   Leaflets 5-23 on a leaf (occasionally fewer and then leaflets thick and leathery). 3
   Leaflets usually 4 (occasionally 6). Leaflets not thick and leathery. ..**428. *Cynometra***

3. Leaflets thick and leathery, often 9-20 x 4.5-9 cm (occasionally larger). Venation normally indistinct. ....**427. *Baikiaea*; 429. *Craibia***
   Leaflets not thick and leathery. 4

4. Leaves paripinnate. ....**424. *Cassia* 426; *Afzelia***
   Leaves imparipinnate. 5

5. Mature petiole and leaf rachis together less than 10 cm long. ....**425. *Dialium***
   Mature petiole and rachis together over 10 cm long. 6

6. Leaflets usually fewer than10 on a leaf. 7
   Leaflets usually more than 10 on a leaf. 8

7. Leaf venation conspicuous and arcuate (see Plates 34,35). ....**421. *Connarus*; 432. *Millettia psilopetala***
   Leaf venation not markedly arcuate. ....**429. *Craibia***

8. Stipules present. Leaves acuminate. ....**430. *Millettia dura*; 431. *M. eetveldeana***
   Stipules absent. Leaves rounded to very shortly acuminate at apex. ......**422. *Cnestis***

9. Thorns present. 10
   Thorns absent. 11

10. Thorns present on branchlets. ....**437. *Acacia***
    Thorns absent from branchlets (except on very young plants and on sucker shoots). ....**436. *Cathormion***

11. Leaflets very small (0.1-1.5(-2) mm wide) Very large trees with big buttresses. ....**434. *Piptadeniastrum*; 435. *Newtonia***
    Leaflets larger. 12

12. Leaflets alternate. 13
    Leaflets opposite or nearly so. 14

13. Leaflets short and rounded, c. 1.2 x 0.6 cm. ....**439. *Tetrapleura***
    Leaflets comparatively large, c. 7 x 3.5 cm. ....**433. *Erythrophleum***

14. Leaflets small, often c. 1.2 x 0.3 cm. ....**436. *Cathormion***
    Leaflets usually over 0.3 cm wide. 15

15. Leaflets with 2 main veins from the base (as seen on upper surface), both veins more or less parallel to the margin, one near the centre and one near the lower margin (see Plate 36). ....**440. *Parkia***
Venation not as above. Each leaflet with one main vein. 16

16. Midrib more or less centrally placed on (or slightly off) the centre of the leaflet. Bark rough. 17
Midrib running diagonally across the leaflet. Bark smooth on forest specimens. 18

17. Slash red, fibrous. ....**442. *Albizia coriaria***
Slash yellow, fibrous. ....**441. *Albizia ferruginea***

18. Leaflets all of about the same size on a pinna or, if of different sizes, then the terminal leaflets smaller than the central. 19
Leaflets of markedly different sizes on some pinnae, the terminal pair being the largest. 21

19. Leaflets with short, but distinct, petiolules (see Plate 36). ....**443. *Albizia glaberrima***
Leaflets sessile or indistinctly stalked. 20

20. Mature rachis more or less glabrous. ....**444. *Albizia gummifera***
Mature rachis densely covered with red-brown hairs. ......**445. *Albizia adianthifolia***

21. Leaflets with short, but distinct, petiolules (see Plate 36). ...**443. *Albizia glaberrima***
Leaflets sessile or indistinctly stalked. 22

22. Stipules rounded. ....**446. *Albizia grandibracteata***
Stipules long and thin. ....**447. *Albizia zygia***

***Connarus longistipitatus* Gilg (421)** Connaraceae

12 m. Understorey tree, branched from near base, with a spreading crown. Bark thin and smooth, light brown. Slash fibrous, pink, turning darker. Young stems more or less glabrous. Leaves imparipinnate, with 5-9 (usually 7) leaflets. Leaflets c. 12 x 5 cm, with 5-8 arcuate main lateral veins on each side of the midrib, apex acuminate, both surfaces more or less glabrous. Petiolules c. 0.5 cm long. Inflorescence terminal, large and conspicuous. Flowers white. Fruit a follicle, c. 2.5 cm long.
OCCURRENCE: U2 and 4. Most abundant in Kalinzu Forest.
NOTE: This species is typically an upper canopy climber.

***Cnestis mildbraedii* Gilg (422)** Connaraceae

SYNONYM: *C. ugandensis* Schellenb.
Small tree to 7 m. Leaves large (to 36 cm long), imparipinnate, with c. 15-31 leaflets. Leaflets with c. 5-11 arcuate main lateral veins on each side of the midrib. Young branches and undersurface of leaflets covered with grey or brown hairs.
OCCURRENCE: U2 and 4. Uncommon, except perhaps in Budongo Forest.
CONSERVATION STATUS: NE (IUCN), Global NT (TOU); National NE.
NOTE: The leaflets are similar to those of *Mildbraediodendron*, but are not gland-dotted.

***Mildbraediodendron excelsum* Harms (423)** Fabaceae (Caesalpinioideae)

Bombo (am); Nabulere (ga); Omuyati (no); Muyati (tn).

50 m. Tall, deciduous, upperstorey or emergent tree with a straight, cylindrical, trunk and a spreading crown with fern-like foliage as seen from the ground. Buttresses large. Bark quite thick, brown, splitting into rectangles (which resemble those of *Entandrophragma utile*, but are a bit smaller). Slash fibrous, layered, yellow (to orange and white), smelling of peas. Leaves pinnate, with 24-38 leaflets. Leaflets c. 5.5 x 2 cm, gland-dotted, covered with yellow hairs when young, more or less glabrous when mature. Fruit spherical, green, c. 5.5 cm diameter, smelling like a pea pod, containing large seeds.
OCCURRENCE: U1, 2 and 4. Common in forests in Bunyoro.
NOTES: Young individuals may be confused with *Cnestis*, but the leaves differ in being gland-dotted. The wood is handsome, very hard, and resistant to decay. It is difficult to work. According to ITU, elephants are very fond of the fruits.

***Cassia mannii* Oliv. (424)** Fabaceae (Caesalpinioideae)
Mutubanka (am); Entanyenya (no).
25 m. Deciduous tree. Bark dark brown, shaggy, with projecting, corky, brown, lenticels. Slash dark red with brown fibres, with a thin orange/yellow outer layer. Leaves paripinnate, with 10-24 leaflets. Leaflets c. 6.5 x 2.5 cm. Petiolule c. 0.5 cm long. Flowers pink, produced in profusion. Fruit to nearly 1 m long.
OCCURRENCE: U1 and 2. Bwamba, Zoka, Budongo, Kalinzu-Maramagambo and Bugoma forests. Rare.
NOTES: There are many indigenous species of *Cassia* and *Senna* (a related genus). The following are yellow-flowered shrubs: (1) *Cassia bicapsularis* L. (with 2-3 pairs of thick fleshy leaflets); (2) *Senna septemtrionalis* (Viv.) H.S. Irwin & Barneby (Syn.: *Cassia floribunda* Cav.) (with 3-4 pairs of leaflets which are not fleshy; common in secondary forest in Kalinzu Forest); (3) *Senna didymobotrya* (Fresen.) H.S. Irwin & Barneby (Syn.: *Cassia didymobotrya* Fresen.) (with 10-20 pairs of leaflets, persistent stipules and very dark-coloured flower buds); and (4) *Senna petersiana* (Bolle) Lock (Syn.: *Cassia petersiana* Bolle) (with 6-12 pairs of leaflets and deciduous stipules - except on uppermost leaves). There are also several introduced trees of *Cassia* and related genera. The yellow-flowered *Cassia spectabilis* DC. is widely planted as a boundary marker in forests and sometimes found in secondary forest. *Cassia siamea* Lam. is planted for fuel and poles. *Cassia javanica* L. subsp. *nodosa* (Buch.-Ham. ex Roxb.) K. Larsen & S.S. Larsen (Syn.: *Cassia javanica* L. var. *agnes* de Wit) is a pink to red-flowered species. *Peltophorum pterocarpum* (DC) Backer ex K. Heyne is a yellow-flowered tree similar to *Cassia* in general appearance.

***Dialium excelsum* Steyaert (425)** Fabaceae (Caesalpinioideae)
Bukirima, Kadindinsimbo (am).
40 m. Deciduous. Trunk straight. Bark light brown, smooth, flaking. Slash layered, pale brown and off-white. Petiole and rachis together only 5-8.5 cm long. Leaves imparipinnate, with 5-9 leaflets. Leaflets small, the uppermost c. 6 x 2 cm, glabrous. Petiolule c. 0.25 cm long. The hairy, brown, axillary buds are conspicuous. Fruit ovoid, almost spherical, brown and indehiscent, c. 1.5 cm long.
OCCURRENCE: U2. Only recorded from Budongo and Bwamba forests. Uncommon.
CONSERVATION STATUS: Global EN (IUCN), LC (TOU); National VU (WCS).

***Afzelia bipindensis* Harms (426)** Fabaceae (Caesalpinioideae)
Mbande (am).
40 m. Trunk straight, with a dark green, thick crown. Bark usually red-brown, scaling. Slash light brown. Leaves paripinnate, with 10-14 leaflets. Leaflets c. 7.5 x 3 cm. Petiolule c. 0.3 cm long. Pod woody, thick, c. 13 cm long.
OCCURRENCE: U2. Local in Bwamba Forest.
CONSERVATION STATUS: Global VU (IUCN), LC (TOU); National VU (WCS).

Plate 34. Connaraceae and Fabaceae (421-428); see also Plate 37

421. *Connarus longistipitatus* 423. *Mildbraediodendron excelsum*
425. *Dialium excelsum* 427. *Baikiaea insignis* 428. *Cynometra alexandri*

Actual sizes: leaves x 2; trunk bases x 80; tree profiles x 800.

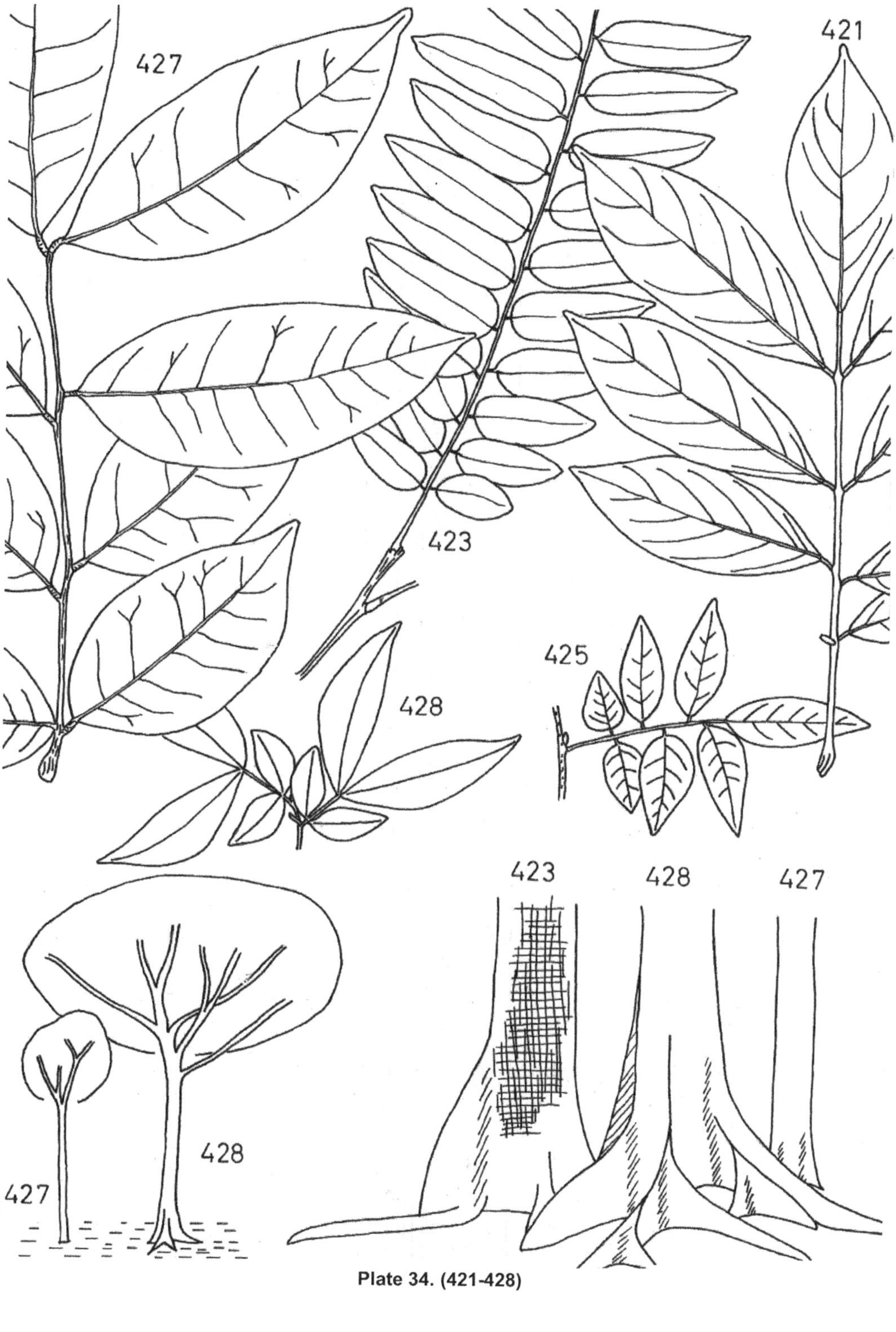

Plate 34. (421-428)

Baikiaea insignis Benth. (427) Fabaceae (Caesalpinioideae)

Mukubampanga, Nkoba (ga).

35 m. Evergreen tree with a straight, cylindrical trunk and a small, dense crown. Buttresses absent. Bark brown, fairly thin, with small vertical fissures. Slash hard, granular (fibrous when young), of rather indeterminate colour (yellow-red, pink, red-brown to brown). Leaves paripinnate or imparipinnate, with c. 2-12 leaflets (usually 5-9). Leaflets c. 17 x 7 cm (but very variable in size), venation usually indistinct on both surfaces, thick and leathery. Petiolules thick, c. 0.8 cm long. Flowers white, with petals up to 12 cm long. Pod thin, brown and fairly woody, up to 40 cm long (often 25 x 6.5 cm).

OCCURRENCE: U2-4. Widely distributed, usually near water, occasionally on dry sites. A dominant tree in the Sango Bay swamp forests and common in swamps in Mabira Forest. Rare elsewhere.

Cynometra alexandri C.H. Wright (428) Fabaceae (Caesalpinioideae)

Kahimbi (am); Muhindi (na); Omuhindi (no); Muhimbi, Uganda ironwood (tn).

50 m. Evergreen tree with a rather crooked trunk, usually branched comparatively low down, with a large spreading crown. Buttresses large, thin and spreading. Bark quite thin and smooth, brown, with numerous prominent lenticels, flaking in both large sheets and smaller pieces. Slash fibrous, light yellow to yellow, turning darker. Young leaves reddish. Leaves paripinnate, with 4 (occasionally 6) sessile leaflets, the uppermost the largest. Leaflets c. 6.5 x 2.25 cm, but larger on young trees. Flowers white, fragrant. Pod thin, containing only a few seeds.

OCCURRENCE: U2. Not found above 1225 m. The dominant tree over large areas of undisturbed parts of Bwamba, Budongo, South Kibale and Maramagambo forests.

CULTIVATION AND PROPAGATION: Slow-growing, requiring shade. Collect pods on the ground beneath mother trees and remove seeds manually after the pods have split open. Sow seeds as soon as possible.

NOTE: The wood is extremely heavy and hard. It is suitable for heavy duty flooring.

Craibia brownii Dunn (429) Fabaceae (Faboideae)

Shitoho (ms).

10 m. Understorey tree with a crooked trunk and dark-coloured crown. Bark thin and smooth, greenish to light-coloured, sometimes with vertical lines. Phellogen green. Slash off-white to yellow, sometimes with some brown colour, granular. Leaves imparipinnate, with (2-)5-9 leaflets, which are alternate or sub-opposite. Leaflets c. 13 x 4 cm, more or less acuminate. Leaf buds characteristic, shiny and globular. Inflorescence attractive. Flowers white tinged with pink or blue. Fruit thin, c. 5-7.5 cm long.

OCCURRENCE: U2 and 3. Uncommon, except in South Kibale Forest, where it is a dominant understorey tree beneath *Cynometra*.

Millettia dura Dunn (430) Fabaceae (Faboideae)

Omutate, Omutete (ki); Kiragara, Murongo (na); Muhakwa (to).

Plate 35. Fabaceae (429-444); see also Plates 36-37

429. *Craibia brownii* 430. *Millettia dura* 432. *Millettia psilopetala*
433. *Erythrophleum suaveolens* 434. *Piptadeniastrum africanum*
435. *Newtonia buchananii* 444. *Albizia gummifera*

Actual sizes: leaves x 2; trunk bases x 80; tree profile x 800.

Plate 35. (429-444)

12 m. Small tree, trunk branching near base, crown spreading and untidy. Bark thin and smooth (occasionally rough at base), green to grey, with vertical columns of lenticels. Phellogen green. Slash soft, white to light yellow, sometimes with vertical brown lines, very rarely exuding yellow latex. Leaves imparipinnate, with c. 15-23 leaflets which are usually opposite. Leaflets c. 9 x 2.75 cm, rather glaucous (bluish or greyish) below, acuminate, midrib with brown hairs. Flowers blue, borne in an attractive inflorescence.
OCCURRENCE: U1 and 2. A forest-edge species, abundant in Kibale Forest. Occasionally an understorey tree in open forest types.
CULTIVATION AND PROPAGATION: Fast-growing. Well adapted to a variety of conditions from shady to open and moist to dry. Can be grown to restore depleted land. Can be used as a cover for slower-growing species. Can be pruned and pollarded to provide poles. Collect ripe pods (brown in colour) from mother trees and remove the seeds manually. Sow the seeds as soon as possible. Stored seeds should be soaked for a few hours in water or else nick their coats slightly before sowing.
NOTE: Widely planted.

Millettia eetveldeana **(Micheli) Hauman (431)** Fabaceae (Faboideae)

Bungbara (am).
15 m. Understorey tree. Very similar in most respects to *Millettia dura*, from which it differs in having glabrous leaves (except there may be a few scattered hairs towards the base of the midrib).
OCCURRENCE: U2. Only known from Bwamba. Under *Cynometra* and on forest edges.

Millettia psilopetala **Harms (432)** Fabaceae (Faboideae)

Omutaate (ki).
Understorey tree to 7 m or (more commonly) a climber. Trunk crooked, with a spreading crown. Young stems with brown hairs. Leaves imparipinnate, with c. 5-9 leaflets, which are alternate or sub-opposite. Leaflets broader than those of other species of *Millettia*, the terminal being c. 13 x 5 cm, with c. 6-11 arcuate main lateral veins on each side of the midrib. Base of petiole and petiolules markedly swollen.
OCCURRENCE: U2. Recorded from Ankole, Kigezi and Tooro (Bwamba Forest). Common in the Kayonza area.
NOTES: The leaves resemble those of *Connarus*. However, the young stems are hairy, rather than being more or less glabrous.

Erythrophleum suaveolens **(Guill. & Perr.) Brenan (433)** Fabaceae (Caesalpinioideae)

Akwir (ac); Aramori (at); Ordeal tree (en); Odiodi (lo); Omumara (no); Missanda, Mumara, Sasswood (tn).
30 m. Trunk thick, often wavy, often with very large branches and a large spreading crown. Trunk flared at base, buttresses usually (but not always) absent. Bark thin, brown to brown-black, with prominent lenticels, often with small vertical fissures, flaking on older trees, general effect smooth. Slash granular, red to red-brown (said to be sometimes closer to yellow), often with white lines and sometimes with orange granules, sometimes in two layers. Leaves bipinnate, with 2-5 pairs of pinnae. Leaflets alternate, c. 8 on each side of the larger pinnae, c. 7 x 3.5 cm (but variable in size). Petiolules 3-5 mm long. Pod woody, thin.
OCCURRENCE: U1-4. Widely distributed, most common in Bunyoro and in lake-shore forests
CONSERVATION STATUS: Global LC (TOU); National VU (WCS).
NOTES: The wood and bark contain alkaloids. Infusions of the bark were formerly used in parts of Africa in trials by ordeal.

Piptadeniastrum africanum **(Hook. f.) Brenan (434)** Fabaceae (Mimosoideae)
Mpewere (ga); Omugeye (no); Agboin. Dahoma (tn).
50 m. Very large, flat-topped, deciduous tree. Trunk often leaning or wavy, branching from relatively low down, foliage in layers. Buttresses large and thin. Bark very thin and smooth, light-coloured, with ring marks, typically with very small lenticels (but lenticels on buttresses occasionally large). Slash granular, light yellow, dry. Leaves bipinnate, with 10-19 pairs of pinnae. Pinnae normally alternate, occasionally opposite. Leaf rachis lacking glands. Leaflets very small, c. 0.5 x 0.09 cm. Pod c. 12-30 x 2-5 cm long.
OCCURRENCE: U2 and 4. Abundant in lake-shore forests, where it is often dominant. Absent from Mabira Forest. Common in South Kibale Forest. Rare in Bunyoro. This is the common, large, spreading tree in Entebbe Botanical Gardens.
CULTIVATION AND PROPAGATION: Slow-growing. Prefers light shade. Collect seeds from ground near mother trees as soon as possible after falling. Remove insect-damaged seeds. Sow as soon as possible. Grow in the nursery for at least 6-12 months before transplanting (as the seedlings grow slowly at first).
NOTES: The timber has been widely used in Uganda, including for the manufacture of railway wagons. The wood has an unpleasant smell when wet.

Newtonia buchananii **(Baker) G.C.C. Gilb. & Boutique (435)** Fabaceae (Mimosoideae)
Mpewere (ga); Omukungu, Omutooyo (ki); Mutole, Mutoyo (na); Muchenche, Muchensi (to).
50 m. Tree very similar to *Piptadeniastrum*, but differing in the following respects. Bark with easily distinguishable lenticels in places. Slash brown, pink or red-brown, usually exuding yellow-brown drops of resin, occasionally producing a red exudate. Rachis with narrow glands between the pinnae. Pinnae opposite. Leaflets c. 0.3 x 0.08 cm.
OCCURRENCE: U2 and 4. Common and sometimes abundant in many forests, e.g. Bwindi, Kalinzu, Kayonza and Kibale. Found up to 2200 m. It is largely replaced in lake-shore forests by *Piptadeniastrum*, but is abundant on the Ssese Islands.
CULTIVATION AND PROPAGATION: Fairly slow-growing at first, speeding up after 2 years. Prefers higher-rainfall areas or damp soils. Can be used to provide light shade in plantations, e.g. of coffee. Preferably collect the fruits from mother trees when they are ripening (brown in colour) and then dry in the sun to release the seeds. Remove insect-damaged seeds. Sow the seeds as soon as possible.

Cathormion altissimum **(Hook. f.) Hutch. & Dandy (436)** Fabaceae (Mimosoideae)
Omuchoole (no).
Flat-topped, deciduous tree to 35 m, but usually less. Spines often present on young plants and sucker shoots. Bark brown, rough. Phellogen pink. Slash thin and fibrous, pale yellow. Leaves bipinnate, with 5-7 pinnae on each side of the rachis. Leaflets 11-25 on each side of the pinna, opposite, c. 1.2 x 0.5 cm. The leaflets are narrower than those of *Tetrapleura* and broader than those of *Newtonia* and *Piptadeniastrum*.
OCCURRENCE: U1 and 2. Bunyoro and W. Nile. On river banks and in swamp forest.

Acacia kirkii **Oliv. (437)** Fabaceae (Mimosoideae)
Mutiti, Muzibi (am); Mukinga (to); Lugando (ko).
16 m. Flat-topped tree. Bark smooth, thin and green. Slash red. Spines present, paired. Leaves bipinnate, with 5-12 pinnae on each side of the rachis. Leaflets 15-25 leaflets on each side of the pinna. Leaflets very small.
OCCURRENCE: U1, 2 and 4. In swamp forest.

Tetrapleura tetraptera (Schumach. & Thonn.) Taub. (439) Fabaceae (Mimosoideae)

Kikangabalimi (am); Munyegenye (ga); Namahumbi (sa).

30 m. Deciduous understorey tree with a straight trunk and either a rounded crown or layered, spreading, branches. Buttresses present on larger trees. Bark of medium thickness, brown, fairly smooth, becoming slightly vertically fissured. Slash more or less two-layered, the outer granular, orange to white (often becoming reddish near the bark), the inner fibrous and white to yellow. Leaves bipinnate with 5-9 pinnae on each side of the rachis. Leaflets 6-12 on each side of the pinna. Leaflets alternate, c. 1.2 x 0.6 cm. Fruit a curved 4-winged pod, smelling of caramel (like *Albizia* fruits). Fruits often found beneath the tree.

OCCURRENCE: U1, 2 and 4. Widespread in lower altitude forests, nowhere abundant.

NOTE: The leaflets differ from those of *Parkia* and *Albizia* in being alternate (not opposite) and in being generally more rounded.

Parkia filicoidea Oliv. (440) Fabaceae (Mimosoideae)

African locust bean (en); Joge (ga); Omujojo (no); Muyenjayenja, Muyenjeyenje (sa).

30 m. Flat-topped tree. Buttresses present, small and rounded. Bark dark-coloured, with vertical lines of lenticels, smooth, becoming fissured and scaling with age. Slash granular, red, exuding an amber-coloured resin, smelling. Leaves bipinnate, with 4-14 pinnae on each side of the rachis. Leaflets 11-30 on each side of the pinna. Leaflets c. 2.5 x 0.8 cm, with two main veins from base (as seen on the upper surface), one vein more or less central and the other near the lower margin. Flowers in a pendulous, club-shaped cluster, red. Pods to 45 cm long, several together dangling from the inflorescence stalk.

OCCURRENCE: U1, 2 and 4. Usually near water.

CULTIVATION AND PROPAGATION: Slow-growing. Tolerant of a wide range of soils. Collect mature pods from mother trees or from the ground just after falling. The germination rate is improved by removing the seed coat. Soak the seeds for 24 hours and plant as soon as possible.

NOTES: Distinguished from *Albizia* by the leaf venation. The tree can be smelt yards away if flowering or dropping leaves. The many uses to which this plant is put in West Africa are described in ITU. Flowers much visited by fruit bats.

Albizia ferruginea (Guill. & Perr.) Benth. (441) Fabaceae (Mimosoideae)

Enyakatoma, Omuchooli (no).

45 m. Large deciduous tree with a straight, cylindrical, trunk and very large crown (said to be second in size only to that of *Klainedoxa* in Budongo Forest). Buttresses sometimes present. Bark thick, brown, flaking in long strips which can be peeled off by hand. Slash very fibrous, yellow. Shoots covered with red-brown hairs. Leaves bipinnate, with 3-9 pinnae on each side of the rachis. Leaflets 10-14 on each side of the pinna. Leaflets c. 1.2 x 0.5 cm, all more or less the same size on a leaf, hairy on the lower surface. Pod c. 15-20 cm long.

OCCURRENCE: U1, 2 and 4. Widely distributed tree of lower altitudes, found in dense forest.

CONSERVATION STATUS: Global VU (IUCN), LC (TOU); National EN (WCS).

NOTE: Distinguished from other species of *Albizia* by the yellow, fibrous, slash.

Plate 36. Fabaceae (pinnae) (436-447); see also Plate 35

436. *Cathormion altissimum* 437. *Acacia kirkii* 439. *Tetrapleura tetraptera* 440. *Parkia filicoidea* 441. *Albizia ferruginea* 442. *Albizia coriaria* 443. *Albizia glaberrima* 444. *Albizia gummifera* 446. *Albizia grandibracteata* 447. *Albizia zygia*

Actual sizes: x 2.

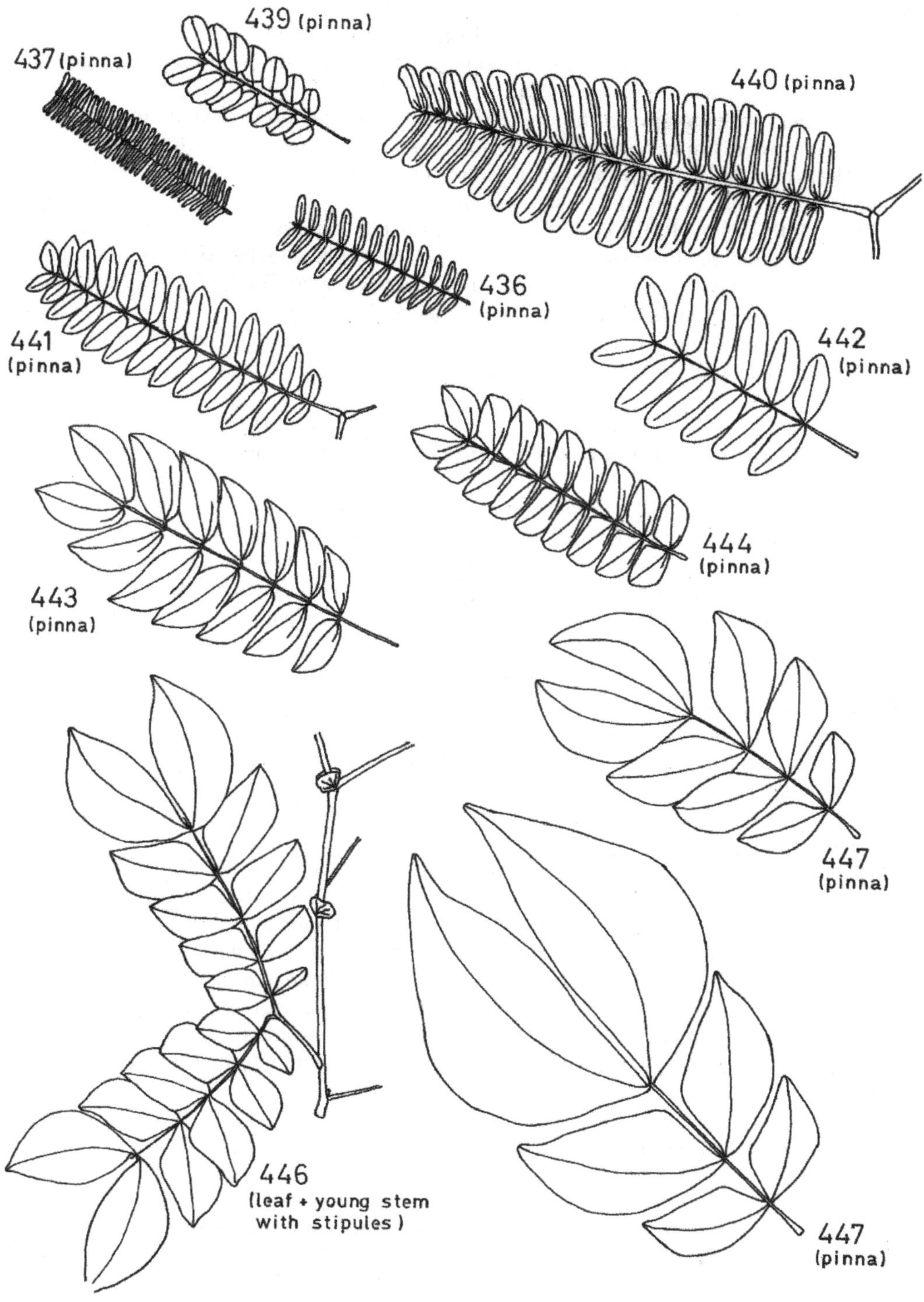

Plate 36. (436-447)

Albizia coriaria Oliv. (442) Fabaceae (Mimosoideae)
Ayekayek, Latoligo (ac); Ober, Omogi (al); Musisiya (am); Etekwa (at); Mugavu (ga, tn, to); Bata, Itek (la); Chesovio, Kumoluho, Kumoluno (ms); Musisa (na, no, to); Omurongo (no); Mubere, Muberi (sa); Musita (so).
35 m. Deciduous tree with layered foliage. Bark rough, dark-coloured, flaking raggedly. Slash fibrous, red. Leaflets similar to those of *Albizia ferruginea*, but less hairy below.
OCCURRENCE: U1-4. Widely distributed. On forest edges and in wooded grassland and farmland (where it can be common). Not in dense forest.
CULTIVATION AND PROPAGATION: Slow-growing. Can grow under a wide range of conditions, including on dry sites and poor soils. Suitable for land reclamation. Germination of fresh-sown seeds takes 1-2 weeks. Retain seedlings in the nursery for up to a year and protect from cattle-browsing. Can be propagated by root sucker induction.
NOTES: Distinguished from other species of *Albizia* by the slash. An important medicinal plant. The wood is burnt to smoke barkcloth. The bark is used for fish poisoning in West Nile and Madi.

Albizia glaberrima (Schumach. & Thonn.) Benth. (443) Fabaceae (Mimosoideae)
White nongo (tn).
30 m. Shape and slash similar to *Albizia gummifera*. Bark similar to *A. gummifera*, but rather darker in colour than other smooth-barked species of *Albizia*. Leaves with 1-4 pinnae on each side of the rachis. Leaflets 3-8 on each side of the pinna. Leaflets with a distinct petiolule, c. 0.1-0.2 cm long. Terminal leaflet c. 3.5 x 1.5 cm (much larger on young plants), glabrous.
OCCURRENCE: U1-4. Widespread in lower altitude forests in western Uganda. Possibly the most abundant species of *Albizia* in Central Kibale Forest. Common in Budongo.

Albizia gummifera (J.F. Gmel.) C.A. Sm. (444) Fabaceae (Mimosoideae)
Omushebeya (ki); Musebere, Mushebera (ko); Seswa, Swessu (ku); Chiruku, Kirongo, Kisubi (ms); Mulera, Mushebeya (na); Red nongo (tn); Mulongo (to).
30 m. Flat-topped, deciduous tree with a cylindrical trunk and layered foliage. Small buttresses occasionally present. Bark thin and smooth, light brown, with ring marks and horizontal lines of lenticels. Slash granular, orange and white, with vertical columns of yellow fibres. Leaves bipinnate, with 3-8 pinnae on each side of the rachis. Leaflets 8-18 on each side of the pinna. Leaflets c. 1.6 x 0.8 cm, all of about the same size on a leaf. Pod c. 10-20 cm long.
OCCURRENCE: U1-4. Widely distributed. Particularly common at higher altitudes, reaching 2400 m. Tends to be in secondary forest and on forest edges at lower altitudes.
NOTES: Distinguished from *Albizia zygia* and *A. grandibracteata* by the more numerous leaflets, all of which are of about the same size on a leaf. However, hybridization is thought to occur between these species and it may be impossible to place some specimens. Distinguished from *A. glaberrima* by the greater number of leaflets. For separation from *Albizia adianthifolia*, see below.

Albizia adianthifolia (Schumach.) W. Wight (445) Fabaceae (Mimosoideae)
Mshebeya, Mulera, Murera (na).

Plate 37. Fabaceae (flower, fruits and seed) (421-446)

421. *Connarus longistipitatus* 423. *Mildbraediodendron excelsum*
425. *Dialium excelsum* 427. *Baikiaea insignis* 428. *Cynometra alexandri*
429. *Craibia brownii* 430. *Millettia dura* 433. *Erythrophleum suaveolens*
434. *Piptadeniastrum africanum* 439. *Tetrapleura tetraptera* 442. *Albizia coriaria*
444. *Albizia gummifera* 446. *Albizia grandibracteata*
Actual sizes: x 2.

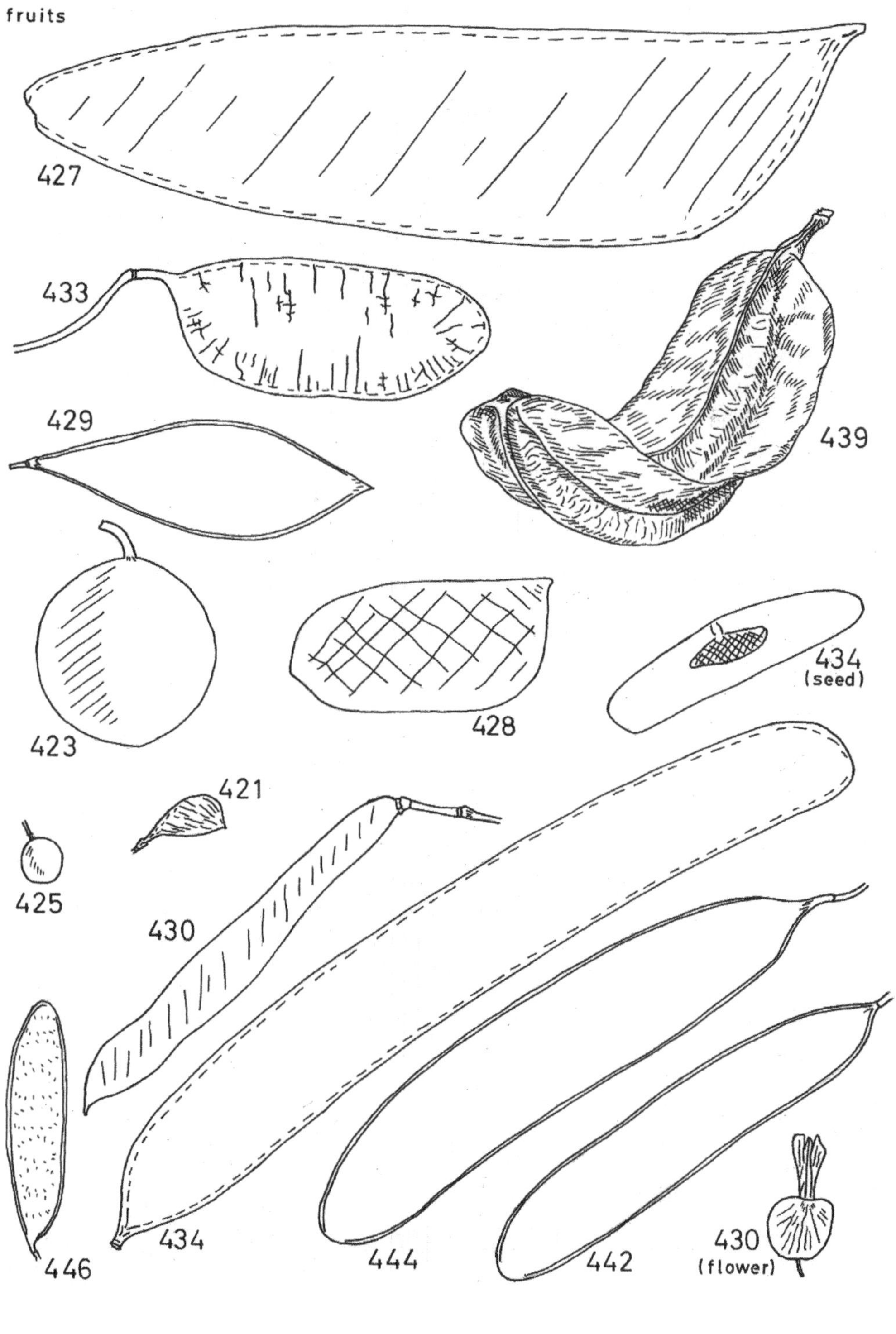

Plate 37. (421-446)

Very similar to *Albizia gummifera* in nearly all respects. It differs in the young shoots, leaves and rachises of the pinnae being densely covered with red-brown hairs, persisting on the rachises.
OCCURRENCE: U2 and 4. Recorded from Ankole, Kigezi and Mengo on forest edges and in woodland. Mainly above 1200 m. Only abundant south of the Equator.

Albizia grandibracteata Taub. (446) Fabaceae (Mimosoideae)

Bulera (am); Nongo (ga, so, to); Omushebeya (ki); Awak, Owak (lo); Enongo, Omulongo, Omurongo (no); Mulongosulwe (sa); Mulongo (so); Red nongo (tn).
30 m. Deciduous tree with a cylindrical trunk and layered foliage. Bark thin and smooth (rough on some forest-edge trees), brown, with vertical columns of lenticels. Slash as for *Albizia gummifera*, but sometimes red near the bark. Leaves bipinnate, with 1-4 pinnae on each side of the rachis. Leaflets 2-6 on each side of the pinna. Leaflets markedly different in size on the same leaf, the terminal being the largest (to 3-7 x 1.5-3 cm on older trees, but much larger on young plants). Stipules rounded. Flowers pink. Pod c. 7-10 cm long.
OCCURRENCE: U1-4. Widespread and common in secondary forest and on forest edges.
CULTIVATION AND PROPAGATION: Moderately fast-growing in well-watered places. Can grow on dry sites. Can be used as a pioneer species on depleted land or as a shade tree over slower-growing species. Collect pods on tree before they split and manually remove the seeds, discarding those that are insect-damaged. Sow as soon as possible. Germination occurs within a few weeks. Can be propagated by root sucker induction.
NOTE: Distinguished from *Albizia zygia* by the rounded stipules and from *A. glaberrima* by the more or less sessile leaflets.

Albizia zygia (DC.) J.F. Macbr. (447) Fabaceae (Mimosoideae)

Bedo (al); Blera (am); Ebata (at); Nongo (ga, sa, so); Ajua (gb); Swessu (ku); Abata-achol (la); Owak (lo); Adzimeli (md); Chiruku, Komosovio (ms); Musebeya (na); Enongo (no); Mulongo (so); Red nongo (tn); Murongo (to).
30 m. Tree with a cylindrical trunk and layered foliage. Buttresses absent. Bark smooth on forest specimens (rough when growing in savanna), with ring marks and vertical columns of lenticels. Slash as for *Albizia gummifera*. Leaves as for *A. grandibracteata*. Stipules thin, not rounded.
OCCURRENCE: U1-4. Forest and wooded grassland. A light-requiring species, found mainly in secondary forest.
CULTIVATION AND PROPAGATION: Moderately fast-growing when older, but growth of seedlings slow. Can grow in many situations, including as a pioneer species on open land. Discard insect-damaged seeds. If using stored seeds, soak in cold water before sowing. Can be propagated by root cuttings and air-layering.
NOTE: Distinguished from *Albizia grandibracteata* by the linear stipules and from *A. glaberrima* by the more or less sessile leaflets.

Part 5

Glossary

aculeate with prickles
acumen abruptly tapering point at leaf apex
acuminate (of a leaf apex) tapering to a slender point. Fig. 5.1
acute an angle of less than 90°; (of a leaf apex) leaf margins meeting at an acute angle. Fig. 5.1
adventitious root root in an unusual position, e.g. arising from a stem
aerial root root arising from above the level of the soil or water. See stilt root
air-layering method of propagating a plant involving wrapping a stem with damp moss to encourage the formation of roots
alternate leaves (as used here) leaves borne singly at each node. Their orientation with respect to one another may vary, e.g. all in the same plane or arranged in a spiral
androgynophore stalk above insertion of petals, carrying the stamens and ovary
arcuate venation main lateral veins of leaf curving around towards the leaf apex. Fig. 5.1
aril layer partially or wholly surrounding a seed
aristate with a long bristle-like point
armed bearing sharp-pointed structures, such as thorns, spines or prickles
asymmetric(al) not divisible by one or more planes into two or more equal parts
attenuate (of a leaf base) tapering gradually over a long distance towards the petiole. Fig. 5.1
auricle rounded ear-like lobe at base of organ
auriculate with auricles
axil upper angle between shoot and leaf, or midrib and vein
axillary situated in an axil
basal lateral veins veins arising at the base of a lamina (at its junction with the petiole). Fig. 5.1
berry fleshy indehiscent simple fruit lacking a stony layer, usually with more than one seed
bi- (a prefix) two or twice
bifoliolate leaf compound leaf having two leaflets. Fig. 5.1
bipinnate leaf twice-pinnate leaf. Fig. 5.1
blade (of leaf or leaflet) = lamina
bole lower branch-free part of trunk
boss knob-like outgrowth, usually on stem or root
branchlet small, usually terminal, branch
bud developing shoot or flower
buttress thin, more or less triangular outgrowth at base of trunk, running from the trunk to a lateral root situated near the soil surface. Some species have root spurs, which differ from buttresses in being broader, more rounded and usually smaller
caducous falling off very early
calyx sepals of a flower, considered as a whole
canopy the uppermost, more or less continuous, stratum of trees in a forest
capsule dry, many-seeded, dehiscent fruit formed from more than one carpel
carpel the basic female reproductive unit of a flowering plant, composed of ovary, style and stigma

catena sequence of different soil profiles found down a slope
climatic climax (of a vegetation type) plant community that is stable under the prevailing climate, especially in the absence of human influence. See primary forest
colleter glandular hair-like structure associated with a petiole or stipule in a leaf axil
compound leaf leaf having more than one blade. See leaf type
compressed flattened
coppicing method of managing a tree involving cutting it regularly at its base to yield poles
cordate (of a leaf base) lamina having two rounded basal lobes, one on each side of the petiole. Fig. 5.1
cordulate subcordulate (somewhat cordate)
coriaceous (of a leaf or leaflet) rather thick and stiff, leathery
corolla petals of a flower (free or fused), considered as a whole
crenate (margin of leaf or leaflet) with rounded teeth. Fig. 5.1
crenulate (margin of leaf or leaflet) with very small rounded teeth
crown (of a tree) top part of a tree carrying all or the great majority of the branches and leaves
cuneate (of a leaf base) wedge-shaped to triangular. Fig. 5.1
cuspidate abruptly tipped, with a short sharp point
cutting a length of stem (stem cutting) or root (root cutting) cut from a plant and placed in soil or water and from which roots and shoots may develop (a method of propagation)
cyme inflorescence in which the central flower opens first, the axillary buds arising from below the central flower
deciduous tree tree that seasonally loses all its leaves
decurrent (of a leaf base) extending downwards from the point of insertion
dehiscent fruit a fruit that opens spontaneously when ripe to display or release the seeds
dentate (of margins) toothed, the teeth directed outward rather than forward
denticulate (of margins) with very small teeth directed outward rather than forward
dichotomous key (for plant identification) a key used to repeatedly choose between two alternatives, each leading to the next pair of alternatives or to the name of the taxon being determined
digitate leaf compound leaf with leaflets spreading out like the fingers of a hand. Fig. 5.1
domatium (plural domatia) cavity, often in the axils of veins on the lower surface of leaves, housing commensal arthropods (e.g. ants)
drupe fleshy indehiscent fruit with (usually a single) seed surrounded by a stony coat
drupelet small individual drupe in a multiple fruit
ellipsoid elliptic in long section and circular in cross-section
elliptic (of a lamina) widest in the middle, narrowing towards both ends. Fig. 5.1
emarginate (of a leaf apex) notched. Fig. 5.1
emergent tree a tree that protrudes above the forest canopy
entire margin (of a lamina) margin unbroken by teeth, crenations or other irregularities. Fig. 5.1
epicormic shoot new shoot growing directly on a trunk or large branch
epidermis outermost layer of cells
epiphyte plant growing on another, but not drawing food or water from it. See strangler
evergreen retaining leaves throughout the year
exfoliation (of bark) the process of the bark becoming detached naturally from a tree
exstipulate lacking stipules
exudate a substance that pours or oozes out of a plant, especially when damaged. See latex, resin, sap
fascicle cluster of branches, leaves or flowers from the same point
feathery (appearance of a tree crown) having very small leaves or leaflets (almost dot-like as seen from the ground); feather-like

ferruginous rust-coloured
fibrous bark bark appearing to possess vertical strands
filiform slender or threadlike; filamentous
fissured bark having long narrow cracks or clefts
flange (on bole of a tree) projecting part of a fluted bole. Flanges can expand downwards into buttresses in some species. See buttress
fluted bole (of a tree) having vertical rounded grooves. A fluted bole is more or less stellate in cross-section (rather than circular)
-foliolate (a suffix) (referring to a leaf) having leaflets. A 5-foliolate leaf is one having five leaflets
follicle dry, single-chambered fruit formed from one carpel, containing two or more seeds
fruit a ripened, fertilized ovary containing seeds
fulvous tawny, dull yellowish-brown
glabrescent becoming (nearly) glabrous
glabrous without hairs
gland (of a plant) a secreting structure on or below the surface of the plant
glaucous dull green, with a bluish or greyish tinge
globose spherical
gregarious (of a type of plant) having a tendency for its individuals to grow close together
habit (1) general appearance of a plant; (2) mode of existence or growth of a plant. See shape
heart-shaped (of a leaf) ovate with a cordate base
imparipinnate leaf pinnate leaf with an unpaired terminal leaflet. Fig. 5.1
indehiscent fruit fruit that is not dehiscent
indumentum epidermal covering
induplicate (of a palm leaf) leaflets V-shaped in cross-section (as seen from above), with the margins higher than the midrib
interpetiolar stipule stipule placed between opposite (or whorled) leaves; typical of the family Rubiaceae
intrapetiolar stipule stipule located in the angle between petiole and stem
laciniate divided into slender lobes or segments
lamina expanded part of leaf or leaflet (= blade)
lanceolate (of a lamina) very narrowly ovate, about 4-6 times as long as wide, broadest in lower half. Fig. 5.1
lateral veins veins arising on either side of a midrib. The longer, more distinct ones are called main lateral veins. Fig. 5.1
latex fluid exuded by a plant, especially when damaged; usually white, but can be cream, yellow, orange or red. Less runny than sap
layer (of plants in a forest) = stratum
leaf thin green organ borne at a node on a stem, divisible in its simplest form into a lamina (the upper expanded portion) and a petiole. The blade may be much modified and the petiole may be lacking
leaf arrangement the arrangement of leaves on a stem, e.g. whether alternate, opposite or whorled
leaf texture the feel of the surface of a leaf (e.g. rough)
leaf type (as used here) type of leaf in the sense of whether simple or compound and, if compound, type of compound. See 5.1
leaflet single division of a compound leaf. See pinnule
lenticel pore in bark, usually raised, usually rounded to elliptic. Lenticels on larger stems can become elongated and more or less linear with time (due to expansion of the stem). See ring mark
lenticellate with lenticels

linear many times longer than wide, margins nearly parallel
lobate lobed
lobe (of a leaf) segment of a leaf separated by more or less rounded indentations
loculus compartment within an ovary containing the ovules or seeds
membranaceous thin and translucent
midrib central or main vein of a leaf or leaflet. Fig. 5.1
mucro short sharp apical point
mucronate leaf leaf narrowing abruptly at apex into a small, short, sharp point
mucronulate ending in a very small sharp point
nerve (of a leaf or stipule) = vein
nick (a seed coat) act of making a shallow cut on a seed (can encourage germination)
node position on a stem where one or more leaves or shoots are borne
non-entire margin (of a lamina) term used for a leaf margin that is broken by teeth, crenations or other such irregularities
obconic(al) (of a lamina) ovate, but widest in the upper (not lower) half (away from the point of attachment)
oblong (of a lamina) longer than broad, with nearly parallel sides, rounded at both ends. Fig. 5.1
obovate (of a lamina) egg-shaped, but widest in the upper (not lower) half (away from the point of attachment
obovoid (of a lamina) inversely ovoid, with the point of attachment at the narrower end
obscure (e.g. of venation) unclear or indistinct
obtuse an angle of more than 90°; (of a leaf apex) leaf margins meeting at an obtuse angle. Fig. 5.1
opposite leaves leaves borne two at each node on opposite sides of a stem
ovary female basal part of the flower immediately enclosing the ovules
ovate (of a lamina) egg-shaped, roughly twice as long as broad, widest in lower half. Fig. 5.1
ovoid (of a lamina) oval-shaped, with the point of attachment at the broader end
ovule grain-like body which, if fertilized, becomes a seed
palmate lobed or compound leaf in which all lobes or leaflets originate from one central point
papyraceous paper-like in texture
paripinnate leaf pinnate leaf, without an unpaired terminal leaflet. Fig. 5.1
pedicel stalk of a flower
peduncle stalk of a group of two or more flowers
peltate leaf leaf with petiole attached to the undersurface of the lamina (rather than at its base or margin)
perianth the outer non-sexual organs of a flower; sepals and petals together
perianth segment one member of the perianth. A term usually used when the calyx and corolla are not or little differentiated
petal one of the segments of a divided corolla, usually white or brightly coloured
petiole stalk of a leaf. Fig. 5.1
petiolule stalk of a leaflet
phellogen layer of tissue under the bark
pilose bearing short, thin, soft hairs
pinna (plural pinnae) primary division of a compound leaf (which may be further divided)
pinnate leaf compound leaf with leaflets arranged along each side of a common rachis. See imparipinnate, paripinnate
pinnule ultimate division of a bipinnate leaf
plumed seed seed adorned with long thin hairs (these aid in its dispersal by wind)

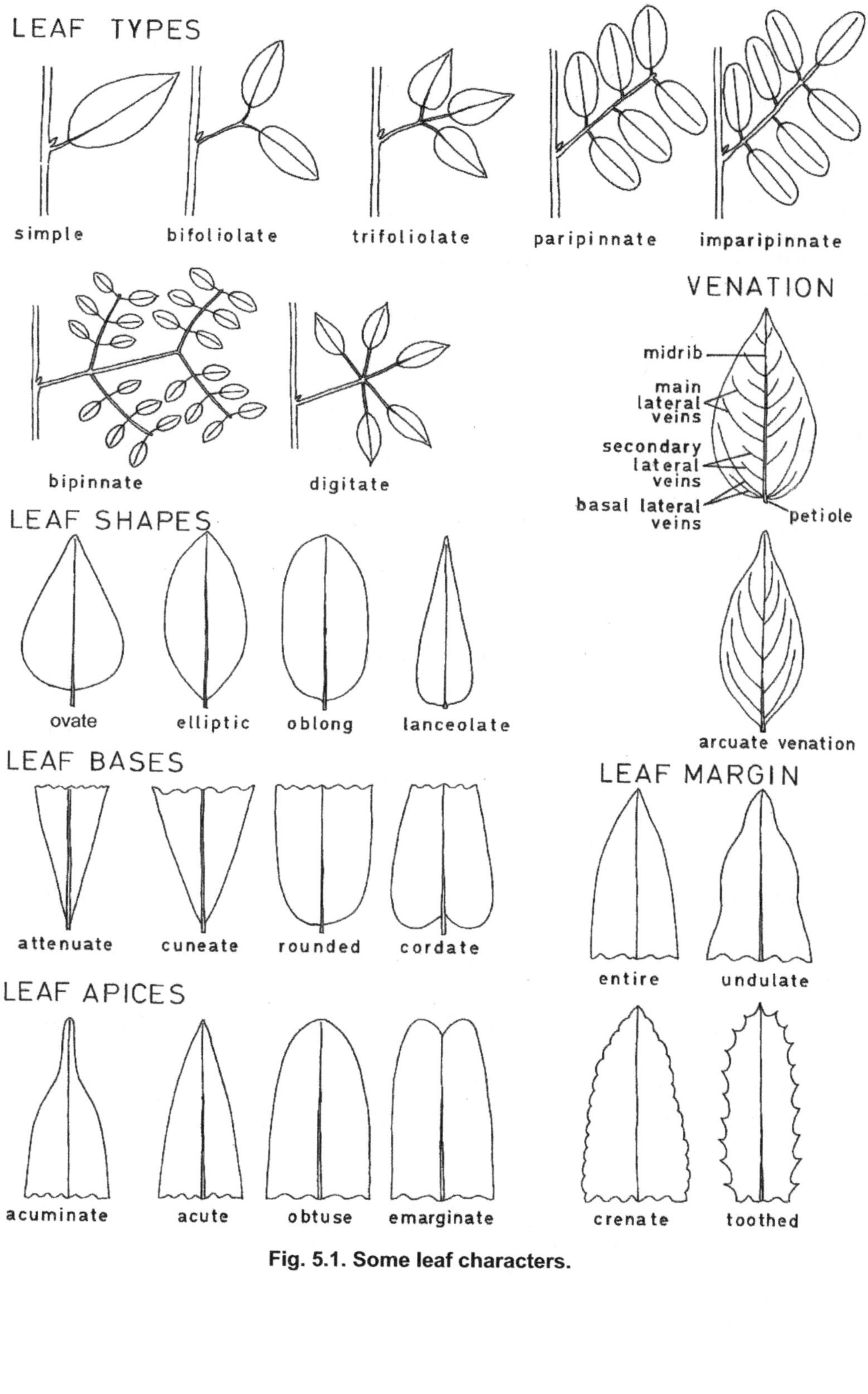

Fig. 5.1. Some leaf characters.

plumose soft-feathered
pneumatophore breathing root produced by some trees that grow in swamps, emerging into the air from below, inverted U-shaped
pod a simple dry dehiscent fruit developed from a single carpel that usually opens along a suture; typical of the family Fabaceae
prickle spiny outgrowth of the epidermis or bark, usually with a broadened base
primary forest (as used here) a forest not or little disturbed. See climatic climax
prop root stilt root
puberulous with dense, soft, very short hairs
pubescent with dense, soft, fine, short hairs
pulvinus (plural pulvini) swelling on petiole or petiolule, usually at the base, sometimes at both ends. Pulvini are the sites of movement of leaves and leaflets
pyriform pear-shaped
raceme inflorescence composed of stalked flowers borne on a common axis (rachis), the lower flowers maturing first
rachis main axis of a compound structure (e.g. compound leaf or inflorescence)
reduplicate (of a palm leaf) ʌ-shaped in cross-section (as seen from above), with the margins lower than the midrib
reniform kidney-shaped
resin semi-solid to solid sticky substance exuded by a plant, especially when damaged, often fragrant, insoluble in water
reticulate having the form of a network
retuse shallow-notched in a round apex
rhombic diamond-shaped; the shape of an equilateral parallelogram that is not a square
ring mark horizontally elongated marking found on smooth or fairly smooth bark. See lenticel
riparian associated with rivers, streams and springs, e.g. of trees found along river and stream banks
rounded (of a leaf base) sides of lamina meeting at base to form a common arc. Fig. 5.1
rugose wrinkled
rugulose somewhat wrinkled
rusty rust-like
sap fluid exuded by a plant, especially when damaged, usually more or less colourless (occasionally coloured, e.g. brown). More runny than latex
scale (on the surface of a plant) tiny, flat, thin and membranous, plate-like body, often attached at its centre
scarification (of a seed) process of scratching or abrading a seed (intended to increase absorption of water and hasten germination)
scrape (1) the process of rubbing over a surface (e.g. the outer side of the bark of a tree) with a sharp or rough instrument; or (2) the layer immediately under thin bark revealed by such rubbing
secondary forest type of forest that has developed as a result of a major past or continuing force, such as by a violent storm or large-scale human disturbance
secondary lateral veins (of leaf or leaflet) the less prominent lateral veins arising on either side of a midrib. Fig. 5.1
seed fertilized, ripened ovule
semi- (a prefix) half or (more loosely) partly
semi-deciduous forest forest containing both deciduous and evergreen trees
sepal one of the segments of the calyx (outermost whorl of floral organs), often green
serrate (of margin of lamina), saw-like, toothed, with teeth more or less apically directed
serrulate toothed along margin with minute sharp teeth that are apically directed
sessile not stalked

setose with bristles

shape (of a tree) the general appearance of a tree, as determined by the form of its trunk, branches, crown, etc. Shape is determined by a combination of age, inherent characteristics and environmental influences. See habit

shrub woody plant, typically smaller than a tree, having several stems arising from ground level

simple leaf an undivided leaf (not divided into leaflets); having a single lamina. Fig. 5.1

simulate the condition or act of resembling something else, e.g. certain arrangements of simple leaves can simulate pinnate leaves

sinuate (of a margin) strongly wavy

slash (1) layer(s) under the bark revealed in a tree trunk by making a shallow cut; (2) the cut made in a trunk to reveal such layer(s); or (3) the process of making this cut

slash exudate exudate produced from a slash. See latex, resin, sap

slash texture texture of the slash; e.g. brittle, fibrous, granular. A fibrous slash can be pulled into long thin pieces. A granular slash breaks up into small, hard, rounded particles when rubbed between the fingers

spatulate spatula-shaped

spine straight, sharply-pointed, woody outgrowth, deep-seated, arising from below the epidermis, representing a modified leaf or stipule

spinulose bearing small spines (spinules)

spinulose-serrate toothed with small spines

spiny margin (of lamina) having stiff teeth, sufficiently sharp to scratch or tear the skin

spiral (of leaf arrangement) situated regularly and spirally around a stem, successive leaves not being at angles of 180° to one other. See alternate leaves

spur (of a root) See buttress

stamen male reproductive organ of a flower, usually consisting of anther and filament

stellate star-shaped, with radiating arms

stilt root stout woody adventitious root arising from a trunk above the level of soil or water, arching down and entering the soil at some distance from the trunk (= prop root)

stipel stipule-like outgrowth occurring (usually in pairs) at the base of a leaflet or a pair of leaflets in some compound leaves, or rarely in simple ones (e.g. in some *Alchornea*)

stipule a small leaf-like appendage to a leaf, typically borne in pairs at the base of a petiole

storey (in forest vegetation) = stratum

strangler a plant that starts life as an epiphyte, sends roots down to the ground, kills the host by 'strangulation' and takes over its place

stratification (1) the layering of plants in vegetation; (2) the process that leads to such layering. Species of plants have inherent tendencies to grow up to given heights when mature. A species described as a canopy species is one that is typically found in the canopy stratum when mature

stratum (plural strata) (in forest vegetation) a layer of plants in a forest. Strata may be more or less well-defined and their number is variable (generally decreasing in number at higher altitudes and under drier climates). Strata can be labelled in various ways, e.g. (from tallest to shortest) emergent, canopy (or upperstorey), second storey, understorey, shrub layer, herbaceous layer. See canopy, emergent tree, understorey, upperstorey

striate with parallel longitudinal grooves

strigose with sharp stiff appressed (pressed flat) hairs

style stalk connecting stigma and ovary

sub- (a prefix) indicating beneath or falling short (less than perfect)

subopposite nearly opposite

subsessile nearly sessile

surface root large root lying close to, or partially above, the surface of the soil

suture (of a carpel) line of union along which a carpel splits
symmetric(al) divisible by one or more planes into similar parts
synonym a scientific name that has been superseded by another (valid) name
taxon a named taxonomic group of any rank, e.g. subspecies, genus, order
tertiary venation (of leaves) third-order vein arrangement
thorn stiff woody, sharply-pointed, modified stem
tomentellous slightly tomentose
tomentose densely covered with short, matted or tangled, soft, woolly hairs
toothed (margin of leaf or leaflet) having short projections with sharp ends. Fig. 5.1. See serrate
translucent allowing the passage of some light, semi-transparent
tri- (a prefix) three or thrice
trifoliolate leaf compound leaf having three leaflets. Fig. 5.1
trigonous obtusely three-angled
trilobate three-lobed
truncate (of apex or base of lamina) square at end (as if cut)
umbel flat-topped or convex inflorescence in which all the pedicels arise from the same point
umbellate in umbels
unarmed lacking stiff, woody, sharply-pointed structures such as thorns, spines and prickles
understorey lower tree layer in forest vegetation. See stratum
undulating (of margin of lamina) waving up and down along its length (but not deeply). Fig. 5.1
unifoliolate (leaf) compound leaf with a single leaflet (the lamina being separated from the petiole by a joint); superficially resembling a simple leaf
upperstorey canopy layer in forest vegetation. See stratum
valve (of a fruit) segment that separates from other segments during dehiscence
vein (of lamina or stipule) vascular strand contained within a lamina or stipule. Fig. 5.1. See basal lateral veins, secondary lateral veins
venation (of lamina or stipule) the arrangement of the veins
vernacular language commonly spoken by a particular people or in a particular place
wavy undulating
whorl (of leaves) group of three or more leaves arising on a stem all at the same node
wing (of a plant structure) flat extension of a stem, fruit or other organ

Part 6

Further Information for the Field Worker

Contact Organizations

Responsible Government Agencies

National Environment Management Authority (NEMA). Plot 17/19/21 Jinja Road, Kampala. Postal address: P.O. Box 2255, Kampala. Enquiries: info@nemaug.org Responsibility: environmental management. Website: https://www.nema.go.ug

National Forestry Authority (NFA). Plot 10/20 Spring Road, P.O. Box 70863, Kampala. Enquiries: info@nfa.org.ug; tel. 0312264035/6. Responsibility: management of Central Forest Reserves; has small nurseries in its forest management zones (forest ranges). Website: https:www.nfa.org.ug

National Forestry Resources Research Institute (NAFORRI), part of the National Agricultural Research Organization – NARO). Kifu, 12 km on Mukono-Kayunga Road. Postal address: P.O. Box 1752, Kampala. Enquiries: naforridir@infocom.co.ug; tel. 0414383028, 0712161161. Responsibility: forestry research. Website: www.naforri.org.ug

National Tree Seed Centre (NTSC, part of NFA). 12 km Kampala-Jinja Road (before Coca Cola Plant), Namanve. Postal address: P.O. Box 23889, Kampala, Uganda. Enquiries: treeseed@nfa.org.uk; tel. 0414286049. Nurseries at Namanve (the main one), Banda (near Kyambogo), Nagojje (Nandagi), Kimada (Jinja) and Karugutu (along Fort Portal to Bundibugyo Road). Relevance: raises tree seedlings for sale. Concentrates mainly on eucalyptus and pines, but has seedlings of some indigenous forest species.

Uganda Wildlife Authority (UWA). Plot 7, Kira Road, Kamwokya. Postal address: P.O. Box 3530, Kampala. Enquiries: info@ugandawildlife.org; tel. 0414355000. Responsibility: wildlife management and protection. Website: https://ugandawildlife.org

Field Stations, Research Institutes, Botanical Gardens and Herbaria

Budongo Conservation Field Station. P.O. Box 362, Masindi, Uganda. Mission: sustainable management of Budongo Forest Reserve. Website: www.budongo.org

Institute of Tropical Forest Conservation (ITFC). Comes under Mbarara University of Science and Technology (MUST). Location: Bwindi Impenetrable National Park (BINP). Postal address: ITFC, P.O. Box 44, Kabale, Uganda. Enquiries: info@itfc.org. Mission: ecological and sociological research; has a herbarium (most collections from BINP and neighbourhood). Website: https://itfc.must.ac.ug

Makerere University Biological Field Station (MUBFS). Comes under the Department of Environmental Management, Makerere University. Location: Kibale National Park. Postal address: College of Agricultural and Environmental Sciences, P.O. Box 7062, Kampala.

Enquiries: pr@caes.mak.ac.ug; tel. 0414542277. Relevance: research and training. Has a herbarium with specimens mainly from the Albertine Rift Region.
Website: caes.mak.ac.ug/makerere-university-biological-field-station

Makerere University Herbarium (MHU). Department of Plant Sciences, Microbiology and Biotechnology, Makerere University, P.O. Box 7062, Kampala. Enquiries: pmb@cns.mak.ac.ug. Relevance: serves as the national herbarium; a small botanic garden is attached, serving mainly university teaching purposes.
Website: http://plantscience.mak.ac.ug

National Biodiversity Data Bank (NBDB). College of Agricultural and Environmental Sciences, P.O. Box 7062, Makerere University, Kampala. Enquiries: nbdb@caes.mak.ac.ug. Relevance: the national repository for biodiversity data; conducts inventories and monitoring; provides data for conservation and sustainable use of resources. The college is developing a botanic garden at its field station at Kabanyolo. Website: www.nbdb.mak.ug

Plant Genetic Resources Centre (PGRC). Part of NARO. Berkely Street, P.O. Box 40, Entebbe. Comprises Entebbe Botanic Gardens and the Uganda National Gene Bank. Tel. 0414320638, 041321070. Mission: conservation of plant genetic resources.
Website: https://www.pgrcuganda.co.ug

Tooro Botanical Gardens. Njara Road, P.O Box, Fort Portal. Relevance: offers seedlings of indigenous forest trees for sale.

Uganda Wildlife Education Centre (UWEC). Plot 56/7 Lugard Avenue, Entebbe. Enquiries: info@uwec.ug; tel. 0784147027, 0705277863. Mission: conservation education. Website: http//uwec.ug

International Conservation Groups

Bioversity International, P.O. Box 24384, Plot 106, Katalima Road, Naguru, Kampala. Enquiries: Bioversity-uganda@cgiar.org. Tel. 0393216106. Relevance: agrobiodiversity. website: https://www.bioversityinternational.org

Wildlife Conservation Society (WCS). Plot 802, Kiwafu Road, Kansanga, Kampala. Postal address: P.O. Box 7487, Kampala. Enquiries: sampindo@wcs.org. Tel. 039200381. Relevance: conservation science and action; landscape-level orientation. Website: https://uganda.wcs.org

World Agroforestry (ICRAF). At NAFORRI compus (see under Government Agencies). Postal address: P.O. Box 26416, Kampala. Enquiries: c.okia@cgiar.org; tel. 04144660647. Relevance: agroforestry. Website: www.worldagroforestry.org

World Wide Fund for Nature (WWF). WWF-Uganda Country Office, Plot 2 Sturrock Road, P.O. Box 8758, Kampala. Enquiries: kampala@wwfuganda.org; tel. 020051800. Relevance: conservation of protected areas; sometimes supports local tree nurseries. Website: wwfuganda.org

Civil Society Organizations

Mbale Coalition Against Poverty. Majanga Road, Mbale. Enquiries: info@mbalecap.org; tel. 0782274713. Relevance: promotes indigenous tree planting. Website: www.mbale-cap.org

NatureUganda. Plot 1, Katalima Crescent, Lower Naguru. Postal address: P.O. Box 27034, Kampala. Enquiries: nature@natureuganda.org; tel. 0414540719. Mission: promotion of the understanding, appreciation and conservation of nature. Website: natureuganda.org

PROMETRA Uganda. Buyijja, Buwama Sub-County, Mpigi District. Enquiries: info@prometra.ug.com; tel. 0772403900, 0750956490. Relevance: emphasizes importance of indigenous knowledge in development; encourages forest conservation and tree planting; has seedlings of indigenous tree species for sale. Website: www.prometraug.com

Tree Talk Plus. Plot 842, Lugolobi Close, Sempagala Zone, Buye, Ntinda, Kampala. Enquiries: info@treetalkplus/org; tel. 0392177128. Promotes ecologically sound land practices and natural resource management; planting of indigenous trees; has seedlings of fast-growing species of indigenous trees suitable for agroforestry available. Website: www.treetalkplus.org

Wildlife Clubs of Uganda. Contact: wildlifeclubsofuganda1@gmail.com; tel. 0787395648. Association of young people's environmental clubs.

On-line Information on Ugandan Forest Tree Species

Useful Trees of East Africa by Rudi Lemmens. This app contains information on 109 useful species of trees of Kenya, Tanzania and Uganda, including about their identification, ecology, habitats, and the products and services that they provide. Some Ugandan forest species are included. Search on your browser for 'Useful Trees of East Africa' or go to the following websites:

https://play.google.com › store › apps › details › id=com.fatslambco.which...
For Android: play.google.com/store/apps/developer?id=Staf+Lemmens
For Apple: https://apps.apple.com/us/app/useful-trees-of-east-africa/id981531635

Part 7

The Indigenous Languages of Uganda

Abbreviations Used (see also Table 7.2)

ITU	The Indigenous Trees of the Uganda Protectorate	Eggeling and Dale 1951
UFT	A Field Guide to Uganda Forest Trees	Hamilton 1981

Indigenous Languages Included

The 42 indigenous languages of Uganda (Ethnologue 2019) fall into two language families, Nilo-Saharan (with a traditional home area in the north) and Niger-Congo (in the south). The various languages included in each of these language families vary in their similarities to one another and can be grouped into linguistic sub-categories accordingly (Table 7.1). Common ancestries can be postulated on the basis of similarities in phonology, morphology and syntax (Barasa 2017). Dialects within the languages remain to be fully explored and also the nature of linguistic change across the boundaries between the languages (Watters 2018; Lesage 2019). It is reported for Kupsapiny that it is losing its vitality and needs revitalization (Kawachi 2010) and the same is true of some of the other languages.

Knowing the geography of the languages and their linguistic relationships is useful for the field botanist, since species commonly have similar names in closely related languages and their use can unlock helpful local information. The home areas of the languages with tree names included in the present work are shown on Fig. 7.1.

- Nilo-Saharan Language Family
    - Central Sudanic: **Lugbara**, **Madi**
    - Eastern Sudanic
        - Western Nilotic (all in Luo sub-group, for which a single index is provided)
            - **Acholi**, **Lango**
            - **Alur**
        - Eastern Nilotic: **Karimojong**, **Teso**
        - Southern Nilotic: **Kupsapiny**
- Niger-Congo Language Family (all in Bantu sub-group)
    - Kongo: **Lukonzo**
    - Lega-Kalanga: **Kwamba**
    - Masaba-Luhya: **Lunyole**, **Lusaamia-Gwe**, **Lumasaba**
    - Nyoro-Ganda
        - **Luganda**, **Lusoga**
        - **Lugwere**
        - Runyakitara sub-group: **Rukiga**, **Runyankore**, **Runyoro**, **Rutooro**
    - Ruanda-Rundi: **Rufumbira**

Table 7.1. Indigenous language families of Uganda and some of their subdivisions. Only the names of families with plant names included in the present book are shown.

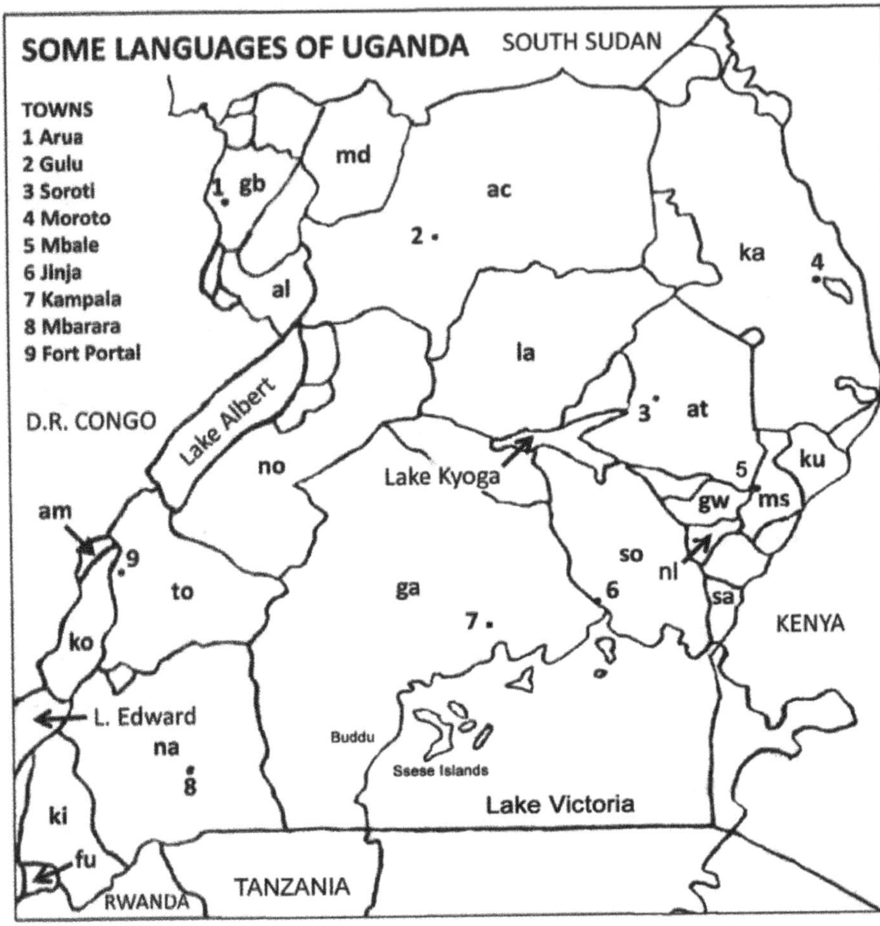

Fig. 7.1. Home areas of some of the indigenous languages of Uganda. See Table 7.2 for key to abbreviations.

Notes on the Languages

Alur. This includes a dialect known as **Jonam**.

Kupsabiny. Also known as **Sebei**.

Luganda. Some names in UFT and ITU are designated Luganda-Buddu or Luganda-Ssese. These names have been combined with Luganda (not geographically specified) in the present work.

Lumasaba. Also known as **Lugisu** (Kawachi 2010). However, Lwangale (2015) treats **Lugisu**, **Lumasaba** and **Lubukusu** as dialects of **Luluhyia**, all three being seen as having a common ancestry.

Luo. This sub-group of Nilo-Saharan languages includes (of the languages in the table) **Acholi**, **Alur** and **Lango**. UFT and ITU describe some of the names that they include as Luo without further specification, while others are assigned to particular Luo languages. All Luo names have been combined into a single index (Part 8) in the present work, though with more detailed affiliation given, where known.

Rufumbira. This is the name given to **Kinyarwanda** (the language of Rwanda), as spoken in the Kisoro District of Uganda.

Rukiga, Runyankore, **Runyoro** and **Rutooro**. These are similar languages that tend to grade into one another in boundary areas. A combined standardized version of them, known as **Runyakitara**, has been devised to facilitate certain activities within them, such as teaching.

Most of the vernacular names in the present work are from UFT, which, in turn, borrowed heavily from ITU. Katende *et al.* 1995 has been another fruitful source. The total numbers of tree names included per language are shown in Table 7.2. We are aware that some names in the sources consulted were initially inaccurately transcribed and that some are probably wrong. As an example of probable errors in transcription, we note that the four Lumasaba names given for *Alangium chinense* (Guronono, Kistono, Lusontono and Lusotono) all likely refer to the same root (-sontono?). We also note that this species appears to have been assigned to different noun classes on different instances of recording (as indicated by the variety of prefixes used, namely gu-, ki- and lu-). At least some Bantu languages have strong capacities for nouns to shift between noun classes (Watters 2018).

Abr	Language	No.	Speakers	Abr	Language	No.	Speakers
ac	Acholi	16	1,500,000	ku	Kupsabiny	69	274,000
al	Alur	17	985,000	la	Lango	20	2,130,000
am	Kwamba	86	42,600	lo	Luo	8	*
at	Ateso	24	2,360,000	md	Madi	14	293,000
en	English	61	-	ms	Lumasaba	120	1,650,000
fu	Rufumbira	2	713,000	na	Runyankore	106	3,420,000
ga	Luganda	144	5,560,000	nl	Lunyole	17	530,000
gb	Lugbara	16	1,100,000	no	Runyoro	74	967,000
gw	Lugwere	7	621,000	sa	Lusaamia-Gwe	44	525,000
ka	Karamojong	8	691,000	so	Lusoga	51	2,960,000
ki	Rukiga	115	2,390,000	tn	Trade names	81	-
ko	Lukonzo	45	893,000	to	Rutooro	105	846,000

Table 7.2. The number of tree names included in this field guide per language. Also shown are the Abbreviations (Abr) used for languages and trade names and the estimated numbers of first-tongue speakers in Uganda in 2014 (Ethnologue 2019). *See note on previous page.

We have failed to find up-to-date published compilations of plant names for most of the languages. Our approach to indexing has been to divide up the names into the individual languages (not done in the indexes of ITU and UFT), because we believe that this will often be helpful for the field worker and provides useful foundations for those wishing to make improvements. Meanwhile, we have sought botanists who are familiar with the names of the trees, both scientifically and in particular languages, and requested them to edit the lists. Those who have done so are acknowledged.

Nouns in Bantu languages are assigned to noun classes, each characterized by particular prefixes (which are generally different for singular and plural forms). These prefixes are often themselves preceded by pre-prefixes, commonly referred to as augments or initial vowels (Watters 2018). As examples, the most frequent noun classes used for the names of plants in Luganda are shown in Table 7.3, together with the names of example species or plant types, written with and without initial vowels.

The editors of the plant lists and other linguistic experts consulted for the present work were divided in opinion as to whether or not initial vowels should be included with the names of plants in Bantu languages, in particular when the names are presented in list form, as here. We decided to follow the styles followed by the experts for each of the languages in cases where these had been consulted, but otherwise have tended not to use initial vowels (since this is the form in which most of the names are presented in UFT).

Noun class prefixes (and initial vowels in brackets)		Examples of tree names (singular forms)		Species/plant type
Singular	**Plural**	**Without initial vowels**	**With initial vowels**	
(o)mu	(e)mi	Muvule	Omuvule	*Milicia excelsa*
(e)ri*	(a)ma*	Bbanda	Amabanda	Bamboo
(e)n	(e)n	Nzo	Enzo	*Vepris nobilis*
(o)lu	(e)n	Lusambya	Olusambya	*Markhamia lutea*
(a)ka	(o)bu	Kaliba	Akaliba	*Musanga cecropioides*

Table 7.3. Noun classes commonly used for the names of plants in Luganda. *This noun class is irregular. The singular forms of many nouns in this noun class, when written with initial vowels, have doubled initial consonants following the initial vowels, as shown by the example given.

Part 8

References

Abbreviations Used for Publications in other Parts of this Field Guide

FTEA	Flora of Tropical East Africa	
ITU	The Indigenous Trees of the Uganda Protectorate	Eggeling and Dale 1951
TOU	Conservation Checklist of the Trees of Uganda	Kalema and Beentje 2012
UFT	A Field Guide to Uganda Forest Trees	Hamilton 1981
WCS	Nationally Threatened Species for Uganda	WCS 2016

Adia, M.M., Anywar, G., Byamukama, R., Kamatenesi-Mugisha, M., Sekagya, Y., Kakudidi, E.K and Kiremire, B.T. (2014). Medicinal plants used in malaria treatment by Prometra herbalists in Uganda. *Journal of Ethnopharmacology* 155, 580-588. http://dx.doi.org/10.1016/j.jep.2014.05.060/

African Plant Database (version 3.4.0) (2019). *Conservatoire et Jardin botaniques de la Ville de Genève and South African National Biodiversity Institute, Pretoria.* http://www.ville-ge.ch/musinfo/bd/cjb/Africa/ [Accessed November 2019]

Akwatulira, F., Gwali, S., Ssegawa, P., Okullo, J.B.L., Tumwebaze, S.B., Mwambo, J.R. and Muchugi, A. (2011). Vegetative propagation of *Warburgia ugandensis* Sprague: an important medicinal tree species in eastern Africa. *Journal of Medicinal Plants Research* 5(30), 6615-6621. http://www.academicjournals.org/JMPR/

Anywar, G., van't Klooster, C.I.E.A., Byamukama, R., Willcox, M., Nalumansi, P.A., de Jong, J., Rwaburindore, P. and Kiremire, B.T. (2016). Medicinal plants used in the treatment and prevention of malaria in Cegere SubCounty, Northern Uganda. *Ethnobotany Research and Applications* 14, 505-516. Available from http://dx.doi.org/10.17348/era.14.0.505-516/

APG IV (2016). An update of the Angiosperm Phylogeny Group classification for the orders and families of flowering plants: APG IV. *Botanical Journal of the Linnean Society* 181, 1-20.

Banana, A.Y., Vogt, N.D., Bahati, J. and Gombya-Ssembajjwe, W. (2007). Decentralized governance and ecological health: why local institutions fail to moderate deforestation in Mpigi District of Uganda. *Scientific Research and Essay* 2, 434-445.

Barasa, D. (2017). *Ateso Grammar: A Descriptive Account of an Eastern Nilotic Language*. Thesis presented for the Degree of Doctor of Philosophy in Linguistics, University of Cape Town, South Africa.

Basamba, T.A., Mayanja, C., Kiiza, B., Nakileza, B., Matsiko, F., Nyende, P., Kukunda, E.B., Tumushabe, A. and Ssekabira, K. (2016). Enhancing adoption of agroforestry in the Eastern Agro-Ecological Zone of Uganda. *International Journal of Ecological Science and Environmental Engineering* 3(1), 20-31.

Beentje, H. 1994. *Kenya Trees, Shrubs and Lianas*. National Museums of Kenya, Nairobi, Kenya.

Beentje, H. 2010. *The Kew Plant Glossary: an Illustrated Dictionary of Plant Terms*. The Board of Trustees of the Royal Botanic Gardens, Kew, UK.

Beentje, H. and Cheek, M. (2003). Glossary. In: Beentje H.J. and Ghazanfar S.A. (eds) *Flora of Tropical East Africa*. A.A. Balkema Publishers, Rotterdam, the Netherlands.

BirdLife International (2008). *Campaign to save Mabira Forest in Uganda from sugarcane plantation for biofuels*. http://www.birdlife.org/ [Accessed November 2019]

Brack, D. (2019). *Forests and Climate Change*. 14th Session of the United Nations Forum on Forests.

Brownstein, G., Johns, C., Fletcher, A., Pritchard, D. and Erskine, P.D. (2015). Ecotones as indicators: boundary properties in wetland-woodland transition zones. *Community Ecology* 16(2), 235-243. doi: 10.1556/168.2015.16.2.11.

Cabrera, O., Benítez, A., Cumbicus, N., Naranjo, C., Ramón, P., Tinitana, F. and Escudero, A. (2019). Geomorphology and altitude effects on the diversity and structure of the vanishing montane forest of Southern Ecuador. *Diversity* 11. doi:10.3390/d11030032.

Christopher, M.D., Uchechukwu, E.E. and Ernest, A.A. (2009). Phytochemical analysis and antimicrobial activity of the bark extracts of *Voacanga africana* Stapf. *Nigerian Journal of Biotechnology* 20, 61–65.

CITES (2017). (Convention on International Trade in Endangered Species of Wild Fauna and Flora). *Appendices I, II and III*. CITES Secretariat, Geneva, Switzerland. https://www.cites.org/eng/app/appendices.php/ [Accessed 2019]

Clark, L.G., Londoño, X. and Ruiz-Sanchez, E. (2015). Bamboo taxonomy and habitat. In: Liese, W. and Köhl, M. (eds), *Bamboo*. Tropical Forestry 10. Springer International Publishing, Switzerland. doi 10.1007/978-3-319-14133-6_1.

Côte, M., Wartmann, F. and Purves R (2018). Introduction: the trouble with forest: definitions, values and boundaries. *Geographica Helvetica* 73, 253-260.

Darbyshire I, Anderson S, Asatryan, A., Byfield, A., Cheek, M., Clubbe, C., Ghrabi, Z., Harris, T., Heatbun, C.D., Kalema, J., Magassouba, S., McCarthy, B., Milliken, W., de Montmollin, B., Nic Lughadha, E., Onana, J.-M., Saïdou, D., Sarbu, A., Shrestha, K. and Radford, E. (2017). Important Plant Areas: revised selection criteria for a global approach to plant conservation. *Biodiversity Conservation*. doi 10.1007/s10531-017-1336-6.

Davenport, T. and Howard, P. (eds) (1996). *Semliki Forest Biodiversity Report*. Forest Department, Kampala, Uganda.

Davenport, T., Howard, P. and Matthews, R. (eds) (1996). *Bwindi Impenetrable National Park Biodiversity Report*. Forest Department, Kampala, Uganda.

Dawkins, H.C. (1951). Graphical field keys of Uganda trees. 1. Forest trees, Mengo District. *East African Agricultural Journal* 17, 1-14.

de Oliveira, T., Crafford, J., Naidoo, N., Mathebula, V., Mulders, J., Maila, D. and Harris, K. (2018). *Forestry and Macroeconomic Accounts of Uganda: the Importance of Linking Ecosystem Services to Macroeconomics*. Uganda Technical Report, Ministry of Water and Environment, Kampala, Uganda.

Deb, J.C., Phinn, S., Butt, N. and McAlpine, C.A. (2018). Climate change impacts on tropical forests: identifying risks for tropical Asia. *Journal of Tropical Forest Science* 30(2), 82-194.

Egeru, A., Okia, C. and de Leeuw, J. (2015). *Trees and Livelihoods in Karamoja, Uganda*. Evidence on Demand. http://dx.doi.org/10.12774/eod_hd.december2014.egeruaetal/

Eggeling, W.J. and Dale, I.R. (1951). *The Indigenous Trees of the Uganda Protectorate*. 2nd edition. Government Printer, Entebbe, Uganda.

Ellison, D. (2018). *Forests and Water*. Background Analytical Study 2, prepared for the thirteenth session of the United Nations Forum on Forests, Global Forest Goals.

Ethnologue, (2019). *Languages of Uganda*. In: Simons, G.F. and Fennig, C.D. (eds). https://ethnologue.com/ [Accessed 2019]

FAO (2010a). *Global Forest Resource Assessment 2010*. Food and Agricultural Organization of the United Nations, Rome, Italy.

FAO (2010b). *Global Forest Resources Assessment: Country Report for Uganda*. Food and Agricultural Organization of the United Nations, Rome, Italy.

FAO (2017). *Analysis of Forests and Climate Change in Eastern Africa*. Forests and Climate Change Working Paper 16, Food and Agriculture Organization of the United Nations, Rome, Italy.

FAO (2018a). *Global Forest Resource Assessment 2020*. Forest Resources Assessment Working Paper 188. Food and Agricultural Organization of the United Nations, Rome, Italy.

FAO (2018b). *The State of the World's Forests: Forest Pathways to Sustainable Development*. Food and Agricultural Organization of the United Nations, Rome, Italy.

FAO (2018c). *Climate Change for Forest Policy-makers - an approach for Integrating Climate Change into National Forest Policy in Support of Sustainable Forest Management* - Version 2.0. FAO Forestry Paper no.181. Rome.

FAO (2018d). *Terms and Definitions*. Global Forest Resources Assessment, Food and Agriculture Organization of the United Nations, Rome, Italy.

Flora of Ethiopia and Eritrea (1989-2009). National Herbarium, Addis Ababa University Ethiopia and Uppsala University, Sweden.

Flora of Tropical East Africa (1952-2012). Edited by C.E. Hubbard, E. Milne-Redhead, R.M. Polhill and W.B. Turrill. Crown Agents, London.

Flora of West Tropical Africa. Vols 1-2 (1954-1963). By J. Hutchinson, J. and J.M. Dalziel. 2nd Edition revised by R.W.J. Keay. Crown Agents. London, UK.

Flora Zambesiaca Vols. 1-4 (1960-) Edited by A.W. Exell and H. Wild. Crown Agents. London, UK.

Flore du Congo, du Rwanda and du Burundi (1958-). Jardin botanique national de Belgique, Brussels, Belgium.

Galabuzi, C., Nabanoga, G.N., Ssegawa, P., Obua, J. and Eilu, G. (2015). Double jeopardy: bark harvest for malaria treatment and poor regeneration threaten tree population in a tropical forest of Uganda. *African Journal of Ecology* 53, 214-222.

Gallery, R.E. (2014). Ecology of Tropical Rain Forests. *Ecology and the Environment.* doi 10.1007/978-1-4614-7612-2_4-1.

Garrity, D.P. (2004). Agroforestry and the achievement of the Millennium Development Goals. *Agroforestry Systems* 61, 5-17.

Government of Uganda (1967). *Atlas of Uganda.* Uganda Government Printery, Entebbe, Uganda.

Government of Uganda (1991). *The Uganda Forestry Policy.* Ministry of Water, Lands and Environment, Kampala, Uganda.

Government of Uganda (2003). *The National Forestry and Tree Planting Act 2003.* Ministry of Water and Environment, Kampala, Uganda.

Hamilton, A.C. (1981). *A Field Guide to Uganda Forest Trees.* Privately published, Kampala, Uganda.

Hamilton, A.C. (1984). *Deforestation in Uganda.* Oxford University Press, Nairobi, Kenya.

Hamilton, A.C. (1989). African Forests. In: Lieth, H. and Werger, M.J.A. (eds) *Tropical Rain Forest Ecosystems: Biogeographical and Ecological Studies* 14B, 155-182. Ecosystems of the World, Elsevier, Amsterdam, the Netherlands.

Hamilton, A.C., Hamilton, N.N., Mukasa, P.N., Ssewanyana Masembe, D., Ssentoogo, C.S.N.K. and Kabuye, C.S. (2016). *Luganda Dictionary and Grammar: Luganda-English and English-Luganda Dictionary with Notes on Luganda Grammar.* Published by Alan Hamilton, Godalming, UK.

Hamilton, A.C., Karamura, D. and Kakudidi, E.K. (2016). History and conservation of wild and cultivated plants in Uganda: forest species and banana varieties as case studies. *Plant Diversity* 38, 23-44.

Harris, J.G. and Harris, M.W. (1994). *Plant Identification Terminology: an Illustrated Glossary.* Spring Lake Publishing, USA.

Hart, T.B., Hart, J.A., Dechamps, R., Fournier, M. and Ataholo, M. (1996). Changes in forest composition over the last 4000 years in the Ituri basin, Zaire. In: van der Maesen, L.J.G., van der Burgt, X.M. and van Medanback de Rooy, J.M. (eds) *The Biodiversity of African Plants.* Kluwer Academic Publishers, Dordrecht, the Netherlands, pp. 541-563.

Hillbrand, A., Borelli, S., Conigliaro, M. and Olivier, A. (2017). *Agroforestry for Landscape Restoration: Exploring the Potential of Agroforestry to Enhance the Sustainability and Resilience of Degraded Landscapes.* Food and Agriculture Organization of the United Nations Rome, Italy.

HLPE (2017). *Sustainable Forestry for Food Security and Nutrition.* A report by The High-Level Panel of Experts on Food Security and Nutrition of the Committee on World Food Security, Rome.

Howard, P. (1991). *Nature Conservation in Uganda's Forest Reserves.* IUCN, Gland, Switzerland.

Howard, P., Davenport, T. and Baltzer, M. (eds) (1996a). *Kalinzu and Maramagambo Forest Reserves Biodiversity Report.* Forest Department, Kampala, Uganda.

Howard, P., Davenport, T. and Dickinson, C. (eds) (1996b). *Kasyoha-Kitomi Forest Reserve Biodiversity Report.* Forest Department, Kampala, Uganda.

Howard, P., Davenport, T. and Matthews, R. (eds) (1996c). *Budongo Forest Reserve Biodiversity Report.* Forest Department, Kampala, Uganda.

Howard, P., Davenport, T. and Matthews, R. (eds) (1996d). *Kibale Forest Biodiversity Report.* Forest Department, Kampala, Uganda.

Hufkens, K., Scheunders, P. and Ceulemans, R. (2009). Ecotones in vegetation ecology: methodologies and definitions revisited. *Ecological Research.* https://doi 10.1007/s11284-009-0584-7/

IUCN (2012). *Red List Categories and Criteria*, version 3.1, Second Edition. https://www.iucnredlist.org/resources/categories-and-criteria

IUCN (2018). *Assessing IUCN's Contribution to Uganda's Forest Landscape Restoration Processes, 2010-2017.* IUCN, Kampala, Uganda.

IUCN (2019). *IUCN Red List of Threatened Species. Version 2019.2.* www.iucnredlist.org/ [Accessed November 2019]

Josephat, M. (2018). Deforestation in Uganda: population increase, forests loss and climate change. *Environmental Risk Assessment and Remediation* 2(2), 46-50.

Kabiru, S., Hassan, S., Hadi, R., Umar, U.A., Musab, I. and Bello, M. (2018). Limiting factors affecting agroforestry adoption in Butta Sub-county, Manafwa District, Uganda. *Asian Journal of Advances in Agricultural Research* 5(3), 1-9.

Kaboggoza, J. (2011). *Forest Plantations and Woodlots in Uganda.* African Forest Forum, Nairobi, Kenya.

Kalema, J. and Bukenya-Ziraba, R. (2005.) Patterns of plant diversity in Uganda. *Biologiske Skrifte* 55, 331-341.

Kalema, J. (2006). The significance of Important Bird Areas for conservation of plants in Uganda. In: Ghazanfar S.A. and Beentje H.J. (eds), *Taxonomy and Ecology of African Plants, their Conservation and Sustainable Use*. Royal Botanic Gardens, Kew, UK, pp. 457-472.

Kalema, J. and Kasenene, J.M. (2007). A comparative study of regeneration under live and dead canopy trees in a tropical rain forest ecosystem of Kibale National Park, Uganda. *African Journal of Ecology* 45 (Suppl. 3), 11–17.

Kalema, J., Namaganda, M., Ssegawa, P., Kabuye, C., Maganyi, O. and Mucuguzi, P. (2010). Status of higher plants in Uganda. In: Kaddu, J.B. and Busuulwa, H. (eds) *Baseline Report on State of Biodiversity in the Nile.* A production of the Wetlands and Biodiversity Conservation Component of the Nile Transboundary Environmental Action Project. Nile Basin Initiative Secretariat, Entebbe, Uganda.

Kalema, J. and Beentje, H. (2012). *Conservation Checklist of the Trees of Uganda.* Kew Publishing, Royal Botanic Gardens, Kew, UK.

Kalema, J., Kiwuka, C., Namaganda, M. and Mulumba, J. (2016). Distribution, habitat and conservation of range-restricted species of the genera *Eleusine* Gaertn. and *Pennisetum* Rich. (Poaceae) in Uganda. *Symbolae Botanicae Upsalienses* 38, 93-100.

Katende, A.B., Birnie, A. and Tengnäs, B. (1995). *Useful Trees and Shrubs for Uganda.* Regional Soil Conservation Unit, SIDA, Nairobi, Kenya.

Kawachi K (2010). An overview of the sociolinguistic situation of Kupsapiny, a Southern Nilotic language of Uganda. *African Study Monographs* 31(3), 127-137.

Kazoora, C. (2001). *Forest Landscape Restoration*. Uganda Country Report, IUCN-EARO and WWF-EARPO.

Kyarikunda, M., Nyamukuru, A., Mulindwa, D. and Tabuti J.R.S. (2017). Agroforestry and management of trees in Bunya County, Mayuge District, Uganda. *International Journal of Forestry Research*, Article ID 3046924. doi.org/10.1155/2017/3046924.

Langdale-Brown, I., Osmaston, H.A. and Wilson, J.G. (1964). *The Vegetation of Uganda and its Bearing on Land-use.* Government Printer, Entebbe, Uganda.

Lee, A.T.K., Carr, J.A., Ahmad, B., Arbainsyah, F.A, Handoko, Y., Harsono, R., Graham, L., Kabangnga, L., Kurniawan, N.P., Keßler, P.J.A., Kuncor, P., Prayunita, D., Priadjati, A., Purwanto, E., Russon, A., Sheil, D., Sylva, N., Wahyudi, A. and Foden, W. (2019). *Reforesting for the Climate of Tomorrow: Recommendations for Strengthening Orangutan Conservation and Climate Change Resilience in Kutai National Park, Indonesia.* IUCN, Gland, Switzerland.

Lesage, J. (2019). *The position of Kam (Central-Eastern Nigeria) within Niger-Congo: and the overestimation of genealogical uniformity in African languages.* ASIMIL, 3 April 2019, Leiden, the Netherlands.

Lewis, S.L. (2006). Tropical forests and the changing Earth system. *Philosophical Transactions of the Royal Society of London B* 361, 195-210.

Liautaud, K., Barbier, M. and Loreau, M. (2019). *Ecotone formation through ecological niche construction: the role of biodiversity and species interactions.* http://dx.doi.org/10.1101/740282/

Liu, C.L.C., Kuchma, O. and Krutovsky, K.V. (2018). Mixed-species versus monocultures in plantation forestry: development, benefits, ecosystem services and perspectives for the future. *Global Ecology and Conservation* 15. https://doi.org/10.1016/j.gecco.2018.e00419/

Lwangale, D.W. (2015). A genealogical reconstruction of Lubukusu, Lumasaba and Lugisu. *Journal of Applied Science* 2(9), 1-26.

Marfo, T.D., Datta, R., Vranovám V. and Ekielski, A. (2019). Ecotone dynamics and stability from soil perspective: forest-agriculture land transition. *Agriculture* 9, 228. doi:10.3390/agriculture9100228.

Marfo, T.D., Resjek, K. and Vranova, V. (2018). Spatial variations in soil properties across ecotones: a short review. *Bulletin of Geography*: Physical Geography Series 14, 71-77. http://dx.doi.org/10.2478/bgeo-2018-0006/

Mau, A.C., Reed, S.C., Wood, T.E. and Cavaleri, M.A. (2018). Temperate and tropical forest canopies are already functioning beyond their thermal thresholds for photosynthesis. *Forests* 9, 47. doi:10.3390/f9010047/

McElhinny, C., Gibbons, P., Brack, C. and Bauhus, J. (2005). Forest and woodland stand complexity: its definition and measurement. *Forest Ecology and Management* 218, 1-24.

Meunier, Q., Lemmens, R. and Morin, A. (2010). *Alternatives to Exotic Species in Uganda: Growth and Cultivation of 85 Indigenous Trees*. Printed by Graphic Consult (U) Ltd., Kampala, Uganda.

Ministry of Water and Environment (2011). *Forest Management Plan for Budongo Central Forest Reserves (Budongo, Siba, Biiso, Kitigo, Busaju and Kaniyo-Pabidi Blocks) for the Period 1st July 2011 – 30th June 2021*. Ministry of Water and Environment. Government of the Republic of Uganda, Kampala, Uganda.

Ministry of Water and Environment (2013). *The National Forest Plan 2011/12-2021/22*. Ministry of Water and Environment, Government of the Republic of Uganda.

Ministry of Water and Environment (2016). *State of Uganda's Forestry*. Ministry of Water and Environment. Government of the Republic of Uganda, Kampala, Uganda.

Ministry of Water and Environment (2017). *Forest Investment Program for Uganda*. Ministry of Water and Environment. Government of the Republic of Uganda, Kampala, Uganda.

Mujawamariya, M., Manishimwe, A., Ntirugulirwa, B., Ziberam E., Ganszkym D., Bahati, E.N., Nyirambangutse, B., Nsabimana, D., Wallin, G. and Uddling, J. (2018). Climate sensitivity of tropical trees along an elevation gradient in Rwanda. *Forests* 9, 647. doi:10.3390/f9100647.

Mulugo, L.W., Galabuzi, C., Nabanoga, G.N., Turyahabwe, N., Eilu, G., Obua, J., Kakudidi, E. and Sibelet, N. (2019). Cultural knowledge of forests and allied tree system management around Mabira Forest Reserve, Uganda. *Journal of Forestry Research*. https://doi.org/10.1007/s11676-019-00961-6/

Nabunya, M. (2017). *Contribution of Agroforestry Practices to Reducing Farmers' Vulnerability to Climate Variability in Rakai district, Uganda*. MSc. Thesis for Degree in Tropical Forestry, Institute of International Forestry and Forest Products, Technische Universität Dresden, Germany.

Nakamura, R. (2019). *United Nations Strategic Plan for Forests 2030*. Workshop among biodiversity-related conventions on the post-2020 global biodiversity framework, 10-12 June 2019, Bern, Switzerland.

Nakkazi, E. (2011). *Ugandans mobilize to save Mabira Forest from sugarcane plantation*. www.theecologist.org/campaigning/wildlife/1057616/ [Accessed 2015]

NatureUganda (2011). *Implementing Collaborative Forest Management in Uganda: Lessons Learnt from Kasyoha Kitomi and Echuya Central Forest Reserve*. NatureUganda, Kampala, Uganda.

Nyeko, P. and Nakabonge, G. (2008). *Occurrence of pests and diseases in tree nurseries and plantations in Uganda*. Sawlog Production Grant Scheme (SPGS), Uganda.

Obua, J., Agea, J.G. and Ogwal J.J. (2010). Status of forests in Uganda. *African Journal of Ecology* 48(4), 853-859.

Ojelel, S., Mucunguzi, P., Katuura, E., Kakudidi, E.K., Namaganda, M. and Kalema, J. (2019). Wild edible plants used by communities in and around selected forest reserves of Teso-Karamoja region, Uganda. *Journal of Ethnobiology and Ethnomedicine* 15(3). https://doi.org/10.1186/s13002-018-0278-8/

Okai B.-D.J. (2012). *Voacanga africana Farming System in the Assin South District: Socio–economic and Soil Nutrient Implications*. MSc Thesis submitted to the Kwame Nkrumah University of Science and Technology, Ghana.

Omino, E. (1996). *A Contribution to the Leaf Anatomy and Taxonomy of Apocynaceae in Africa: a Monograph of Pleiocarpinae*. Thesis, Landbouwuniversiteit, Wageningen, the Netherlands. Backhuys Publishers, Wageningen, Netherlands.

POWO 2019. *Plants of the World Online*. Facilitated by the Royal Botanic Gardens, Kew, UK. http://www.plantsoftheworldonline.org/ [Accessed November 2019]

Sassen, M. and Sheil, D. (2013). Human impacts on forest structure and species richness on the edges of a protected mountain forest in Uganda. *Forest Ecology and Management* 307, 206-218.

Schulte, I., Streck, C. and Roe, S. (2019). *Protecting and Restoring Forests: a Story of Large Commitments yet Limited Progress*. New York Declaration on Forests Five-Year Assessment Report. Climate Focus (coordinator and editor). http://forestdeclaration.org/

Secretariat of the Convention on Biological Diversity (2001). *Global Biodiversity Outlook*. Montreal, Canada.

Spracklen, D.V., Baker, J.C.A., Garcia-Carreras, L. and Marsham, J. (2018). The effects of tropical vegetation on rainfall. *Annual Review of Environment and Resources* 43, 14.1-14.26. Available from https://doi.org/10.1146/annurev-environ102017-030136/

Ssegawa, P. and Kasenene, J.M. (2007a). Medicinal plant diversity and uses in the Sango bay area, Southern Uganda. *Journal of Ethnopharmacology* 113, 521-540.

Ssegawa, P. and Kasenene, J.M. (2007b). Plants for malaria treatment in Southern Uganda: traditional use, preference and ecological viability. *Journal of Ethnobiology* 27, 110-131.

Tang, K.H.D. (2019). Impacts of climate change on tropical rainforests' adaptive capacity and ecological plasticity. *Climate Change Facts, Impacts and Solutions* 1, 1-5.

Tenywa, G. (2005). BIDCO gets Bugala Island forest land. *New Vision*, 14 September 2005, Kampala, Uganda.

Tenywa, G. (2013). Oil palm growing threatens Bugala Island forest cover. *New Vision*, 17 April 2013. Kampala, Uganda.

Tor-Anyiin, T.A. and Yakumbur, D.T. (2012). Phytochemical screening and antimicrobial activity of stem bark extracts of *Antidesma venosum*. *Journal of Natural Products and Plant Resources* 2(3), 427-430.

TROPICOS (2019). Missouri Botanical Garden. http://www.tropicos.org/ [Accessed November 2019]

Tugume, P., Kakudidi, E.K., Buyinza, M., Namaalwa, J., Kamatenesi, M., Mucunguzi, P. and Kalema, J. (2016). Ethnobotanical survey of medicinal plant species used by communities around Mabira Central Forest Reserve, Uganda. *Journal of Ethnobiology and Ethnomedicine* 12,5. DOI 10.1186/s13002-015-0077-4.

Tumushabe, G. and Mugyenyi, O. (2017). *Forest Governance Training Manual*. Advocates Commission for Development and Environment, Kampala, Uganda.

van Goor, W. and Snoep, M. (2019). *The Contribution of Forests to Climate Change Mitigation: a Synthesis of Current Research and Understanding*. REDD+ Business Initiative and Greenchoice, Wageningen, The Netherlands.

van Schaik, A. and Tickell, O. (2015*). UN, Banks and Oil Palm Giants Feast on the Stolen Land of Uganda's Disposessed.* [Accessed 2015.] www.theecologist,org/News/news_analysis/2759987/

Veit, P. (2010). *Land for Private Investors and Economic Development: Uganda*. World Resources Institute, Washington, USA.

Watters, J.R. (2018). East Benue-Congo. In: Watters, J.R. (ed.) *East Benue-Congo: Nouns, Pronouns, and Verbs*, 1–25. Berlin: Language Science Press. doi 10.5281/zenodo.1314306.

WCS (2016). *Nationally Threatened Species for Uganda*. Wildlife Conservation Society, New York, USA.

White, F. (1983). *The Vegetation of Africa*. UNESCO, Paris.

Part 9

Indexes of Vernacular and Trade Names

Names of Trees by Language

The numbers given are those of species in Part 4 or (for species not allocated their own numbers in Part 4) those of the species in which they are mentioned or of the next numbered species. See Part 7 for more information on these languages, including the relationships between them, also for a map showing those parts of Uganda that are their traditional home areas (Fig. 7.1).

ACHOLI (ac)

See Luo.

ALUR (al)

See Luo.

ATESO (at)

Names from ITU and UFT, edited by Nelson Omagor.

ENGLISH (en)

Names from ITU and UFT, reformatted but not otherwise edited.

KARAMJONG (ka)

Names from ITU and UFT, not further edited.

KUPSABINY (ku)

Names from ITU and UFT, not further edited.

KWAMBA (am)

Names from ITU and UFT, not further edited.

LANGO (la)

See Luo.

LUGANDA (ga)

Names from ITU and UFT, edited by Olivia Wanyana Maganyi. Names recorded as used in the Ssese Islands in ITU and UFT are not distinguished in this list, awaiting further research on Luganda dialects. Names of trees marked with an asterisk (*) are given in a list of Luganda plant names in Hamilton (2016), which benefitted from editing by Christine Kabuye. Shifts between noun classes in Luganda can carry meaning. For instance, for fruit trees, it is common for the prefix mu- to be used for the names of the trees (e.g. *musaali*) and the prefix n- for the fruits (e.g. *nsaali*). Some names have obvious meanings, thus *mubajjangabo* is a type of tree useful for carving into shields (*-bajja* = carve; *engabo* = shield).

LUGBARA (gb)

Names from ITU and UFT, not further edited.

LUGWERE (gw)

Names from ITU and UFT, not further edited.

LUKONZO (ko)

Names from ITU and UFT, not further edited.

LUMASABA (ms)

Names from ITU and UFT, not further edited.

LUNYOLE (nl)

Names from ITU and UFT, not further edited.

LUO (lo), including ACHOLI (ac), ALUR (al) and LANGO (la)

Names from ITU and UFT. Acholi, Lango and Luo edited by Perpetra Akite. The names of these closely related languages are combined into a single index to allow for inclusion of Luo (undifferentiated) names. The names are assigned to particular languages where this has been recorded.

LUSAAMIA-GWE (sa)

Names from ITU and UFT, not further edited.

LUSOGA (so)

Names from ITU and UFT, edited by Ben Kirunda.

MADI (md)

Names from ITU and UFT, not further edited.

RUFUMBIRA (fu)

Names from ITU and UFT, edited by Samuel Nsutiyayesu.

RUKIGA (ki)

Names from ITU and UFT, edited by Robert Barigyira and Dennis Babaasa.

RUNYANKORE (na)

Names from ITU and UFT, not further edited.

RUNYORO (no)

Names from ITU and UFT, edited by Patrick Mucunguzi.

RUTOORO (to)

Names from ITU and UFT, not further edited.

TRADE NAMES (tn)

Names from ITU and UFT, not further edited.

Part 10

Index of Scientific Names

Names in bold type are valid names; those in regular type are no longer valid. The latter are mainly synonyms.

The numbers are those given to species in Part 4 or (for species not allocated their own numbers in Part 4) those of the species in which they are mentioned or of the next numbered species. The numbers or number ranges for families are the numbers of the species that belong to them.

A star (*) indicates that there is a description of the taxon inserted among the descriptions. A hash sign (#) indicates that there is a key. The numbers associated with these are the numbers of the next numbered species following the description or key. The location of these descriptions or keys are shown on Table 4.1.

Printed and bound by CPI Group (UK) Ltd, Croydon, CR0 4YY
16/09/2026
14963183-0003